T0328374

The Future of Soil Carbon

The Future of Soil Carbon

Its Conservation and Formation

Edited by

CARLOS GARCIA

CEBAS-CSIC, Campus Universitario de Espinardo, Murcia, Spain

PAOLO NANNIPIERI

University of Firenze, Firemze, Italy

TERESA HERNANDEZ

CEBAS-CSIC, Campus Universitario de Espinardo, Murcia, Spain

Academic Press is an imprint of Elsevier
125 London Wall, London EC2Y 5AS, United Kingdom
525 B Street, Suite 1800, San Diego, CA 92101-4495, United States
50 Hampshire Street, 5th Floor, Cambridge, MA 02139, United States
The Boulevard, Langford Lane, Kidlington, Oxford OX5 1GB, United Kingdom

Library of Congress Cataloging-in-Publication Data
A catalog record for this book is available from the Library of Congress

British Library Cataloguing-in-Publication Data
A catalogue record for this book is available from the British Library

ISBN: 978-0-12-811687-6

For information on all Academic Press publications visit our website at
https://www.elsevier.com/books-and-journals

Publisher: Candice Janco
Acquisition Editor: Anneka Hess
Editorial Project Manager: Carly Demetre
Production Project Manager: Maria Bernard
Designer: Victoria Pearson

Typeset by Thomson Digital

CONTENTS

LIST OF CONTRIBUTORS

Judith Ascher-Jenull
University of Innsbruck, Innsbruck, Austria

Felipe Bastida
CEBAS-CSIC, Campus Universitario de Espinardo, Murcia, Spain

Nanthi S. Bolan
University of Newcastle, Callaghan, NSW, Australia

Gordon J. Churchman
University of Adelaide, Urrbrae, SA, Australia

Vincenza Cozzolino
Interdepartmental Research Center on Nuclear Magnetic Resonance for the Environment, Agro-Food and New Materials (CERMANU); University of Napoli Federico II, Portici, Italy

Serena Doni
National Research Council—Institute of Ecosystem Study (CNR-ISE), Pisa, Italy

Marios Drosos
Interdepartmental Research Center on Nuclear Magnetic Resonance for the Environment, Agro-Food and New Materials (CERMANU), University of Napoli Federico II, Portici, Italy

María Gómez-Brandón
University of Innsbruck, Innsbruck, Austria

Carlos García
CEBAS-CSIC, Campus Universitario de Espinardo, Murcia, Spain

Gabriel Gascó
Universidad Politécnica de Madrid, Madrid, Spain

Teresa Hernández
CEBAS-CSIC, Campus Universitario de Espinardo, Murcia, Spain

Heribert Insam
University of Innsbruck, Innsbruck, Austria

Matthias Kästner
Helmholtz Centre for Environmental Research—UFZ, Leipzig, Germany

Ana M. Méndez
Universidad Politécnica de Madrid, Madrid, Spain

Cristina Macci
National Research Council—Institute of Ecosystem Study (CNR-ISE), Pisa, Italy

Fernando T. Maestre
Universidad Rey Juan Carlos, Móstoles, Spain

Sanchita Mandal
University of South Australia, Mawson Lakes, SA, Australia

Grazia Masciandaro
National Research Council—Institute of Ecosystem Study (CNR-ISE), Pisa, Italy

Anja Miltner
Helmholtz Centre for Environmental Research—UFZ, Leipzig, Germany

Paolo Nannipieri
University of Firenze, Firenze, Italy

Eleonora Peruzzi
National Research Council—Institute of Ecosystem Study (CNR-ISE), Pisa, Italy

Alessandro Piccolo
Interdepartmental Research Center on Nuclear Magnetic Resonance for the Environment, Agro-Food and New Materials (CERMANU); University of Napoli Federico II, Portici, Italy

César Plaza
Instituto de Ciencias Agrarias, Madrid; Universidad Rey Juan Carlos, Móstoles, Spain; Northern Arizona University, Flagstaff, AZ, United States

Cornelia Rumpel
Institute of Ecology and Environment Paris, Thiverval-Grignon, France

Binoy Sarkar
University of South Australia, Mawson Lakes, SA, Australia; The University of Sheffield, Sheffield, United Kingdom

Balaji Seshadri
University of Newcastle, Callaghan, NSW, Australia

Mandeep Singh
University of South Australia, Mawson Lakes, SA, Australia

Riccardo Spaccini
Interdepartmental Research Center on Nuclear Magnetic Resonance for the Environment, Agro-Food and New Materials (CERMANU); University of Napoli Federico II, Portici, Italy

Donald Sparks
University of Delaware, Newark, DE, United States

Giovanni Vinci
Interdepartmental Research Center on Nuclear Magnetic Resonance for the Environment, Agro-Food and New Materials (CERMANU), University of Napoli Federico II, Portici, Italy

Yilu Xu
University of Newcastle, Callaghan, NSW, Australia

Claudio Zaccone
University of Foggia, Foggia, Italy

PREFACE

It is well established that soil is a natural resource that we need to protect and conserve. However, it is not always adequately managed due to the scarcity of appropriate regulations, and also to the attitude of the majority of our society, which does not demand to administrations the necessary actions for soil protection, minimizing the possible risks of certain actions leading to soil degradation. As soil scientists, we must assume our responsibility for convincing institutions and society about the importance of soil, which is fundamental for the life in the planet. Soil should be considered as a living system being capable of performing key functions from ecological and human perspectives. The decrease in soil functionality or the soil loss can generate negative consequences on the functionality, and thus the production, of agricultural and forest systems as well as the functionality of terrestrial ecosystems. The organic C content of soil plays a key role in soil functionality affecting physical, chemical, and biological properties, including biodiversity of soil. This book tries to highlight the importance of organic C in soils, and discusses what can be expected in the future, considering the changes in both land use and climate change.

The book "The Future of Soil Carbon: Its Conservation and Formation," has a total of nine chapters and it will contribute to increasing our knowledge on soil C and its future dynamics. It offers to the readers (students, researchers, academics, etc.) a consistent vision of the complex problems and processes concerning the soil organic C. The chapter by Masciandaro et al. shows the importance of soil management practices for soil C storage, as affected by climate changes, and it describes the ecosystem services of soil. The essential ecosystem services that organic C provides in dryland ecosystems is highlighted in the chapter by Plaza et al., whereas the chapter by Bolan et al. shows how the interactions between the clay minerals and organic C can stabilize and protect organic C in soil against the degradation by soil microbial communities. Piccolo et al. discuss the recent advances in the formation and chemical structure of soil humus. The degradation of plant residues to form soil organic matter by producing microbial necromasses is discussed in the chapter by Kaestner and Miltner, whereas Xu et al. show that the activity, composition, and biomass of microbial communities are related to soil organic C dynamics, and this relationship depends on the spatial accessibility of organic substrates to soil microorganisms. Insam et al. discuss the importance of organic wastes as organic C and nutrient resources, and their contribution to soil fertility. The chapter by Bastida et al. considers the losses of C provoked by erosion processes, and the possibility of increasing soil organic C by adding organic residues or by using afforestation practices. The last chapter by book's editors analyzes the importance of soil C to guarantee a sustainable and productive use of soil, and discusses the future of soil C.

Carlos Garcia
Paolo Nannipieri
Teresa Hernandez

CHAPTER 1

Soil Carbon in the World: Ecosystem Services Linked to Soil Carbon in Forest and Agricultural Soils

Grazia Masciandaro, Cristina Macci, Eleonora Peruzzi, Serena Doni
National Research Council—Institute of Ecosystem Study (CNR-ISE), Pisa, Italy

Chapter Outline

SOIL ORGANIC CARBON IN THE WORLD

Monitoring and modeling spatial distribution of soil organic carbon (SOC), for better understanding and quantifying global carbon (C) cycles, are presently global challenges that involve environmental scientists, socioeconomists, and policy makers for the coming decades (Galati et al., 2016).

SOC is a key component of any terrestrial ecosystem linked to ecosystem services, such as provisional (e.g., food, fuel, fiber), regulating (e.g., climate and greenhouse gas regulation), cultural (e.g., recreation, ecotourism), and supporting (e.g., weathering, soil formation, nutrient cycling) services (Adhikari and Hartemink, 2016).

The C stored in soils is nearly three times that in the aboveground biomass and approximately double that in the atmosphere. Under natural conditions

The Future of Soil Carbon
http://dx.doi.org/10.1016/B978-0-12-811687-6.00001-8

the content of organic C in soil is constant; the rate of decomposition is equal to the rate of supply of organic C, generally from plants (Stevenson, 1986). The SOC balance can be disturbed markedly by human activities, including deforestation and agricultural use of land. There is also a decline in SOC content when grassland is transformed into cropland in the tropics and subtropics (Hartemink et al., 2008) or when savannahs are burned (du Preez et al., 2011). SOC is also lost through misuse or deterioration of land (soil erosion, salinization, alkalization, and soil degradation), and because of increasing nonagricultural use of land (urbanization and highway construction) (Hutyra et al., 2011).

Small relative changes in SOC content and composition can have profound effects on many processes that occur within the system. For example, small increases in organic carbon loss from soils could greatly enhance carbon dioxide concentrations in the atmosphere, potentially creating a positive feedback on the climate cycle (Smith and Fang, 2010).

Despite its importance, the size of the global SOC stock and its distribution in space and among land use/land cover classes is not well known (Jandl et al., 2014), hence possible C emissions from soils due to changes in land use and land cover.

Guo and Gifford (2002) found an average reduction in SOC stocks of 42% when land was converted from forest to cropland. The decrease was even greater when pasture was converted to cropland (59%). Variations in SOC stocks when land use change is in the opposite direction were also important, with an average increase of 53% when cropland was converted to secondary forest but only 19% on average when cropland was converted to pasture.

Improving soil C projections is considered a high priority to track the changes in SOC and thus, to identify where the world soils are close to reaching critical thresholds of sustainability and places under which environmental conditions are likely to provoke maximal change. The Harmonized World Soil Database (HWSD) (Fisher et al., 2008) provides one of the most recent and coherent global data sets of SOC, giving a total stock of 1417 Pg C under partially disturbed conditions when bulk densities for organic soils are derived from a pedotransfer function (Hiederer and Köchy, 2011).

Although there is large uncertainty associated with estimating the mean soil C content of a biome or soil taxonomic order, with coefficients of variation in the order of 65%, recent estimates for the global soil C reservoir converge (Table 1.1).

The spatial distribution of mean SOC content is shown in Fig. 1.1 (Batjes, 2016). Based on analyses of the WISE30sec database, about the 30% of the total SOC stock to 2 m depth is held in the Northern Circumpolar

Table 1.1 Global soil organic C (SOC) estimates (Pg) in soils

Source	0–1 m depth	0–2 m depth
Post et al. (1982)	1395	
Batjes (1996)	1462–1548	2376–2456
Kasting (1998)	1580	
Jobbágy and Jackson (2000)	1502	1993
Robert (2001)	1500	2456
Hiederer and Köchy (2011)	1417	
Govers et al. (2012)	1400–1600	1990–2460
Todd-Brown et al. (2013)	1260 (890–1660)	
Batjes (2016)	1408	2060

Region (Hugelius et al., 2014). These soil C reserves are particularly vulnerable to loss with climate change (Schuur et al., 2008). Models of permafrost soil C have only recently been integrated into Earth system models (ESMs) (Koven et al., 2011) and further improvements in the representation of thermokarst dynamics, peat accumulation, and soil hydrology are needed to reduce uncertainties related to climate–carbon feedbacks in northern biomes.

De Brogniez et al. (2015) created a topsoil organic carbon map using Generalized Additive Models (GAM) for the area of the European Union (EU) (Fig. 1.2). In this study, topography and land use were recognized as key indicators to assess SOC stock and its distribution.

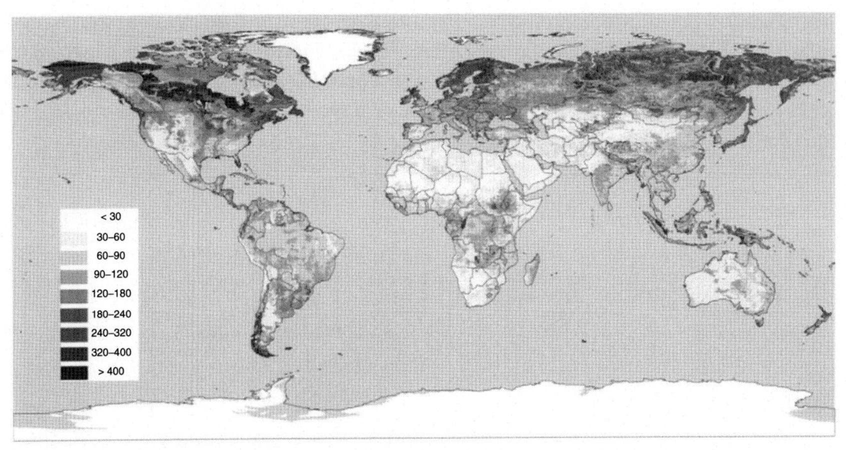

Figure 1.1 ***Soil organic carbon content to 1 m depth (Mg C ha^{-1}).*** *(Modified from Batjes, N.H., 2016. Harmonized soil property values for broad-scale modelling (WISE30sec) with estimates of global soil carbon stocks. Geoderma 269 (C), 61–68. Available from http://doi.org/10.1016/j.geoderma.2016.01.034).*

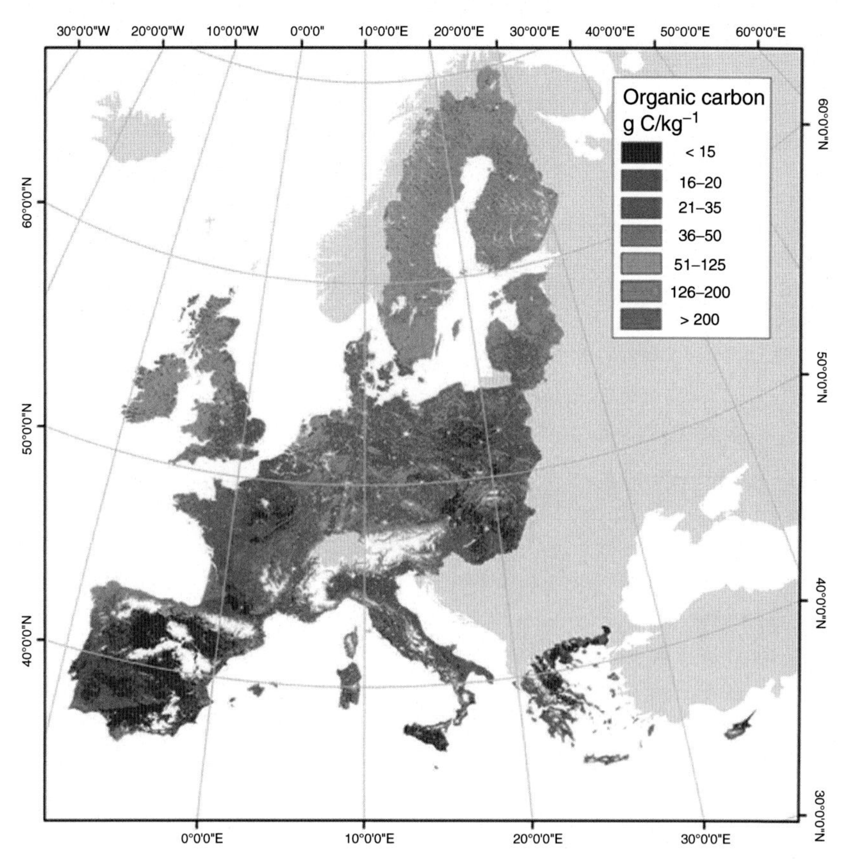

Figure 1.2 ***Map of predicted topsoil organic C content (g C kg^{-1}).*** *(Modified from De Brogniez, D., Ballabio, C., Stevens, A., Jones, R.J.A., Montanarella, L., van Wesemael, B., 2015. A map of the topsoil organic carbon content of Europe generated by a generalized additive model. Eur. J. Soil Sci. 66, 121–134. Available from http://dx.doi.org/10.1111/ejss.12193).*

The smallest organic carbon (OC) concentrations were mainly located in Mediterranean soils and also in some soils of France, Germany, Poland, the Czech Republic, Slovakia, and Hungary, covered by croplands or permanent crops subjected to intensive management practices, which increase the mineralization of soil organic matter and therefore reduce OC contents (Lal, 2002).

In addition, Novara et al. (2014, 2013) also highlighted that SOC levels in Mediterranean soils increase when the soils are no longer cultivated. This suggests that soil cultivation can play a major role in affecting SOC content under Mediterranean conditions. However, SOC dynamics also depend on other factors, including climate, soil type and texture, soil moisture, temperature regimes, lithology, morphology, land use history and management (Fantappiè et al., 2011, 2010; Pisante et al., 2015).

Climate, use, and management are very influential in the C variability in Spanish soils (Muñoz-Rojas et al., 2012; Parras-Alcántara et al., 2013; Ruiz Sinoga et al., 2012), particularly in semiarid regions, which are characterized by low levels of soil organic matter content (10 g kg^{-1}) (Acosta-Martínez et al., 2003).

ECOSYSTEM SERVICES: AN INTRODUCTION

Ecosystems provide a wide array of goods and services to people, called ecosystem services. As extensively discussed by Gómez-Baggethun et al. (2010), the modern concept of ecosystem services started in the late 1970s to highlight the societal dependence on ecosystems, stating the ecosystem functions as services, so as to increase public awareness about biodiversity conservation (De Groot, 1987; Ehrlich and Ehrlich, 1981; Westman, 1977). Then, in 1990s, several influential papers about ecosystem services (Costanza and Daly, 1992; Daily, 1997; De Groot, 1992) and methods to estimate their economic values (Costanza et al., 1997) laid the foundation for a new field of research between ecological, economic, and social sciences, which can effectively support decision makers about environmental management and policy.

According to the Millennium Ecosystem Assessment, ecosystem services are divided into four basic categories (MEA, 2005):

1. Provisioning services: goods, such as food or freshwater that ecosystems provide and humans consume or use;
2. Regulatory services: services, such as flood reduction and water purification that healthy natural systems, such as wetlands, can provide;
3. Cultural services: intangible benefits, such as aesthetic enjoyment or religious inspiration that nature often provides;
4. Supporting services: basic processes and functions, such as soil formation and nutrient cycling that are critical to the provision of the first three types of ecosystem services

Other recent and important studies dealing with this subject include "Valuing the Protection of Ecological Systems and Services" (Thompson et al., 2009), "The Economics of Ecosystems and Biodiversity: Mainstreaming the Economics of Nature" (TEEB, 2010), and *Natural Capital: Theory and Practice of Mapping Ecosystem Services* (Kareiva et al., 2011).

SOIL ECOSYSTEM SERVICES

Soil is a species-rich habitat of terrestrial ecosystems carrying out many important functions (e.g., biomass production, maintaining nutrient balance, chemical recycling, and water storage) (Blum, 2005). Jónsson and DavíÐsdóttir (2016) compared different frameworks to evaluate soil ecosystem services, identifying as one of the most comprehensive frameworks that were proposed by Dominati et al. (2010) and by Dominati (2013) (Table 1.2), where ecosystem services provided by soil are linked together to the natural capital soil resource also considering the relative properties.

Lal (2016) discussed the relationships between quality, health, and ecosystem services in soil. Soil health, defined as "capacity of soil to function as a vital living system to sustain biological productivity, maintain environment quality and promote plant, animal and human health" by Doran and Jones (1996), represents "the engine of economic development," influencing both human being and environment, as detailed in Table 1.3.

Keesstra et al. (2016) and Lal (2016) have recently pinpointed the important connection between maintaining ecosystem services and soil management so as to reach the sustainable development goals (SDG) proposed by United Nations in 2015 (UN, 2015; Table 1.4). Soil functionality and maintaining and restoring ecosystem services are in fact clearly linked to SDGs 1–4, 6–9, 11–13, 15–16.

SOIL ORGANIC C PROPERTIES

The SOC pool size depends on the balance between formation of soil organic matter from decomposition of organic residues, mainly plant litter materials, and its mineralization to inorganic carbon by soil organisms. During the decomposition of organic residues, formation of humus, chemically complex and often associated with soil minerals, occurs (Blanco-Moure et al., 2016).

The amount of plant litter, its composition and its properties, as well as land-use intensity, environmental conditions, and soil properties are controlling factors for the formation of soil organic matter (SOM) and humification processes in terrestrial ecosystems (Blanco-Moure et al., 2016). Plant roots also play a major role in controlling SOM dynamics through their effects on aggregation and the movement of C to deeper soil layers. Carbon input from plant to soil through rhizodeposition is one of the major sources of available C for microorganisms (Luo et al., 2014). Exudates from living roots can stimulate a quick response of soil microbes with acceleration

Table 1.2 Ecosystem services provided by soil

Service	Goods or Services	Definition
Provisioning services	Food, wood, and fiber	Agroecosystem's first purpose is to produce food and grow crops for several purposes. Soils physically support plants and supply them with nutrients and water.
	Raw materials	Soils and vegetation can be source of raw materials (top soils, peat, turf, sand, clay minerals, biomedical and medicinal resources, genetic resources, ornamental resources)
	Support for human infrastructure and animals	Soils represent the physical base on which human infrastructures and animals stand
Supporting and regulating services	Flood mitigation	Soils have the capacity to absorb and store water, thereby regulating water flows and mitigating flooding
	Filtering of nutrients and contaminants	Soils can absorb and retain nutrients and contaminants and avoid their release in water bodies
	Carbon storage and greenhouses gases regulation	Soils have the ability to store C and regulate their production of GHG
	Detoxification and the recycling of wastes	Soils can adsorbs nutrients and contaminants and avoid their release in water bodies
	Carbon storage and GHG regulation	Soils have the ability to store carbon and regulate GHG production
	Detoxification and waste recycling	Soils can adsorb and degrade harmful compounds
	Pest and disease regulation	By providing habitat to beneficial species, soils and vegetation can control pests (crops, animals, and humans) and harmful disease vectors (viruses, bacteria), thus providing biological control
Cultural services	Recreation/ ecotourism	Natural and managed landscapes can be used for pleasure and relaxation
	Aesthetics	Appreciation of the beauty of natural managed landscapes
	Heritage values	Memories in the landscape from past cultural ties
	Spiritual values	Sacred places
	Cultural identity/ inspiration	Natural and cultivated landscapes provide a sense of cultural identity

Source: Modified from Dominati, E.J., Patterson, M., Mackay, A., 2010. A framework for classifying and quantifying the natural capital and ecosystem services of soils. Ecol. Econ. 69, 1858–1868 and Dominati, E.J., 2013. Natural capital and ecosystem services of soils. In: Dymond J.R. (Ed.), Ecosystem Services in New Zealand – Conditions and Trends. Manaaki Whenua Press, Lincoln, New Zealand.

Table 1.3 Ecosystem services for human well-being and environmental preservation provided by healthy soils

Ecosystem services	Goods and functions provisioned by healthy soils
Biogeochemical cycles	Element transformations
	Nutrient retention
	Nutrient diffusion
Hydrogeological cycle	Pollutant decontamination
	Water filtration
	Water storage
Moderating climate	Carbon sequestration
	Greenhouse gases regulation
	Resilience against extreme events
Biodiversity	Habitat for organisms
	Energy source
	Waste transformation
Good provisions	Food production
	Pharmaceuticals
	Raw materials
Other services	Physical support
	Archive of climate, human, and planetary history

Source: Modified from Lal, R., 2016. Soil health and carbon management. Food Energy Secur. 5 (4), 212–222. Available from http://doi.org/10.1002/fes3.96.

of native SOC mineralization, the so-called rhizosphere priming effect, which also depends on nutrient availability and substrate quality (Murphy et al., 2015).

On the basis of the turnover times, SOM can be divided into two pools: (1) rapid and medium turnover fractions, and (2) more recalcitrant forms that turn over slowly. The quickly degradable fraction is produced by soil enzyme-catalyzed depolymerization of fresh organic matter and it is comprised of low molecular weight chemicals that are often water soluble and thus more accessible to microbial assimilation as energy, C, and nutrient sources. This pool, with mean residence times (MRTs) ranging from months to decades, accounts for 40% of total SOM (Paul et al., 2015).

On the other hand, the fraction resistant to decomposition represents a long-term reservoir of energy that serves to sustain the system in the longer term and it contributes to soil aggregate formation and erosion resistance (Grandy and Neff, 2008). This fraction, with MRTs of hundreds of years, is largely represented by humic substances and other organic macromolecules, which are intrinsically resistant to microbial attack and physically protected by association with mineral surfaces or trapped within mineral aggregates

Table 1.4 Sustainable development goals (SDG)

1	End each poverty form everywhere.
2	End hunger, achieve food security and improve nutrition, and promote sustainable agriculture.
3	Ensure healthy lives and promote well-being for all at all ages.
4	Ensure inclusive and equitable quality education and promote lifelong learning opportunities for all.
5	Achieve gender equality and empower all women and girls.
6	Ensure availability and sustainable management of water and sanitation for all.
7	Ensure access to affordable, reliable, sustainable, and modern energy for all.
8	Promote sustained, inclusive, and sustainable economic growth, full and productive employment and decent work for all.
9	Build resilient infrastructure, promote inclusive and sustainable industrialization, and foster innovation.
10	Reduce inequality within and among countries.
11	Make cities and human settlements inclusive, safe, resilient, and sustainable.
12	Ensure sustainable consumption and production patterns.
13	Take urgent action to combat climate change and its impacts.
14	Conserve and sustainable use of the oceans, seas, and marine resources for sustainable development.
15	Protect, restore, and promote sustainable use of terrestrial ecosystems, sustainably manage forests, combat desertification, and halt and reverse land degradation and halt biodiversity loss.
16	Promote peaceful and inclusive societies for sustainable development, provide access to justice for all and build effective, accountable and inclusive institutions at all levels.
17	Strengthen the means of implementation and revitalize the global partnership for sustainable development.

Source: From UN, 2015. https://sustainabledevelopment.un.org/sdgs.

or clay layers (Mikutta et al., 2006). Recent studies have shown that in the early stages of litter decomposition, a surprisingly large amount of litter-C is sequestered in mineral soils (Rubino et al., 2010), because dissolved organic matter (DOM) produced during litter decomposition can rapidly become associated with silt and clay fractions in the top soil mineral layer (Kaiser and Kalbitz, 2012). However, compounds produced during the microbial transformation of plant litter are likely to contribute more to stabilized SOM than do original plant litter compounds (Mambelli et al., 2011). Indeed, improved analytical techniques have established that mineral-bound organic matter (OM) is predominately derived from microbial products (Miltner et al., 2011), with a layering of OM on clay particles with proteins and polysaccharides from microbial residues involved in the mineral–OM stabilization (Cotrufo et al., 2013; Kleber et al., 2007; Ladd et al., 1996). The quantity of microbial-derived OM formed through this pathway depends on the amount of OM produced during decomposition as well as the capacity of mineral surfaces to protect OM from further decomposition (Six et al., 2006).

High efficiency of SOM formation does not necessarily correspond to high persistence of the formed organic C in soil. In the past the chemical composition of SOM was thought to determine its turnover, with greater stabilization of more than less-complex biopolymers due to slower rates of decomposition (Derenne and Largeau, 2001). However, recent work emphasizes that microbial access to SOM, rather than chemical composition, controls its turnover (Dungait et al., 2012; Schmidt et al., 2011). Microbial access to SOM is restricted by C association with mineral surfaces and by spatial isolation within soil aggregates (Jastrow et al., 2007). The importance of mineral surfaces for SOM stabilization has been recognized for decades, with increases in the content of reactive mineral phases corresponding to increases in the potential for organic C accumulation (Ladd et al., 1996; Six et al., 2002). However, as mineral surfaces in a soil become saturated with C, C decomposition rates increase, and the rate of SOC storage per unit of C input declines, thus resulting in an asymptotic response of SOC stocks to increasing C inputs (Heitkamp et al., 2012). In the conceptual soil C saturation model by Six et al. (2002), OC associated with silt and clay particles, that microaggregrate-protected, that biochemically protected, and that unprotected will all reach their own saturated levels. Additionally, soil cations, ionic strength, and soil pH can regulate OM adsorption and precipitation (Schneider et al., 2010). Soil organic matter associated with silt and clay particles was correlated with cation exchange capacity (Guibert et al., 1999). Amounts of Fe/Al oxyhydroxides and oxides were correlated to OC concentration (Kleber et al., 2005) and higher hydroxyl density on Fe/Al oxides than on phyllosilicates adsorbed more OM (Gu et al., 1994).

The content of humic substances content has widely been recognized as an important fraction of SOM playing an important role in soil quality and being influenced by agricultural practices or adverse climate conditions (Stevenson, 1986). Humic substances are able to bind extracellular enzymes (humic–enzyme complexes) and preserve them from proteolysis and chemical degradation (Nannipieri et al., 1996). Humic-bound enzymes extracted from soils can be separated by isoelectric focusing (IEF) without modifying the enzyme activity and the molecular structure of the complexes (Benítez et al., 2000; Ceccanti et al., 2008).

As suggested by other studies (Ceccanti et al., 2008; Doni et al., 2014), the relationship between enzyme activity and humic carbon content might reflect the potential for enzyme immobilization in soil and, therefore, the potential for soil resilience (Benítez et al., 2004; Ceccanti and Masciandaro, 2003). Because they have been found in a great variety of natural and managed soils humic-enzyme complexes they are supposed to constitute structural components of SOM that significantly contribute to empowering the biological barrier at the protection of the final and irreversible soil degradation.

García et al. (2016) found that if the soil contains a more structured humus, with a greater proportion of carboxylic, phenolic, and alcohol groups, it has a higher capacity of enzyme immobilization.

For the EU, forests are an important ecosystem in terms of recreation, biodiversity, timber, and carbon storage (Hansen et al., 2013). They cover about 177 million ha (42.3% of the total land area) of the EU27 territory and provide living space for ca. 4 million people (forestry and forest-based industries) (Forest Europe, UNECE and FAO, 2011).

It is estimated that about 13 million ha of forestlands are converted to other land uses every year (FAO, 2010). The way of forest management practices have a high impact on soil-related forest ecosystems (Lal, 1996) especially with regard to its biodiversity (Torras and Saura, 2008), water quality (Stott et al., 2001), and the related ecosystem services (Chazdon, 2008).

Conversion of natural to agroecosystems leads to depletion of the SOC pool because of a multitude of interacting factors including lower input of biomass C, a higher rate of decomposition caused by alterations in soil moisture and temperature regimes, and vulnerability of soils under agroecosystems to accelerated erosion and other degradation processes. Vulnerability to decomposition and depletion of SOC is affected by complex factors that are not clearly understood (Conant et al., 2011; Falloon et al., 2011; Tuomi et al., 2008).

Therefore, soils of agroecosystems are presently sources of major GHGs (Powlson et al., 2011). The magnitude of emission from soils may increase with expansion of agriculture into ecologically sensitive ecoregions, such as tropical savannahs (Noellemeyer et al., 2008), rain forests (Cerri et al., 2007), and peatlands (Jungkunst et al., 2012) and especially tropical peatlands (Jauhiainen et al., 2011).

Achieving a sustainable equilibrium among agricultural practices and environment is the main goal for agronomic research nowadays (Keesstra et al., 2016; Smith et al., 2015). Land management should tend to a sustainable agriculture to guarantee or improve crop yields and soil quality, while limiting greenhouse gases emissions (IPCC, 2013; Lal, 2000).

SOIL ECOSYSTEM SERVICES LINKED TO SOIL CARBON

Soil carbon is one of the main properties with a great influence on soil health and quality. It is, in fact, critical for the biophysical processes operating in the soil, thus conferring fertility to the soil beyond that provided by the weathered parent material (Clothier et al., 2011).

Birgé et al. (2016) provide a useful adaptive management for soil ecosystem services, identifying the main ecosystem services provided by soil mainly associated with soil carbon and OM soil content:

1. Soil stability
2. Primary production
3. Carbon sequestration
4. Water-holding capacity
5. Biodiversity
6. Nutrient cycles and soil fertility

Soil Stability

Aggregate stability can affect a wide range of soil properties, including organic C stabilization, soil porosity, water infiltration, aeration, compactibility, water retention, hydraulic conductivity, resistance to erosion by water and overland flow. Maintaining high stability of soil aggregate is essential for preserving soil productivity, minimizing soil erosion and degradation, and thus minimizing environmental pollution as well (Six et al., 2000).

OM promotes the formation of soil aggregates (Verchot et al., 2011) and thus influences soil structure and stability, through mechanisms, such as promoting the binding soil mineral particles, reducing aggregate wettability, and influencing the mechanical strength of soil aggregates, which is the measure for the coherence of interparticle bonds (Zinn et al., 2007). OM stabilization in soil aggregates is the principal mechanism for long-term sequestration of organic C in SOM (Mohammadi and Motaghian, 2011). The hypothesis that soil erosion by water may result in a source of atmospheric C is largely based on the assumption that water erosion accelerates mineralization of SOC mainly owing to the breakdown of aggregates. According to Lal (2003) it could represent a significant source of atmospheric CO_2 of 0.8–1.2 Pg C year^{-1}.

While increases in SOM are generally associated with increases in organic C-rich macroaggregates, long-term sequestration depends on stabilization of organic C in microaggregates (Six et al., 2000). Larger aggregate fraction was the most reactive to changes in land use, and aggregate disruption induced by tillage has been correlated to organic C losses, worsening of soil physical properties and soil quality (Bronick and Lal, 2005; Six and Paustian, 2014). Ayoubi et al. (2012) found that the highest percent of two large aggregates (i.e., 2.00–4.75 and 0.25–2.00 mm) were in the natural forest soil and the highest percent of microaggregates (0.053–0.25 mm) were in the cultivated and disturbed forest soils. Mikha and Rice (2004) showed that the aggregates larger than 2.00 mm and those of 0.25–2.00 mm were higher in the no-till soil than in the conventional tiled soil. However, the highest percentages of the smaller aggregates (i.e., 0.053–0.25 and 0.02–0.053 mm) had were in the conventionally tilled soil. Many long-term cropping systems resulted in reduction of aggregation and stability of macroaggregates (Mikha and Rice, 2004). The reduction of large aggregates in the cultivated soil is probably due to the degradation of young and relatively unstable OM, including fungal hyphae and plant roots of large aggregates whereas old and more stable OM was in small aggregates (Bronick and Lal, 2005; Wright and Hons, 2005).

An increase in SOC content as the result of high surface residue cover and lack of soil disturbance is typically accompanied by an increase in near-surface soil aggregation, as well as by an increase in water infiltration (Quintero and Comerford, 2013). SOC accumulation with the concomitant increase in soil biological activity can improve soil aggregation and create macropores capable of channeling water quickly through soil profile (Shipitalo et al., 2000).

Vegetation regulates soil erosion and thereby provides a major contribution to Mediterranean agro-forestry system's sustainability (Olesen et al., 2011). Soil erosion is detrimental to agricultural ecosystems due to loss of SOM, nutrients, and tillage, but also detrimental to neighboring ecosystems due to the transport and deposition of sediment, nutrients, and agrochemicals (see Chapter 9 by García et al. (2016) in this book).

An et al. (2009) showed that soil quality parameters were extremely affected by deforestation. The decrease of soil quality in differently cultivated areas also shows the importance of maintaining the natural vegetation on the eroded land. Mitigation of soil erosion can be obtained by appropriate land use, for example, afforestation and a natural succession on eroded land. An et al. (2010) reported that soil aggregate stability was extremely high under forest soils compared to bare fallow land and cropland soils.

In cultivated soils, soil structure is destroyed by plow and the stabilizing effects of root fibers become insignificant because the roots are shredded by the tillage and the subsequent microbial decomposition (An et al., 2010). As pore space increased due to the mechanical cultivation, the air exchange increased the oxygen availability for the microbial decay of OMs. This factor, coupled with the accelerated erosion, can rapidly deplete the SOM in the plow layer and weakened the soil aggregate stability (Six et al., 2000). Pasture soils showed a higher aggregate stability than cultivated soils due to the extensive rooting of grasses, higher SOC content, permanent plant coverage, and higher soil conservation (Bronick and Lal, 2005). Balabane and Plante (2004) reported that the greater stable macroaggregates in the pasture soils may be due to the higher amount of microbial biomass, plants residue, plants root, polysaccharides, and humic materials in the macroaggregates of these soils. Gol (2009) reported significantly larger water stable aggregates, SOC content, and total nitrogen in forest and pasture soils than in cultivated soils.

Primary Production

Primary production, classified as supporting service by MEA (2005), is the assimilation or accumulation of energy and nutrients by organisms, and thus it represents the ultimate source of energy in all ecosystems (Birgé et al., 2016). Vitousek et al. (1986) estimated that the quantity of the terrestrial net primary productivity (NPP) taken by humans worldwide, both directly (consumption) or indirectly (loss), was greater than 15%. The support, provision, and regulation ecosystem services (MEA, 2005) may be decreased by NPP seasonality, NPP reductions and C exported out of the system (Caride et al., 2012). The net primary productivity (NPP) of an ecosystem, one of the fundamental parameters describing its functioning, is the rate of formation of biomass used to create organic structures in plants, including woody, leaf and root tissues, but also root exudates and volatile organic compounds. Hence, NPP is an important factor determining the amount of the organic material available to higher trophic levels. It can also indicate the turnover rate of C and nutrient cycles in the ecosystem, and the potential response times to disturbance. For example, NPP can provide information about CO_2 sinks and sources (Sallaba et al., 2015). The allocation of NPP between different tissues and products is also an important for describing forest ecosystem ecology (Malhi et al., 2011). NPP is also an important indicator for biodiversity, species composition, and ecosystem services.

Peterson and Lajtha (2013) investigated the relationships between aboveground structure and function in a forest, specifically aboveground net primary productivity (ANPP) and stand composition, and soil C dynamics, finding expected positive correlation between ANPP, C inputs and exports. However, they did not find a relationship between ANPP and stand composition and SOM stores, C content of the soil, or DOC pools, suggesting that the biogeochemical processes controlling C storage and lability in soil may relate to longer-term variability in aboveground inputs that result from a heterogeneous and evolving forest stand.

Emmett et al. (2016) characterized the delivery of ecosystem service provided by an agricultural catchment by estimating the relationships among metrics of ecosystem service supply and identifying a single productivity gradient, with agricultural production and C storage at opposing ends of the gradient.

In a study about interannual variability of carbon dioxide exchange and relative controlling factors in agroecosystems, Suyker and Verma (2012) examined 8 years of results of measurements in an irrigated and rain-fed maize–soybean rotation cropping system. The net biome production indicated that the irrigated maize–soybean rotation was initially a moderate source of carbon; however, the system appears to be approaching near C neutral at the end of investigated period.

C Sequestration

In 2010 forests were estimated to cover about 28% of the total land area of the world, representing the major reserve of terrestrial C stock (FAO and ITPS, 2015). Schlesinger (1990) estimated that soil C accumulation of 0.4 Pg C $year^{-1}$ occurs mainly in forest soils. A report by DOE (1999) estimated the potential of C sequestration of world forests at 1–3 Pg C $year^{-1}$ out of the total potential of 5.7–10.1 Pg C $year^{-1}$ in all biomes of the world. Similar estimates were reported by Trexler (1998) and Watson et al. (1996). The total global potential of C sequestration in forest biomes was estimated at 196 Tg C $year^{-1}$ in 2010 and 748 Tg C $year^{-1}$ by 2040 (IPCC, 2000). Several factors can affect the amount of SOC in forest soils. For example, important climatic factors, such as precipitation, potential evapotranspiration (PET), and the ratio between PET and annual precipitation (PET ratio) have a pronounced effect on SOC concentration (Lal, 2005). For any given rate of annual precipitation,

SOC storage increases by decreasing PET ratio. There are numerous other soil and landscape factors that also affect SOC stock in forests (Wilcox et al., 2002). Prichard et al. (2000) observed a strong effect of slope on the SOC stock of a subalpine American forest. Landscape position can impact SOC stock because of its influence on soil water regime. The SOC stock also depends on soil texture and aggregation and cation exchange capacity. Coarser soils had lower total SOC concentration with respect to silt loam and sandy loam soils (Lal, 2005). In addition, forest soil C stock is affected by both natural and anthropogenic factors (Larionova et al., 2002). A natural disturbance can be a destructive event with drastic perturbation of an ecosystem, such as drought, fire, wind, insects, and diseases. Severe natural disturbance is followed by changes in soil moisture and temperature regimes, and succession of forest species with differences in quantity and quality of biomass returned to soil, can impact SOC stock in soil (Overby et al., 2003). Changes in the canopy cover due to fire and other natural disturbances can affect soil erosion thus reducing SOC stock (Elliot, 2003). SOC in forests is also affected by anthropogenic factors, such as forest management activities, deforestation, afforestation of agricultural soils, and subsequent management of forest plantations. Even though forest management is less intensive than cropland management, there are many management systems that may increase or enhance SOC stock in forests. Management options that maintain a continuous canopy cover and mimic regular natural forest disturbance are the best combination for high wood yield and C storage (Thornley and Cannell, 2000). Other management systems that may impact the SOC stock include harvesting and site preparation, fertilization and liming, soil drainage and planting of adapted species with high NPP, and more belowground biomass production (Hoover, 2003).

Land use and particularly its changes can strongly affect the C stocks and fluxes between soils and the atmosphere (Bolin and Sukumar, 2000). In the last years, the area of natural vegetation cover has gradually decreased in favor of agricultural land or urbanization (Field and Raupach, 2004). Under natural conditions, SOM contents are constant in the long term due to balanced input of organic residues and microbial decomposition. The quantity and quality of organic inputs and their rate of decomposition are determined by the combined interaction of soil properties, climate, and land use (EPA, 2006). Contents of SOM can be ranked as forests > grasslands > arable lands. Since the beginning of agriculture, 750 million ha of forests have been converted to agricultural land, causing a loss of 121 Pg C from biomass and soils worldwide (Kasang, 2004).

In a meta-analysis concerning 119 publications of 453 paired or chronosequential sites in 36 countries where forests (tropical, temperate, and boreal) were converted to agricultural land the SOC stocks decreased in 98% of the sites by an average of 31%, 41%, and 52% in boreal, tropical, and temperate regions, respectively (Wei et al., 2014). This meta-analysis considered the SOC in the upper 0–30 cm and the calculation kept into account changes in soil bulk density after land-use modification. In addition, the decrease in SOC stocks varied significantly according to forest type, cultivation stage, climate and soil factors. In a meta-analysis concerning 74 publications across tropical and temperate zones, Guo and Gifford (2002) showed a decline in soil C stocks after conversion from native forest to plantation (−13%) and native forest to crop (−42%). On the other hand, soil C stocks increased after conversions from native forest to pasture (+8%), crop to plantation (+18%), and crop to secondary forest (+53%). In this study, soil depth varied from less than 30 cm to more than 100 cm and changes in bulk density with land-use change were not taken into account. Similar results were also observed by Don et al. (2011) in a meta-analysis concerning 385 studies on land-use changes in the tropics. SOC decreased when primary forest was converted to perennial crops (−30%), cropland (−25%)and grassland (−12%). Instead, SOC increased when cropland was converted to grassland (+26%), it was afforested (+29%), or it was subjected to cropland fallow (+32%). Secondary forests stored 9% less SOC than primary forests. Relative changes were equally high in the subsoil as in the surface soil (Don et al., 2011). In this study, SOC stocks were corrected to an equivalent soil mass and the average sampling depth was 32 cm.

In agricultural cropping systems, the highest amount of C was stored in soil (Grego and Lagomarsino, 2008). Since approximately 12% of the soil C stock is present in cultivated soil and agricultural soils occupy about 35% of the global land surface, soil management is an important tool for climate change mitigation through C sequestration (Söderström et al., 2014).

Mechanisms responsible for SOM stabilization in agricultural soils include: physical protection of partially decomposed plant residues in inter- and intraaggregate; chemical protection due to SOM associations with clay and silt particles; biochemical protection of SOM due to a molecular structure, which is slowly metabolized by soil microorganisms; differences in activity and composition of soil biological communities that affect SOM sequestration and turnover (Paul et al. 2015).Whether the plant residues and their associated biota are free or protected within soil aggregates controls their short-term breakdown (Fortuna et al., 2003). The incorporation of plant and microbial-derived C into aggregates protects plant residues, associated microbial C, and microbial products for periods of weeks to decades depending on soil conditions (e.g., clay content and water availability) and management (e.g., tillage and crop rotation) (Paul et al., 2015).

The loss of organic C from agricultural soils on a global scale has been a topic of considerable debate, and according to Lal (2005) the C flux from soil to the atmosphere is estimated to be 0.8–1.2 Pg C $year^{-1}$, whereas C flux from soil to the ocean is 0.6 Pg C $year^{-1}$. The C losses can depend on: decrease in biomass (above- and belowground) returned to the soil, high decomposability of crop residues due to lignin content and differences in C:N ratio, change in soil moisture and temperature regimes, which influence the OM decomposition rate, tillage-induced perturbations, increase in soil erosion, decrease in soil aggregation and reduction in physical protection of the SOM. In general, low crop yields, high soil C contents, and high soil OM decomposition rates enhance the C losses from agricultural soils. Therefore, agricultural soils and in particular eroded agricultural soils usually contain lower SOC stock than their potential capacity.

However, there are practices other than land-use changes that potentially can slow down or reverse this trend. Management practices favoring C sequestration should increase organic C inputs and/or decrease organic C outputs. Carbon sequestration can also take place through a decrease in soil disturbance, such as tillage as already mentioned. Therefore, reducing soil disturbance include reduced or zero tillage systems, set-aside land, and the growth of perennial crops. Switching from conventional arable agriculture to other land uses with higher carbon inputs or reduced disturbance (e.g., bioenergy crop production, natural regeneration, conversion to grassland) will increase soil carbon stocks. A meta-analysis of data concerning 29 Spanish publications suggested that some forms of conservation agriculture (i.e., no tillage and implementing cover crops) can positively affect SOC content (Gonzalez-Sanchez et al., 2012). Among conservation agricultural practices, such as reduction in tillage, use of crop rotations, and retention of crop residues, the greatest reduction in emissions from farming activities occurred with tillage activities (Govaerts et al., 2009).

Measures for increasing soil carbon inputs include the preferential use of crop residues, animal manure, compost, sewage sludge, improved rotations with higher C inputs to soil, and also fertilization/irrigation/livestock management to enhance productivity. The C sequestration potential in agricultural soils has been estimated globally, emphasizing the role of temperate croplands (Freibauer et al., 2004).

The conventional farming includes management regimes with mineral fertilizer and/or pesticide application, while organic farming includes management types with organic fertilizer and no pesticides. The benefits of organic farming to the environment include less contamination by fertilizers and pesticides, the increase in biodiversity (Rey Benayas and Bullock, 2012), the enhancement of soil C sequestration and nutrients (Pimentel et al., 2005), the enhancement of pest control (Crowder et al., 2010), and the conservation of the genetic diversity of local varieties of domestic plants and animals (Jarvis et al., 2008).

As already mentioned, afforestation of agricultural land can reverse some degradation processes and lead to an increase in SOC sequestration (Ross et al., 2002). During reforestation, C stocks in the organic soil layer can increase linearly at a rate of 0.36 Mg C ha^{-1} $year^{-1}$ due to the increase in litter inputs by forest vegetation, and to low decomposition rates, particularly those of coniferous litter.

However, in a world with an increasing demand for food, any removal of land from agriculture inevitably adds to the pressure for land clearance for increased food production elsewhere. In the case of C sequestration when, for example, arable lands are converted to forest or grassland, it is necessary to consider the possible indirect impact of removing land from food production. If it leads to clearance of seminatural with a large SOC content plus C in vegetation, the climate change benefit is negated.

Water-Holding Capacity

Water cycling is crucial for hydrologic service of an ecosystem (Novick et al., 2004). The amount of water that is available to plants depends on the quantity of water that is able to infiltrate into soil and the quantity of water that soil is able to hold onto. The speed at which water can infiltrate soil is the infiltration rate or hydraulic conductivity. Typically, sandy soils have much faster infiltration rates and conductivity rates than clay soils. However, clay soils, due to higher matric potential and smaller pore size, generally hold significantly more water by weight than sandy soils.

The OM content can increase water infiltration rates (Brown and Cotton, 2011) and water-holding capacity (Brown and Cotton, 2011) in soil. Brown and Cotton (2011) showed that changes in water-holding capacity as a result of increases in soil C were much more pronounced for coarser soils while smaller to no change occurred in finer textured soils.

The water infiltration into soil depends on soil structural stability and porosity. In this context the soil OM content plays an important role since it affects soil aggregate stability positively, providing sorption sites. Transmission of water occurs through the soil macropores, where biota and sorption sites are generally located, thus having a pivotal role in the composition, growth, and activity of soil microbial community (Parr et al., 1981). Agriculture modifies the composition and root structure of plant communities, the production of litter, the extent and timing of plant cover and the composition of the soil biotic communities, all of which influence water infiltration and retention in the soil (Power, 2010).

Both crop yields and management practices affect both the quantity and quality of water in an agricultural landscape. Practices that maximize plant cover, such as minimum tillage, polycultures, or agroforestry systems are likely to decrease runoff and increase infiltration (Power, 2010). Irrigation practices also influence runoff, sedimentation, and groundwater levels in the landscape (Power, 2010).

The effect of organic amendment applications on water content and water-holding capacity of soil can be important in agricultural areas, as soil fertility status is strongly related to water availability. In general, addition of OM to soil increases the water holding capacity, because water is held by adhesive and cohesive forces within soil and an increase in the pore space will lead to an increase in water holding capacity of soil (Reicosky, 2003). Hudson (1994) showed that for a 1% increase in OM content, soil water-holding capacity increased by 3.7%; as a consequence, less irrigation water is needed to irrigate the same crop when the OM content of the soil increases (FAO, 2005; Verheijen et al., 2010).

Perennial vegetation in natural ecosystems, such as forests, can regulate the capture, infiltration, retention, and flow of water across the landscape. In forest soils, litter layers protect soil from raindrop splashes by intercepting rainfall, preventing surface sealing and crusting of soil (Sayer, 2006). In addition, these ecosystems typically support high soil infiltration rates because large soil pores are formed by root systems and soil fauna, so surface runoff is a rare process.

Biodiversity

Soil biodiversity depends on plants, directly, through the influence of living plants and, indirectly, by afterlife effects of dead plants, as primary producers provide energy and nutrients and create soil microhabitats. Soil biota severely affects soil ecosystem C, in that it is the prime agent of decomposition of soil organic C and hence of organic C loss by respiration. In addition, soil biota can promote plant growth and consequently C inputs to soil by enabling plants to access nutrients and/or by protecting plants from adverse biotic (e.g., pathogens) and abiotic (e.g., drought stress) conditions (Lynch, 1990; Pinton et al., 2001). Finally, soil biota can help to stabilize soil organic C by promoting aggregate formation (De Deyn, 2013).

There is a growing interest in determining soil microbial diversity (Nannipieri et al., 2014). The level of the diversity of soil microbial communities is supposed to be critical for maintaining soil health and quality because a broad range of microorganisms is involved in essential soil functions. Nowadays there is a particular interest in the relation between microbial diversity, simply defined as the number of species present in the system, and functions in soil (Nannipieri et al., 2003). This is part of a more general concern to conserve biodiversity and its role in maintaining a functional biosphere.

Determining the composition of soil microbial communities is a difficult task, mainly because of their huge phenotypic and genotypic diversity. According to Torsvik and Ovreas (2002), more than 10^9 bacterial cells per g soil can inhibit the top layers, and most of these cells are generally unculturable. However, in the last decades, the development of molecular techniques has been permitted to determine uncultured organisms. Soil metagenomics (extraction from soil and characterization of all microbial DNA) is the approach used to identify microbial genes that are involved in soil processes especially after the development of next generation sequencing techniques (Nannipieri et al., 2014; Schloter et al., 2017). Generally amplicon sequencing is the approach used, based on the subsequent sequencing analysis of PCR amplified marker genes extracted from soil (Schloter et al., 2017). However, detection of genes does not mean that they are active and for determining their expression is needed to use both metatranscriptomic and proteomic approaches (Nannipieri et al., 2014). Nowadays, methods capable of extracting both DNA and RNA from soil can be used to link microbial biodiversity and functionality. However, it is well established that gene function is expressed by the relative encoded proteins and this makes soil proteomics (characterization of all soil proteins) the best approach to determine gene expression in soil and thus determines microbial activity (Benndorf et al., 2007; Johnson-Rollings et al., 2014; Maron et al., 2007; Masciandaro et al., 2008; Nannipieri, 2006; Renella et al., 2014; Wilmes and Bond, 2006). Therefore, proteomics is an ideal approach to characterize functional genomics (Benndorf et al., 2007). The combination of metagenomic, metatranscriptomic, and proteomic may provide the link between microbial community composition and soil function. Functional diversity is an aspect of the overall microbial diversity in soil, and encompasses a range of microbial activities (Nannipieri et al., 2003).

Soil microorganisms are critical for the maintenance of functions in both natural and managed soils, because they are involved in several key processes, such as decomposition of OM, soil structure formation, the cycling of carbon, nitrogen, phosphorus, and sulfur, and toxin removal. Moreover, microorganisms are fundamental in promoting plant growth and in suppressing soil-borne plant diseases (Garbeva et al., 2004).

Microbiologists have studied the impact of microbial diversity on the stability of ecosystem function since the 1960s (Harrison et al., 1968), but only recently the interest has been directed toward the role of microbial diversity on ecological functions and resilience to disturbances in soil ecosystems (Nannipieri et al., 2003). It is well established that activity, biomass, and composition of soil microbial communities are sensitive to a wide range of factors, such as soil overexploitation, decline in soil OM, soil pollution, compaction, erosion, and desertification. Soil OM serves as the basis of all soil life thus when inputs of OM to soil are reduced, soil biomass, and soil biodiversity will decline as a consequence (Jeffery et al., 2010). However, a relationship does not always exist between microbial diversity and content and decomposition of OM. Nannipieri et al. (2003) reported that a reduction in soil microbial diversity may have little effect on overall processes due to the redundancy of soil functionality.

Soil microbiota are not only sensitive to the input of organic substrates but also to physical and chemical disturbances (Bending et al., 2002). Soil tillage and application of fertilizers and pesticides (Culman et al., 2010), amendment with chitin (Hallmann et al., 1999), compost (Schönfeld et al., 2003) or manure, can affect the composition of soil microbial communities. The physicochemical properties of soil, soil particle size distribution, the presence and age of specific plant species, and crop rotations are key determinative factors (Garbeva et al., 2004).

The conversion of natural forest to a cultivation system can change physical, chemical, and biological soil properties, and often these changes are associated with the reduction of OM content, deterioration of soil structure, and the decrease in microbial biomass and activity (Bossio et al., 2005) and changes in the composition of microbial communities, which may lead to reduction in microbial diversity with the formation of a less functionally stable microbial community (Chaer et al., 2009). For example, Bossio et al. (2005) showed a lower soil organic C content, microbial biomass, and cellulases and chitinase activities in an agricultural soil with respect to the respective forest soil. Islam and Weil (2000) observed that the values of most of the measured biological properties were significantly lower in the cultivated soils than in the natural forest, reforested, and grass land soils. The microbial biomass C (MBC) was about 40% lower in soils under cultivation than in soils under natural forest and the effect was more pronounced (69% reduction) on active MBC (AMBC) (Islam and Weil, 2000). Enhanced microbial activities in soils under natural forest, reforestation, and grass depend on greater levels of available organic C. Thus, reforestation of degraded lands not only increased MBC and the active MBC contents, but also increased the labile fraction of organic C. As a result, soil microbial communities under natural forest, reforestation, and grass were more biologically active and less stressed than those in the cultivated soils (Eleftheriadis and Turrión, 2014).

Nusslein and Tiedje (1999) showed that changes in the composition of soil bacterial community correlated with a change from forest to pasture vegetation in a tropical soil. The G + C content of DNA from the pasture soil was significantly higher than that of the forest soil. Both α- and β-proteobacteria dominated in the pasture soil, whereas fibrobacter types were dominant in the forest soil. Bossio et al. (2005) showed that different farming regimes, that is, organic, low-input, and conventional, influence soil phospholipid fatty acids (PLFA) profiles. In particular, monounsaturated fatty acids increased with organic input in organic and low-input systems. Steenwerth et al. (2002) evaluated soil microbial community composition for nine land uses, including irrigated and nonirrigated agricultural sites, non-native annual grassland and relict, and never-tilled or old field perennial grassland. Four soils from eastern Washington State with contrasting soil management (no-till and conventional till) and environmental conditions were analyzed by PLFA and denaturing gradient gel electrophoresis (DGGE) (Ibekwe et al., 2002). No-till soil practices improved biological properties, particularly promoted high microbial biomass, as determined by PLFA analysis, and greater diversity of ammonia-oxidizing bacteria. Bastida et al. (2016) observed that deforestation induced a long-term loss of enzyme activity and bacterial biomass, but increased the bacterial activity as estimated by metaproteomics. Protein abundances analysis revealed that active proteobacteria was higher in forested than deforested soil. Moreover, the amount of cyanobacterial proteins was significantly higher in deforested (7.3%) with respect to forested (0.9%). Interestingly, cyanobacterial proteins involved in C fixation (phycocyanins and photosystem proteins, ribulose 1,5-bisphosphate carboxylase) were only identified in deforested. This result suggests that cyanobacteria have a significant role in the ecosystem functioning and biotic C fixation when soil is deforested.

Agricultural activities are the main negative environmental impacts worldwide (Kiers et al., 2008). Agriculture is the main cause of deforestation (FAO, 2010) and the main threat to bird species; it accounts for about 12% of total anthropogenic of greenhouse gases emissions, and strongly impacts soil C and nutrients cycles (Rey Benayas and Bullock, 2012). Despite, agriculture produces more important crops, the relative practices have environmental impacts on a wide range of ecosystem services, including water quality, environmental pollution, nutrient cycling, soil retention, C sequestration, and biodiversity conservation. In turn, ecosystem services affect agricultural productivity. Biodiversity conservation is an important service, even though it is difficult to pin down precisely (Helm and Hepburn, 2012).

Nutrient Cycling and Soil Fertility

The interactions between the soil mineral matrix, plants, and microbes are responsible for both building and decomposing OM, and thus for the preservation and availability of nutrients (Cotrufo et al., 2013). The increased availability of N through biological nitrogen fixation and P supply by arbuscular mycorrhizal fungi (Gyuricza et al., 2010), are well-studied examples of nutrient cycling processes. In addition, litter decomposition and mineralization is an ecosystem process, resulting in the breakdown of organic materials into their constituents by which nutrients are made available for nutrient cycling.

The two nutrients that most limit biological production in natural and agricultural ecosystems are N and P. After C, N is the most abundant nutrient in all forms of life, because it is contained in proteins, nucleic acids, and other compounds (Galloway et al., 2008). According to Bai et al. (2010), N deposition can directly affect biodiversity loss and can promote the decline in ecosystem resilience. Nitrogen as nitrate is mobile in soil and subjected to leaching losses if not taken up by plants and/or lost through denitrification; other N losses are represented by ammonia volatilization. Plant N introduced into soil as plant debris or rhizodeposition can be microbially mineralized to ammonium or transformed in a slowly metabolized organic N (Stevenson, 1986). In general, nitrate leaching under grassland is greatly reduced as compared to cropland (Kunrath et al., 2015).

As it concerns soil P, soil erosion is one of the most important processes leading to P loss because of the strong association of P with soil particles. In soil permanently covered, little P can be lost through this process (McLauchlan, 2006), while harvesting removal may lead to P depletion in the long-term unfertilized grassland soils (Pätzhold et al., 2013).

Nutrient management has been extensively studied, with the aim of identifying and proposing management practices for improving nutrient use efficiency and productivity and reducing potentially harmful losses to the environment (Van Groenigen et al., 2010;Venterea et al., 2011). Several studies have investigated nutrient cycling in different ecosystems like forests (Ro en et al. 2010), arable agriculture (Sandhu et al., 2010), and permanent grassland (Hamel et al., 2007). For example, the conversion of forestland to cropland, grazing land, and settlements has often resulted in soil degradation and nutrient losses (Dinesh et al., 2003). Indeed, agroecosystems are subjected to removal of nutrient-rich biomass during harvesting and to the increase of decomposition rates by increasing the frequency of tillage and irrigation. The nutrient losses in agroecosystems contrast with the situation of unmanaged, undisturbed ecosystems in which nutrient cycles tend to be more nearly closed, with inputs approximately equal to outputs. On the other hand, many high-intensity farming systems do not retain soil structure and fertility through biological processes, and nutrient inputs are maintained through tillage (OM mineralization) and additions of chemical and organic fertilizers (Matson et al., 1997). It is well known that management practices, such as compost applications, are being encouraged as a means to improve soil fertility, thereby buffering against the impacts of increasing climate variability (Ng et al., 2015).

To maintain ecosystem services, soil nutrient pools can be intentionally managed to supply crops at the right time, while minimizing nutrient losses by reducing soluble inorganic N and P pools (Drinkwater and Snapp, 2007). For example, cover cropping or intercropping can maintain soil fertility by enhancing plant and microbial assimilation of N and reducing standing pools of nitrate. Other good management practices include diversifying nutrient sources, legume intensification for biological N fixation and P solubilizing properties, and diversifying rotations. Integrated management of biogeochemical processes that regulate the cycling of nutrients and C can reduce the need for surplus nutrient additions in agriculture (Drinkwater and Snapp, 2007).

As already mentioned, the management of SOC in agroecosystems can affect the productive capacity of land as a final ecosystem service by improving the growth conditions for crops and therefore yields, and by increasing nutrient use efficiency that may affect the amount of fertilizer input required for optimal plant growth (Pan et al., 2010). Glenk et al. (2017) investigate farm gross margin effects of management measures aimed at enhancing SOC stocks to maintain soil fertility while providing important ecosystem services. They used a farm-level model to simulate the farm gross margin effects of selected SOC management measures for arable farms in Scotland (UK) and Aragon (Spain). Tillage, fertilizer management, and crop rotations (with legumes) were management measures with positive effects on farm gross margins. Residue management can have a negative effect on farm gross margins. Results of the sensitivity analysis indicate that effects of SOC management on farm gross margins are more sensitive to a change in crop yields than to changes in input costs (Glenk et al., 2017).

CONCLUSIONS

Soil C can protect the environment from global changes. Indeed small relative changes in soil abundance content and composition can affect Earth ecosystem processes. Organic C loss from soil generally increases carbon dioxide emissions. On the other hand, the formation and conservation of soil organic C improves soil stability, primary production, C sequestration, water-holding capacity, biodiversity, nutrient cycles, and soil fertility, and therefore allows the maintaining of ecosystem services essential to human well being.

OM can bind clay and silt components to form soil colloidal organo-mineral aggregates, which improve soil structure, facilitate water infiltration, help hold water, and protect organic C from mineralization. OM also provides cohesive strength to soil thus improving the resistance of soil to erosion. In addition, OM provides nutrients and physical protection for extracellular enzymes, thus sustaining the biochemical processes carried out by soil organisms and plant roots. The availability of C and nutrients in soil, and especially in the rhizosphere, strongly affects activity, biomass, and composition of microbial communities. Soil management can significantly affect the relative balance of these soil processes. When soils are converted to agriculture, C inputs from plants and activity and biomass of microbial communities decline with shifts in microbial diversity; in addition, cultivation breaks up existing soil aggregates, leaving C within aggregates more vulnerable to decomposition.

The irrational land-use change and wrong management options are adverse to maintaining of ecosystem services, mainly resulting in soil C losses; however, suitable measures, such as the use of crop residues, animal manure, compost, and, sewage sludge, can increase soil C storage.

Despite its importance, the size of the global soil organic C stock and its distribution in space and among land-use/land-cover classes is poorly known, hence possible carbon dioxide emissions result from soil due to changes in land use and land cover.

Future research should address processes, such as formation, and conservation of soil OM sink, and the relative new findings should be integrated in decision support tools for policy makers and disseminated and adopted by land managers at various levels. A better understanding of soil organic C stocks and fluxes could allow soil scientists and ecologists to monitor soil status and to predict ecosystem behavior toward climate changes. Moreover, these findings will give a solid base to assist decision makers and land managers to choose the more sustainable land-use and management options. As recommended by Schmidt et al. (2011), new research should be addressed to: (1) study the processes driving SOM stabilization and destabilization; (2) develop new soil models representing the mechanisms driving soil response to global change; and (3) connect the different research communities that are involved in studying SOM cycles and terrestrial ecology.

On the basis of the continuous and higher demand of ecosystem services arising from climate change and population increase, a rational and aware management of soil C will represent one of the key strategies to reach the UN sustainable development goals in a more effective way.

REFERENCES

Acosta-Martínez, V., Klose, S., Zobeck, T.M., 2003. Enzyme activities in semiarid soils under conservation reserve program, native rangeland, and cropland. J. Plant Nutr. Soil Sci. 166, 699–707.

Adhikari, K., Hartemink, A.E., 2016. Linking soils to ecosystem services: a global review. Geoderma 262, 101–111. doi: 10.1016/j.geoderma.2015.08.009.

An, S.S., Mentler, A., Acosta-Martinez, V., Blum, W.E.H., 2009. Soil microbial parameters and stability of soil aggregate fractions under different plant communities of grassland soils on the Loess Plateau, China. Biologia 64 (3), 424–427.

An, S., Mentler, A., Mayer, H., Blum, W.E.H., 2010. Soil aggregation, aggregate stability, organic carbon and nitrogen in different soil aggregate fractions under forest and shrub vegetation on the Loess Plateau. China Catena 81 (3), 226–233. doi: 10.1016/j.catena.2010.04.002.

Ayoubi, S., Karchegani, P.M., Mosaddeghi, M.R., Honarjoo, N., 2012. Soil aggregation and organic carbon as affected by typography and land use change in western Iran. Soil Till. Res. 121, 18–26. doi: 10.1016/j.still.2012.01.011.

Bai, Y., Wu, J., Clark, C.M., Naeem, S., Pan, Q., Huang, J., et al., 2010. Tradeoffs and thresholds in the effects of nitrogen addition on biodiversity and ecosystem functioning: evidence from inner Mongolia Grasslands. Global Change Biol. 16 (1), 358–372. doi: 10.1111/j.1365-2486.2009.01950.x.

Balabane, M., Plante, A.F., 2004. Aggregation and carbon storage in silty soil using physical fractionation techniques. Eur. J. Soil Sci. 55, 415–427. doi: 10.1111/j.1365-2389.2004.00608.x.

Bastida, F., Torres, I.F., Moreno, J.L., Baldrian, P., Ondoño, S., Ruiz-Navarro, A., et al., 2016. The active microbial diversity drives ecosystem multifunctionality and is physiologically related to carbon availability in Mediterranean semi-arid soils. Mol. Ecol. 25 (18), 4660–4673. doi: 10.1111/mec.13783.

Batjes, N.H., 1996. Total carbon and nitrogen in the soils of the world. Eur. J. Soil Sci. 47, 151–163.

Batjes, N.H., 2016. Harmonized soil property values for broad-scale modelling (WISE30sec) with estimates of global soil carbon stocks. Geoderma 269 (C), 61–68. doi: 10.1016/j.geoderma.2016.01.034.

Bending, G.D., Turner, M.K., Jones, J.E., 2002. Interactions between crop residue and soil organic matter quality and the functional diversity of soil microbial communities. Soil Biol. Biochem. 34, 1073–1082.

Benítez, E., Melgar, R., Nogales, R., 2004. Estimating soil resilience to a toxic organic waste by measuring enzyme activities. Soil Biol. Biochem. 36, 1615–1623.

Benítez, E., Nogales, R., Masciandaro, G., Ceccanti, B., 2000. Isolation by isoelectric focusing of humic urease complexes from earthworm (*Eisenia fetida*)-processed sewage sludges. Biol. Fert. Soils 31, 489–493. doi: 10.1007/s003740000197.

Benndorf, D., Balcke, G.U., Harms, H., von Bergen, M., 2007. Functional metaproteome analysis of protein extracts from contaminated soil and groundwater. ISME J. 1, 224–234.

Birgé, H.E., Bevans, R.A., Allen, C.R., Angeler, D.G., Baer, S.G., Wall, D.H., 2016. Adaptive management for soil ecosystem services. J. Environ. Manage. 183 (2), 371–378. doi: 10.1016/j.jenvman.2016.06.024.

Blanco-Moure, N., Gracia, R., Bielsa, A.C., Lopez, M.V., 2016. Soil organic matter fractions as affected by tillage and soil texture under semiarid Mediterranean conditions. Soil Till. Res. 155, 381–389.

Blum, W.E.H., 2005. Functions of soil for society and the environment. Rev. Environ. Sci. Biotechnol. 4 (3), 75–79. doi: 10.1007/s11157-005-2236-x.

Bolin, B., Sukumar, R., 2000. Global perspective. In: Watson, R.T., Noble, I.R., Bolin, B., Ravindranath, N.H., Verardo, D.J., Doken, D.J. (Eds.), Land Use, Land-Use Change, and Forestry. Cambridge University Press, Cambridge, pp. 23–51.

Bossio, D.A., Girvan, M.S., Verchot, L., Bullimore, J., Borelli, T., Albrecht, A., et al., 2005. Soil microbial community response to land use change in an agricultural landscape of western Kenya. Microb. Ecol. 49, 50–62.

Bronick, G.J., Lal, R., 2005. Manuring and rotation effect on soil organic carbon concentration for different aggregate size fractions on two soils northeastern Ohio, USA. Soil Till. Res. 81, 239–252.

Brown, S., Cotton, M., 2011. Changes in soil properties and carbon content following compost application: results of on-farm sampling. Compost Sci. Util. 19, 88–97.

Caride, C., Piñeiro, G., Paruelo, J.M., 2012. How does agricultural management modify ecosystem services in the argentine Pampas? The effects on soil C dynamics. Agr. Ecosyst. Environ. 154, 23–33. doi: 10.1016/j.agee.2011.05.031.

Ceccanti, B., Doni, S., Macci, C., Cercignani, G., Masciandaro, G., 2008. Characterization of stable humic-enzyme complexes of different soil ecosystems through analytical isoelectric focussing technique (IEF). Soil Biol. Biochem. 40 (9), 2174–2177.

Ceccanti, B., Masciandaro, G., 2003. Stable humus-enzyme nucleus: the last barrier against soil desertification. In: Lobo, M.C., Ibanez, J.J. (Eds.), Preserving Soil Quality and Soil Biodiversity: The Role of Surrogate Indicators. CSICIMIA, Madrid, Spain, pp. 77–82.

Cerri, C.E.P., Easter, M., Paustian, K., Killan, K., Coleman, K., Bernoux, M., et al., 2007. Simulating SOC changes in 11 land use change chronosequences from the Brazilian Amazon with RothC and Century models. Agric. Ecosyst. Environ. 122, 46–57.

Chaer, G., Fernandes, M., Myrold, D., Bottomley, P., 2009. Comparative resistance and resilience of soil microbial communities and enzyme activities in adjacent native forest and agricultural soils. Microb. Ecol. 58, 414–424.
Chazdon, R.L., 2008. Beyond deforestation: restoring forests and ecosystem services on degraded lands. Science 320, 1458–1460.
Clothier, B.E., Hall, A.J., Deurer, M., Green, S.R., Mackay, A.D., 2011. Soil ecosystem services: Sustaining returns on investment into natural capital. Sustaining Soil Productivity in Response to Global Climate Change: Science, Policy, and Ethics. John Wiley & Sons Inc, pp. 117–139. doi: 10.1002/9780470960257.
Conant, R.T., Ryan, M.G., Agren, G.I., Birge, H.E., Davidson, E.A., Eliasson, P.E., et al., 2011. Temperature and soil organic matter decomposition rates synthesis of current knowledge and a way forward. Global Change Biol. 17, 3392–3404.
Costanza, R., Daly, H., 1992. Natural capital and sustainable development. Conserv. Biol. 6, 37–46.
Costanza, R., d'Arge, R., de Groot, R., Farber, S., Grasso, M., Hannon, B., et al., 1997. The value of the world's ecosystem services and natural capital. Nature 387 (6630), 253–260. doi: 10.1038/387253a0.
Cotrufo, M.F., Wallenstein, M.D., Boot, C., Denef, K., Paul, E., 2013. The microbial efficiency-matrix stabilization (MEMS) framework integrates plant litter decomposition with soil organic matter stabilization: do labile plant inputs form stable organic matter? Glob. Change Biol. 19, 988–995.
Crowder, D.W., Northfield, T.D., Strand, M.R., Snyder, W.E., 2010. Organic agriculture promotes evenness and natural pest control. Nature 466, 109–112.
Culman, S.W., DuPont, S.T., Glover, J.D., Buckley, D.H., Fick, G.W., Ferris, H., et al., 2010. Long-term impacts of high-input annual cropping and unfertilized perennial grass production on soil properties and belowground food webs in Kansas, USA. Agric. Ecosyst. Environ. 137, 13–24.
Daily, G.C., 1997. Nature's Services: Societal Dependence on Natural Ecosystems. Island Press, Washington, DC.
De Brogniez, D., Ballabio, C., Stevens, A., Jones, R.J.A., Montanarella, L., van Wesemael, B., 2015. A map of the topsoil organic carbon content of Europe generated by a generalized additive model. Eur. J. Soil Sci. 66, 121–134. doi: 10.1111/ejss.12193.
De Deyn, G.B., 2013. Ecosystem Carbon and Soil Biodiversity. In: Lal, R., Lorez, K., Hüttl, R.F., Schneider, B.U., Von Braun, J. (Eds.), Ecosystem Services and Carbon Sequestration in the Biosphere. Springer Science + Business Media, Dordrech, pp. 131–153.
De Groot, R.S., 1987. Environmental functions as a unifying concept for ecology and economics. Environmentalist 7 (2), 105–109.
De Groot, R.S., 1992. In: Groningen, Wolters-Noordhoff (Ed.), Functions of Nature: Evaluation of Nature in Environmental Planning, Management, and Decision Makingdoi: 10.1017/S0032247400023779.
Derenne, S., Largeau, C., 2001. A review of some important families of refractory macromolecules: composition, origin, and fate in soils and sediments. Soil Sci. 166 (11), 833–847.
Dinesh, R., Ghoshal Chaudhuri, S., Ganeshamurthy, A.N., Dey, C., 2003. Changes in soil microbial indices and their relationships following deforestation and cultivation in wet tropical forests. Appl. Soil Ecol. 24, 17–26.
DOE, 1999. Carbon sequestration: research and development. AUS Department of Energy Report Office of Science and Office of Fossil Energy. National Technical Information Service, Springfield, VA.
Dominati, E.J., 2013. Natural capital and ecosystem services of soils. In: Dymond, J.R. (Ed.), Ecosystem Services in New Zealand: Conditions and Trends. Manaaki Whenua Press, Lincoln, New Zealand.
Dominati, E.J., Patterson, M., Mackay, A., 2010. A framework for classifying and quantifying the natural capital and ecosystem services of soils. Ecol. Econ. 69, 1858–1868.

Don, Schumacher, A., Freibauer, J.A., 2011. Impact of tropical land-use change on soil organic carbon stocks: a meta-analysis. Global Change Biol. 17, 1658–1670.

Doni, S., Macci, C., Peruzzi, E., Ceccanti, B., Masciandaro, G., 2014. Factors controlling carbon metabolism and humification in different soil agroecosystems. Sci. World J. 2014doi: 10.1155/2014/416074.

Doran, J.W., Jones, A.J. (Eds.), 1996. Methods for Assessing Soil Quality. Soil Science Society of America, Madison, Wisconsin, Special Publication 49: Soil Science Society of America Special Publication.

Drinkwater, L.E., Snapp, S.S., 2007. Nutrients in agroecosystems: re-thinking the management paradigm. Adv. Agron. 92, 163–186. doi: 10.1016/S0065-2113(04)92003-2.

Dungait, J.A.J., Hopkins, D.W., Gregory, A.S., Whitmore, A.P., 2012. Soil organic matter turnover is governed by accessibility not recalcitrance. Global Change Biol. 18, 1781–1796.

du Preez, C.C., van Huyssteen, C.W., Mnkeni, P.N.S., 2011. Land use and soil organic matter in South Africa 1: a review on spatial variability and the influence of rangeland stock production. S. Afr. J. Sci. 107 (5/6), 1–8. doi: 10.4102/sajs.v107i5/6.354.

Ehrlich, P.R., Ehrlich, A.H., 1981. Extinction: The Causes and Consequences of the Disappearance of Species. Random House, New York.

Eleftheriadis, A., Turrión, M.B., 2014. Soil microbiological properties affected by land use, management, and time since deforestations and crop establishment. Eur. J. Soil Biol. 62, 138–144.

Elliot, W.J., 2003. Soil erosion in forest ecosystems and carbon dynamics. In: Kimble, J.M., Heath, L.S., Birdsey, R.A., Lal, R. (Eds.), The Potential of US Forest Soils to Sequester Carbon and Mitigate the Greenhouse Effect. CRC Press, Boca Raton, FL, pp. 175–190.

Emmett, B.A., Cooper, D., Smart, S., Jackson, B., Thomas, A., Cosby, B., et al., 2016. Spatial patterns and environmental constraints on ecosystem services at a catchment scale. Sci. Total Environ. 572 (C), 586–1600. doi: 10.1016/j.scitotenv.2016.04.004.

EPA, 2006. Land use, land use change and forestry. Available from: http://epa.gov./climatechange/emissions/downloads06/06LULUCF.pdf.

Falloon, P., Jones, C.D., Ades, M., Paul, K., 2011. Direct soil moisture controls of future global soil carbon changes: an important source of uncertainty. Global Biogeochem. Cycles 25, 1–14.

Fantappiè, M., L'Abate, G., Costantini, E.A.C., 2010. Factors influencing soil organic carbon stock variations in Italy during the last three decades. In: Zdruli, P. (Ed.), Land Degradation and Desertification: Assessment, Mitigation and Remediationpp. 435–465. doi: 10.1007/978-90-481-8657-0_34.

Fantappiè, M., L'Abate, G., Costantini, E.A.C., 2011. The influence of climate change on the soil organic carbon content in Italy from 1961 to 2008. Geomorphology 135, 343–352, http://dx.doi.org/10.1016/j.geomorph.2011.02.006.

FAO, 2005. The Importance of Soil Organic Matter Key to Drought-Resistant Soil. FAO Soil Bulletin 80, Rome, Italy.

FAO, 2010. Global Forest Resources Assessment 2010 Main Report. FAO, Rome, Italy.

FAO and ITPS, 2015. Status of the World's Soil Resources (SWSR): Main Report. Food and Agriculture Organization of the United Nations and Intergovernmental Technical Panel on Soils, Rome, Italy.

Field, C.B., Raupach, M.R., 2004. The Global Carbon Cycle: Integrating Humans, Climate and the Natural World. Island Press, Washington, DC.

Fisher, G., Nachtergaele, F., Prieler, S., van Velthuizen, H.T., Verelst, L., Wiberg, D., 2008. Global Agro-ecological Zones Assessment for Agriculture. IIASA, Laxenburg, Austria and FAO, Rome, Italy.

Forest Europe, UNECE, and FAO, State of Europe's Forests 2011: Status and Trends in Sustainable Forest Management in Europe, 2011. Ministerial Conference on the Protection of Forests, Norway. www.foresteurope.org. ISBN 978-82-92980-05-7.

Fortuna, A.M., Paul, E.A., Harwood, R.R., 2003. The effects of compost and crop rotations on carbon turnover and the particulate organic matter fraction. Soil Sci. 168, 434–444.
Freibauer, A., Rounsevell, M.D.A., Smith, P., Verhagen, J., 2004. Carbon sequestration in the agricultural soils of Europe. Geoderma 122, 1–23.
Galati, A., Crescimanno, M., Gristina, L., Keesstra, S., Novara, A., 2016. Actual provision as an alternative criterion to improve the efficiency of payments for ecosystem services for C sequestration in semiarid vineyards. Agric. Syst. 144, 58–64.
Galloway, J.N., Townsend, A.R., Erisman, J.W., Bekunda, M., Cai, Z., Freney, J.R., et al., 2008. Transformation of the nitrogen cycle: recent trends, questions, and potential solutions. Science 320, 889–892.
Garbeva, P., van Veen, J.A., van Elsas, J.D., 2004. Microbial diversity in soil: selection of microbial populations by plant and soil type and implications for disease suppressiveness. Annu. Rev. Phytopathol. 42, 243–270.
García, A.C., de Souza, L.G., Pereira, M.G., Castro, R.N., García-Mina, J.M., Zonta, E., et al., 2016. Structure-property-function relationship in humic substances to explain the biological activity in plants. Sci. Rep. 6, 20798.
Glenk, K., Shrestha, S., Topp, C.F.E., Sánchez, B., Iglesias, A., Dibari, C., et al., 2017. A farm level approach to explore farm gross margin effects of soil organic carbon management. Agr. Syst. 151, 33–46.
Gol, C., 2009. The effects of land use change on soil properties and organic carbon at Dagdami river catchment in Turkey. J. Environ. Biol. 30, 825–830.
Gómez-Baggethun, E., de Groot, R., Lomas, P.L., Montes, C., 2010. The history of ecosystem services in economic theory and practice: from early notions to markets and payment schemes. Ecol. Econ. 69 (6), 1209–1218.
Gonzalez-Sanchez, E.J., Ordonez-Fernandez, R., Carbonell-Bojollo, R., Veroz-Gonzalez, O., Gil-Ribes, J.A., 2012. Meta-analysis on atmospheric carbon capture in Spain through the use of conservation agriculture. Soil Till. Res. 122, 52–60.
Govaerts, B., Verhulst, N., Castellanos-Navarrete, A., Sayre, K.D., Dixon, J., Dendooven, L., 2009. Conservation agriculture and soil carbon sequestration: between myth and farmer reality. Crit. Rev. Plant Sci. 28, 97–122.
Govers, G., Merckx, R., van Ost, L., van Wesemael, B., 2012. Managing Soil Organic Carbon for Global Benefits: ASTAP Technical Report. Global Environmental Facility, Washington, DC, p. 70 http://www.thegef.org/gef/sites/thegef.org/files/publication/STAP-SOC-Report-lowres-1.pdf.
Grandy, A.S., Neff, J.C., 2008. Molecular soil C dynamics downstream: the biochemical decomposition sequence and its effects on soil organic matter structure and function. Sci. Total Environ. 404, 297–307.
Grego, S., Lagomarsino, A., 2008. Soil organic matter in the sustainable agriculture: source or sink of carbon? Marinari, S., Caporali, S. (Eds.), Soil Carbon Sequestration Under Organic Farming in the Mediterranean Environment, 3, Transworld Research Network, Kerala, pp. 39–51.
Gu, B.H., Schmitt, J., Chen, Z., Liang, L., McCarthy, J.F., 1994. Adsorption and desorption of natural organic matter on iron oxide: mechanisms and models. Environ. Sci. Technol. 28, 38–46.
Guibert, H., Fallavier, P., Romero, J.J., 1999. Carbon content in soil particle size and consequence on cation exchange capacity of alfisols. Commun. Soil Sci. Plan. 30 (17–18), 2521–2537.
Guo, L.B., Gifford, R.M., 2002. Soil carbon stocks and land use change: a meta analysis. Global Change Biol. 8 (4), 345–360.
Gyuricza, V., de Boulois, H.D., Declerck, S., 2010. Effect of potassium and phosphorus on the transport of radiocesium by arbuscular mycorrhizal fungi. J. Environ. Radioact. 101, 482–487.

Hallmann, J., Rodriguez-Kabana, R., Kloepper, J.W., 1999. Chitin-mediated changes in bacterial communities of soil rhizosphere and roots of cotton in relation to nematode control. Soil Biol. Biochem. 31, 551–560.

Hamel, C., Schellenberg, M.P., Hanson, K., Wang, H., 2007. Evaluation of the "bait-lamina test" to assess soil microfauna feeding activity in mixed grassland. Appl. Soil Ecol. 36, 199–204.

Hansen, M.C., Potapov, P.V., Moore, R., Hancher, M., Turubanova, S.A., Tyukavina, A., 2013. High-resolution global maps of 21st-century forest cover change. Science 342, 850–853.

Harrison, N.G., Allan, J.D., Colwell, R.K., Futuyma, D.J., Howell, J., 1968. The relationship between species diversity and stability: an experimental approach with protozoa and bacteria. Ecology 49, 1091–1101.

Hartemink, A.E., Veldkamp, T., Bai, Z., 2008. Land cover change and soil fertility decline in tropical regions. Turk. J. Agric. For. 32, 195–213.

Heitkamp, F., Wendland, M., Offenberger, K., Gerold, G., 2012. Implications of input estimation, residue quality and carbon saturation on the predictive power of the Rothamsted Carbon Model. Geoderma 170, 168–175.

Helm, D., Hepburn, C., 2012. The economic analysis of biodiversity: an assessment. Oxford Rev. Econ. Pol. 28 (1), 1–21.

Hiederer, R., Köchy, M., 2011. Global soil organic carbon estimates and the harmonized world soil database. JRC Scientific and Technical Reports. Joint Research Centre, Ispra, Italy, 68528/EUR 25225 EN, Joint Research Centre.

Hoover, C.M., 2003. Soil carbon sequestration and forest management: challenges and opportunities. In: Kimble, J.M., Heath, L.S., Birdsey, R.A., Lal, R. (Eds.), The Potential of U.S. Forest Soils to Sequester Carbon and Mitigate the Greenhouse Effect. CRC Press, Boca Raton, FL, pp. 211–238.

Hudson, B.D., 1994. Soil organic matter available water capacity. J. Soil Water Conserv. 2, 189–194.

Hugelius, G., Strauss, J., Zubrzycki, S., Harden, J.W., Schuur, E.A.G., Ping, C.L., et al., 2014. Estimated stocks of circumpolar permafrost carbon with quantified uncertainty ranges and identified data gaps. Biogeosciences 11, 6573–6593.

Hutyra, L., Yoon, B., Alberti, M., 2011. Terrestrial carbon stocks across a gradient of urbanization: a study of the Seattle, WA Region. Glob. Change Biol. 17 (2), 783–797.

Ibekwe, A.M., Kennedy, A.C., Frohne, P.S., Papriernik, S.K., Yang, C.H., 2002. Microbial diversity along a transect of agronomic zones. FEMS Microb. Ecol. 39, 183–191.

IPCC, 2000. Land use, land use change and forestry: special report. Inter-Governmental Panel on Climate Change. Cambridge University Press, Cambridge, UK, pp. 127-180.

IPCC, 2013. Climate change the physical science basis. Contribution of Working Group I to the Fifth Assessment Report of the Intergovernmental Panel on Climate Change. Cambridge University Press, Cambridge, United Kingdom and New York, NY, USA.

Islam, K.R., Weil, R.R., 2000. Land use effects on soil quality in a tropical forest ecosystem of Bangladesh. Agr. Ecosyst. Environ. 79, 9–16.

Jandl, R., Rodeghiero, M., Martinez, C., Cotrufo, M.F., Bampa, F., van Wesemael, B., et al., 2014. Current status, uncertainty and future needs in soil organic carbon monitoring. Sci. Total Environ. 468–469, 376–383.

Jarvis, D.I., Brown, A.H.D., Cuong, P.H., Collado-Panduro, L., Latournerie-Moreno, L., Gyawali, S., et al., 2008. A global perspective of the richness and evenness of traditional crop-variety diversity maintained by farming communities. Proc. Natl. Acad. Sci. USA 105, 5326–5331.

Jastrow, J.D., Amonette, J.E., Bailey, V.L., 2007. Mechanisms controlling soil carbon turnover and their potential application for enhancing carbon sequestration. Climatic Change 80, 5–23.

Jauhiainen, J., Hooijer, A., Page, S.E., 2011. Carbon dioxide emissions from an Acacia plantation on peatland in Sumatra. Indo. Biogeosci. Disc. 8, 8269–8302.

Jeffery, S., Gardi, C., Jones, A., Montanarella, L., Marmo, L., Miko, L., et al., 2010. Threats to soil biodiversity. European Atlas of Soil Biodiversity. European Commission Publications Office of the European Union, Luxembourg, pp 52-65.

Jobbágy, E.G., Jackson, R.B., 2000. The vertical distribution of soil organic carbon and its relation to climate and vegetation. Ecol. Appl. 10, 423–436.

Johnson-Rollings, A.S., Wright, H., Masciandaro, G., Macci, C., Doni, S., Calvo-Bado, L.A., 2014. Exploring the functional soil-microbe interface and exoenzymes through soil metaexoproteomics. ISME J. 8, 2148–2150.

Jónsson, J.Ö.G., DavíÐsdóttir, B., 2016. Classification and valuation of soil ecosystem services. Agr. Syst. 145, 24–38.

Jungkunst, H.F., Krüger, J.P., Heitkamp, F., Erasmi, S., Glatzel, S., Fiedler, S., et al., 2012. Accounting more precisely for peat and other soil carbon sources. In: Lal, R., Lorenz, K., Hüttl, R.F., Schneider, B.U., Von Braun, J. (Eds.), Recarbonization of the Biosphere: Ecosystems and the Global Carbon Cycle. Springer, Dordrecht, The Netherlands, p. 127- L157.

Kaiser, K., Kalbitz, K., 2012. Cycling downwards-dissolved organic matter in soils. Soil Biol. Biochem. 52, 29–32.

Kareiva, P.M., Tallis, H., Ricketts, T.H., Daily, G.C., Polasky, S., 2011. Natural Capital: Theory and Practice of Mapping Ecosystem Services. Oxford University Press, Oxford, doi: 0.1093/acprof:oso/9780199588992.001.0001.

Kasang, D., 2004. Climate change and the greenhouse effect. Available from: http://lbs.hh.schule.de/welcome.phtml?unten=/klima/greenhouse/.

Kasting J.F., The carbon cycle, climate and the long-term effects of fossil fuel burnings, Consequences: The Nature and Implications of Environmental Change, 1998, Vol. 4, No. 1, Washington, DC. http://www.atmo.arizona.edu/students/courselinks/fall07/atmo551a/pdf/CarbonCycle.html.

Keesstra, S.D., Bouma, J., Wallinga, J., Tittonell, P., Smith, P., Cerdà, A., et al., 2016. The significance of soils and soil science towards realization of the United Nations Sustainable Development Goals. Soil 2 (2), 111–128. doi: 10.5194/soil-2-111-2016-supplement.

Kiers, E.T., Leakey, R.R.B., Izac, A.M., Heinemann, J.A., Rosenthal, E., Nathan, D., et al., 2008. Ecology – agriculture at a cross roads. Science 320, 320–321.

Kleber, M., Mikutta, R., Torn, M.S., Jahn, R., 2005. Poorly crystalline mineral phases protect organic matter in acid subsoil horizons. Eur. J. Soil Sci. 56, 717–725.

Kleber, M., Sollins, P., Sutton, R., 2007. A conceptual model of organo-mineral interactions in soils: self-assembly of organic molecular fragments into zonal structures on mineral surfaces. Biogeochemistry 85, 9–24.

Koven, C.D., Ringeval, B., Friedlingstein, P., Ciais, P., Cadule, P., Khvorostyanov, D., et al., 2011. Permafrost carbon-climate feedbacks accelerate global warming. Proc. Natl. Acad. Sci. USA 108, 14769–14774. doi: 10.1073/pnas.1103910108, 2011.

Kunrath, T.R., de Berranger, C., Charrier, X., Gastal, F., de Faccio Carvalho, P.C., Lemaire, G., et al., 2015. How much do sod-based rotations reduce nitrate leaching in a cereal cropping system? Agr. Water Manage. 150, 46–56. doi: 10.1016/j.agwat.2014.11.015.

Ladd, J.N., Foster, R., Nannipieri, P., Oades, J.M., 1996. Soil structure and biological activity. In: Stotzky, G., Bollag, J.-M. (Eds.), Soil Biochemistry. Marcel Dekker, New York, pp. 23–78, Vol. 9.

Lal, R., 1996. Deforestation and land-use effects on soil degradation and rehabilitation in western Nigeria. I. Soil physical and hydrological properties. Land Degrad. Dev. 7, 19–45.

Lal, R., 2000. Soil conservation and restoration to sequester carbon and mitigate the greenhouse effect. In: Rubio, J.L., Asins, S., Andreu, V., de Paz, J.M., Gimeno, E. (Eds.), Key Notes. 3rd International Congress European Society for Soil Conservation: Man and Soil at the Third Millennium, Valencia, pp. 5–20.

Lal, R., 2002. Introduction. In: Kimble, J.M., Lal., R., Follett, R.F. (Eds.), Agriculture Practices and Policies for Carbon Sequestration in Soil. CRC Press, Boca Raton, FL, pp. 3–5.
Lal, R., 2003. Soil erosion and the global carbon budget. Environ. Int. 29, 437–450.
Lal, R., 2005. Forest soils and carbon sequestration. Forest Ecol. Manage. 220, 242–258.
Lal, R., 2016. Soil health and carbon management. Food Energy Sec. 5 (4), 212–222, http://doi.org/10.1002/fes3.96.
Larionova, A.A., Rozanova, L.N., Evdokimov, I.V., Ermolaev, A.M., 2002. Carbon budget in natural and anthropogenic forest-steppe ecosystems. Pochvovedenie 2, 177–185.
Luo, Y., Xueyong, Z., Olof, A., Yangchun, Z., Wenda, H., 2014. Artificial root exudates and soil organic carbon mineralization in a degraded sandy grassland in northern China. J. Arid Land 6, 423–431.
Lynch, J.M., 1990. The Rhizosphere. John Wiley & Sons, New York.
Malhi, Y., Doughty, C., Galbraith, D., 2011. The allocation of ecosystem net primary productivity in tropical forests. Philos. Trans. R. Soc. B 366, 3225–3245.
Mambelli, S., Bird, J.A., Gleixner, G., Dawson, T.E., Torn, M.S., 2011. Relative contribution of foliar and fine root pine litter to the molecular composition of soil organic matter after in situ degradation. Organ. Geochem. 42, 1099–1108.
Maron, P.A., Ranjard, L., Mougel, C., Lemanceau, P., 2007. Metaproteomics: a new approach for studying functional microbial ecology. Microbial Ecol. 53, 486–493.
Masciandaro, G., Macci, C., Doni, S., Maserti, B.E., Calvo-Bado, L.A., Ceccanti, B., Wellington, E., 2008. Comparison of extraction methods for recovery of extracellular boxx-glucosidase in two different forest soils. Soil Biol. Biochem. 40, 2156–2161.
Matson, P., Parton, W., Power, A., Swift, M., 1997. Agricultural intensification and ecosystem properties. Science 277, 504–509.
McLauchlan, K., 2006. The nature and longevity of agricultural impacts on soil carbon and nutrients: a review. Ecosystems 9, 1364–1382.
MEA, 2005. Millennium ecosystem assessment. Ecosystem and Human Well-being: A Framework for Assessment. Island Press.
Mikha, M.M., Rice, C.W., 2004. Tillage and manure effects on soil and aggregate-associated carbon and nitrogen. Soil Sci. Soc. Am. J. 68, 809–816.
Mikutta, R., Klebber, M., Torn, M., Jahn, R., 2006. Stabilization of soil organic matter: association with minerals or chemical recalcitrance? Biogeochemistry 77, 25–56.
Miltner, A., Bombach, P., Schmidt-Brücken, B., Kastner, M., 2011. SOM genesis: microbial biomass as a significant source. Biogeochemistry 111, 41–45.
Mohammadi, J., Motaghian, M., 2011. Spatial prediction of soil aggregate stability and aggregate-associated organic carbon content at the catchment scale using geostatistical techniques. Pedosphere 21, 389–399.
Muñoz-Rojas, M., Jordán, A., Zavala, L.M., De la Rosa, D., Abd-Elmabod, S.K., Anaya-Romero, M., 2012. Organic carbon stocks in Mediterranean soil types under different land uses (Southern Spain). Solid Earth 3, 375–386.
Murphy, C.J., Baggs, E., Morley, N., Muro, D., Paterson, E., 2015. Rhizosphere priming can promote mobilisation of N-rich compounds from soil organic matter. Soil Biol. Biochem. 81, 236–243.
Nannipieri, P., 2006. Role of stabilized enzymes in microbial ecology and enzyme extraction from soil with potential applications in soil proteomics. In: Nannipieri, P., Smalla, K. (Eds.), Soil Biology Volume 8: Nucelic Acids and Proteins in Soil. Springer Verlag, Heidelberg, Germany, pp. 75–94.
Nannipieri, P., Asher, J., Ceccherini, M.T., Landi, L., Pietramellara, G., Renella, G., 2003. Microbial diversity and soil functions. Eur. J. Soil Sci. 54, 655–670.
Nannipieri, P., Pietramellara, G., Renella, G., 2014. Omics in Soil Science. Caster Academic Press, Norfolk, UK.
Nannipieri, P., Sequi, P., Fusi, P., 1996. Humus and enzyme activity. In: Piccolo, A. (Ed.), Humic Substances in Terrestrial Ecosystems. Elsevier, Amsterdam, pp. 293–328.

Ng, E.L., Patti, A.F., Rose, M.T., Schefe, C.R., Smernik, R.J., Cavagnaro, T.R., 2015. Do organic inputs alter resistance and resilience of soil microbial community to drying? Soil Biol. Biochem. 81 (C), 58–66, http://doi.org/10.1016/j.soilbio.2014.10.028.
Noellemeyer, E., Frank, F., Alvarez, C., Morazzo, G., Quiroga, A., 2008. Carbon contents and aggregation related to soil physical and biological properties under a land-use sequence in the semiarid region of Central Argentina. Soil Till. Res. 99, 179–190.
Novara, A., Gristina, L., Kuzyakov, Y., Schillaci, C., Laudicina, V.A., La Mantia, T., 2013. Turnover and availability of soil organic carbon under different Mediterranean land-uses as estimated by 13 C natural abundance. Eur. J. Soil Sci. 64, 466–475, http://dx.doi.org/10.1111/ejss.12038.
Novara, A., La Mantia, T., Rühl, J., Badalucco, L., Kuzyakov, Y., Gristina, L., et al., 2014. Dynamics of soil organic carbon pools after agricultural abandonment. Geoderma 235–236, 191–198, http://dx.doi.org/10.1016/j.geoderma.2014.07.015.
Novick, K.A., Stoy, P.C., Katul, G.G., Ellsworth, D.S., Siqueira, M.B.S., Juang, J., et al., 2004. Carbon dioxide and water vapor exchange in a warm temperate grassland. Oecologia 138, 259–274.
Nusslein, K., Tiedje, J.M., 1999. Soil bacterial community shift correlated with change from forest to pasture vegetation in a tropical soil. Appl. Environ. Microb. 65, 3622–3626.
Olesen, J.E., Trnka, M., Kersebaum, K.C., Skjelvåg, a.O., Seguin, B., Peltonen-Sainio, P., et al., 2011. Impacts and adaptation of European crop production systems to climate change. Eur. J. Agron. 34, 96–112, http://dx.doi.org/10.1016/j.eja.2010.11.003.
Overby, S.T., Hart, S.C., Neary, D.G., 2003. Impacts of natural disturbance on soil carbon dynamics in forest ecosystems. In: Kimble, J.M., Heath, L.S., Birdsey, R.A., Lal, R. (Eds.), The Potential of U.S. Forest Soils to Sequester Carbon and Mitigate the Greenhouse Effect. CRC Press, Boca Raton, FL, pp. 159–172.
Pan, G., Xu, X., Smith, P., Pan, W., Lal, R., 2010. An increase in topsoil SOC stock of China's croplands between 1985 and 2006 revealed by soil monitoring. Agr. Ecosyst. Environ. 136, 133–138.
Parr, J.F., Gardner, W.R., Elliot, L.F., 1981. Water Potential Relations in Soil Microbiology: Proceedings of a Symposium, SSSA Special Publication Number 9. Soil Science Society of America, Madison, WI, USA, p. 151.
Parras-Alcántara, L., Díaz-Jaimes, L., Lozano-García, B., 2013. Organic farming affects C and N in soils under olive groves in Mediterranean areas. Land Degrad. Dev. 26, 800–806, http://dx.doi.org/10.1002/ldr.2231.
Pätzhold, S., Hejcman, M., Barej, J., Schellberg, J., 2013. Soil phosphorus fractions after seven decades of fertilizer application in the Rengen grassland experiment. J. Plant Nutr. Soil Sci. 176, 910–920.
Paul, E.A., Kravchenko, A., Grandy, A.S., Morris, S., 2015. Soil organic matter dynamics: controls and management for sustainable ecosystem functioning. In: Hamilton, S.K., Doll, J.E., Robertson, G.P. (Eds.), The Ecology of Agricultural Landscapes: Long-Term Research on the Path to Sustainability. Oxford University Press, New York, USA, pp. 104–134.
Peterson, F.S., Lajtha, K.J., 2013. Linking aboveground net primary productivity to soil carbon and dissolved organic carbon in complex terrain. J. Geophys. Res.: Biogeosci. 118 (3), 1225–1236, http://doi.org/10.1002/jgrg.20097.
Pimentel, D., Hepperly, P., Hanson, J., Douds, D., Seidel, R., 2005. Environmental, energetic, and economic comparisons of organic and conventional farming systems. BioScience 55, 573–582.
Pinton, R., Varanini, Z., Nannipieri, P., 2001. The Rhizosphere. Biochemistry and Organic Substances at the Soil-Plant Interface. Marcel Dekker, New York.
Pisante, M., Stagnari, F., Acutis, M., Bindi, M., Brilli, L., Di Stefano, V., et al., 2015. Conservation agriculture and climate change. Conserv. Agric., 579–620.

Post, W.M., Emanuel, W.R., Zinke, P.J., Stangenberger, A.G., 1982. Soil carbon pools and world life zones. Nature 298, 156–159.

Power, A.G., 2010. Ecosystem services and agriculture: tradeoffs and synergies. Philos. Trans. R. Soc. B 365, 2959–2971. doi: 10.1098/rstb.2010.0143.

Powlson, D.S., Whitmore, A.P., Goulding, K.W.T., 2011. Soil carbon sequestration to mitigate climate change: a critical re-examination to identify the true and the false. Eur. J. Soil Sci. 62, 42–55.

Prichard, S.J., Peterson, D.L., Hammer, R.D., 2000. Carbon distribution in sub-alpine forests and meadows of the Olympic Mountains, Washington. Soil Sci. Soc. Am. J. 64, 1834–1845.

Quintero, M., Comerford, N.B., 2013. Effects of conservation tillage on total and aggregated soil organic carbon in the Andes. Open J. Soil Sci. 3, 361–373.

Reicosky, D.C., 2003. Tillage-induced CO_2 emissions and carbon sequestration: effect of secondary tillage and compaction. In: Garcia-Torres, L., Benites, J., Martinez-Vilela, A., Holgado-Cabrera, A. (Eds.), Conservation Agriculture. Kluwer Academic Publisher, Dordrecht, The Netherlands, pp. 291–300.

Renella, G., Giagnoni, L., Arenella, M., Nannipieri, P., 2014. Soil proteomics. In: Nannipieri, P., Pietramellara, G., Renella, G. (Eds.), Omics in Soil Science. Caster Academic Press, Norfolk, UK, pp. 95–125.

Rey Benayas, J.M., Bullock, J.M., 2012. Restoration of biodiversity and ecosystem services on agricultural land. Ecosystems 15 (6), 883–899. doi: 10.1007/s10021-012-9552-0.

Robert, M., 2001. Soil Carbon Sequestration for Improved Land Management: World Soil Resources Reports 96. Food and Agriculture Organization of the United Nations, Rome, p. 58. ftp://ftp.fao.org/agl/agll/docs/wsrr96e.pdf.

Ross, D.J., Tate, K.R., Scott, N.A., Wilde, R.H., Rodda, N.J., Townsend, J.A., 2002. Afforestation of pastures with *Pinus radiata* influences soil carbon and nitrogen pools and mineralisation and microbial properties. Aust. J. Soil Res. 40 (8), 1303–1318. doi: 10.1071/SR;1; 02020.

Ro en, A., Sobczyk, Ł., Liszka, K., Weiner, J., 2010. Soil faunal activity as measured by the bait-lamina test in monocultures of 14 tree species in the Siemianice common-garden experiment. Poland Appl. Soil Ecol. 45 (3), 160–167. doi: 10.1016/j.apsoil.2010.03.008.

Rubino, M., Dungait, J.A.J., Evershed, R.P., Bertolini, T., De Angelis, P., D'Onofrio, A., et al., 2010. Carbon input belowground is the major C flux contributing to leaf litter mass loss: evidences from a13C labelled-leaf litter experiment. Soil Biol. Biochem. 42 (7), 1009–1016. doi: 10.1016/j.soilbio.2010.02.018.

Ruiz Sinoga, J.D., Pariente, S., Romero Diaz, A., Martinez Murillo, J.F., 2012. Variability of relationships between soil organic carbon and some soil properties in Mediterranean rangelands under different climatic conditions (South of Spain). Catena 94, 17–25. doi: 10.1016/j.catena.2011.06.004.

Sallaba, F., Lehsten, D., Seaquist, J., Sykes, M.T., 2015. A rapid NPP meta-model for current and future climate and CO2 scenarios in Europe. Ecol. Model. 302, 29–41. doi: 10.1016/j.ecolmodel.2015.01.026.

Sandhu, H.S., Wratten, S.D., Cullen, R., 2010. The role of supporting ecosystem services in conventional and organic arable farmland. Ecol. Complex. 7 (3), 302–310. doi: 10.1016/j.ecocom.2010.04.006.

Sayer, E.J., 2006. Using experimental manipulation to assess the roles of leaf litter in the functioning of forest ecosystems. Biol. Rev. 81 (1), 1–31. doi: 10.1017/S1464793105006846.

Schlesinger, W.H., 1990. Evidence from chronosequence studies for a low carbon storage potential of soils. Nature 348, 232–234.

Schloter, A., Jacquiod, S., Vestergaard, G., Schulz, S., Schloter, M., 2017. Analysis of soil microbial communities based on amplicon sequencing of marker genes. Biol. Fert. Soils 53 (5), 485–489.

Schmidt, M.W.I., Torn, M., Abiven, S., Dittmar, T., Guggenberger, G., Janssens, I.A., et al., 2011. Persistence of soil organic matter as an ecosystem property. Nature 478, 49–56.

Schneider, M.P.W., Scheel, T., Mikutta, R., van Hees, P., Kaiser, K., Kalbitz, K., 2010. Sorptive stabilization of organic matter by amorphous Al hydroxide. Geochim. Cosmochim. Ac. 74 (5), 1606–1619. doi: 10.1016/j.gca.2009.12.017.

Schönfeld, J., Gelsomino, A., van Overbeek, L.S., Gorissen, A., Smalla, K., van Elsas, J.D., 2003. Effects of compost addition and simulated solarisation on the fate of *Ralstonia solanacearum* biovar 2 and indigenous bacteria in soil. Fems Microbiol. Ecol. 43 (1), 63–74. doi: 10.1111/j.1574-6941.2003.tb01046.x.

Schuur, E.A.G., Bockheim, J., Canadell, J.G., Euskirchen, E., Field, C.B., Goryachkin, S.V., et al., 2008. Vulnerability of permafrost carbon to climate change: Implications for the global carbon cycle. Bioscience 58 (8), 701–714. doi: 10.1641/B580807.

Shipitalo, M.J., Dick, W.A., Edwards, W.M., 2000. Conservation tillage and macropore factors that affect water movement and the fate of chemicals. Soil Till. Res. 53 (3–4), 167–183. doi: 10.1016/S0167-1987(99)00104-X.

Six, J., Conant, R.T., Paul, E.A., Paustian, K., 2002. Stabilization mechanisms of soil organic matter: implications for C-saturation of soils. Plant Soil 241 (2), 155–176.

Six, J., Elliott, E.T., Paustian, K., 2000. Soil macroaggregate turnover and microaggregate formation: a mechanism for C sequestration under no-tillage agriculture. Soil Biol. Biochem. 32, 2099–2103.

Six, J., Frey, S.D., Thiet, R.K., Batten, K.M., 2006. Bacterial and fungal contributions to carbon sequestration in agroecosystems. Soil Sci. Soc. Am. J. 70 (2), 555–569.

Six, J., Paustian, K., 2014. Aggregate-associated soil organic matter as an ecosystem property and a measurement tool. Soil Biol. Biochem. 68, A4–A9. doi: 10.1016/j.soilbio.2013.06.014.

Smith, P., Cotrufo, M.F., Rumpel, C., Paustian, K., Kuikman, P.J., Elliott, J.A., et al., 2015. Biogeochemical cycles and biodiversity as key drivers of ecosystem services provided by soils. Soil 1, 665–685.

Smith, P., Fang, C.M., 2010. A warm response by soils. Nature 464, 499–500.

Söderström, B., Hedlund, K., Jackson, L.E., Kätterer, T., Lugato, E., Thomsen, I.K., et al., 2014. What are the effects of agricultural management on soil organic carbon (SOC) stocks? Environ. Evidence 3 (2).

Steenwerth, K.L., Jackson, L.E., Calderon, F.J., Stromberg, M.R., Scow, K.M., 2002. Soil microbial community composition and land use history in cultivated and grassland ecosystems of coastal California. Soil Biol. Biochem. 34 (11), 1599–1611. doi: 10.1016/S0038-0717(02)00144-X.

Stevenson, F.J., 1986. Cycles of Soil. Carbon, Nitrogen, Phosphorus, Sulphur, Micronutrients. John Wiley & Sons, New York, p. 1.

Stott, T., Leeks, G., Marks, S., Sawyer, A., 2001. Environmentally sensitive plot-scale timber harvesting: impacts on suspended sediment, bedload and bank erosion dynamics. J. Environ. Manage. 63 (1), 3–25. doi: 10.1006/jema.2001.0459.

Suyker, A.E., Verma, S.B., 2012. Gross primary production and ecosystem respiration of irrigated and rainfed maize–soybean cropping systems over 8 years. Agr. Forest Meteorol. 165, 12–24. doi: 10.1016/j.agrformet.2012.05.021.

TEEB, 2010. The Economics of Ecosystem and Biodiversity (TEEB): Ecological and Economic Foundations. Earthscan, London.

Thompson, B.H., Paradise, R.E., McCarty, P.L., Segerson, K., Ascher, W.L., McKenna, D.C., et al., 2009. Valuing the protection of ecological systems and services: A report of the EPA Science Advisory Board. EPA Science Advisory Board. U.S. Environmental Protection Agency, Washington, DC.

Thornley, J., Cannell, M., 2000. Managing forests for wood yield and carbon storage: a theoretical study. Tree Physiol. 20 (7), 477–484.

Todd-Brown, K.E.O., Randerson, J.T., Post, W.M., Hoffman, F.M., Tarnocai, C., Schuur, E.A.G., et al., 2013. Causes of variation in soil carbon simulations from CMIP5 Earth system models and comparison with observations. Biogeosciences 10 (3), 1717–1736. doi: 10.5194/bg-10-1717-2013.

Torras, O., Saura, S., 2008. Effects of silvicultural treatments on forest biodiversity indicators in the Mediterranean. Forest Ecol. Manage. 255 (8–9), 3322–3330. doi: 10.1016/j.foreco.2008.02.013.

Torsvik, V., Ovreas, L., 2002. Microbial diversity and function in soil: from genes to ecosystems. Curr. Opin. Microbiol. 5 (3), 240–245. doi: 10.1016/S1369-5274(02)00324-7.

Trexler, M., 1998. Forestry as a climate change mitigation option. In: Herozog, H.J. (Ed.), In: Proceedings of the Stakeholders Workshop on Carbon Sequestration. Massachusetts Institute of Technology Energy Laboratory, MIT EL 98-002.

Tuomi, M., Vanhala, P., Karhu, K., Fritze, H., Liski, J., 2008. Heterotrophic soil respiration: comparison of different models describing its temperature dependence. Ecol. Model. 211 (1–2), 182–190. doi: 10.1016/j.ecolmodel.2007.09.003.

UN, 2015. General Assembly Resolution 70/1 Transforming our world: the 2030 Agenda for Sustainable Development A/RES/70/1 (25 September 2015). http://www.un.org/ga/search/view_doc.asp?symbol=A/RES/70/1&Lang=E.

Van Groenigen, J.W., Velthof, G.L., Oenema, O., Van Groenigen, K.J., Van Kessel, C., 2010. Towards an agronomic assessment of N_2O emissions: a case study for arable crops. Eur. J. Soil Sci. 61 (6), 903–913. doi: 10.1111/j.1365-2389.2009.01217.x.

Venterea, R.T., Maharjan, B., Dolan, M.S., 2011. Fertilizer source and tillage effects on yield-scaled nitrous oxide emissions in a corn cropping system. J. Environ. Qual. 40 (5), 1521–1531. doi: 10.2134/jeq2011.0039.

Verchot, L.V., Dutaur, L., Shepherd, K.D., Albrecht, A., 2011. Organic matter stabilization in soil aggregates: understanding the biogeochemical mechanisms that determine the fate of carbon inputs in soils. Geoderma 161 (3–4), 182–193. doi: 10.1016/j.geoderma.2010.12.017.

Verheijen, F., Jeffery, S., Bastos, A.C., van der Velde, M., Diafas, I., 2010. Biochar application to soils: A critical scientific review of effects on soil properties, processes and functions. Scientific and Technical Reports. European Commission, Joint Research Centre, Institute for Environment and Sustainability, Ispra (Italy), p. 166.

Vitousek, P.M., Ehrlich, P., Ehrlich, A., Matson, P.M., 1986. Human appropriation of the products of photosynthesis. BioScience 36, 368–373.

Watson, R.T., Zinyowera, M.C., Moss, R.H., 1996. Climate change 1995: impacts. Adaptations and Mitigation of Climate Change: Scientific-Technical Analyses. Cambridge University Press.

Wei, X., Shao, M., Gale, W., Li, L., 2014. Global pattern of soil carbon losses due to the conversion of forests to agricultural land. Sci. Rep.-UK 4, 4062. doi: 10.1038/srep04062.

Westman, W., 1977. How much are nature's services worth? Science 197, 960–964.

Wilcox, C.S., Dominguez, J., Parmelee, R.W., McCartney, D.A., 2002. Soil carbon and nitrogen dynamics in *Lumbricus terrestris. L. middens* in four arable, a pasture, and a forest ecosystems. Biol. Fert. Soils 36 (1), 26–34. doi: 10.1007/s00374-002-0497-x.

Wilmes, P., Bond, P.L., 2006. Metaproteomics: studying functional gene expression in microbial ecosystems. Trends Microbiol. 14 (2), 92–97. doi: 10.1016/j.tim.2005.12.006.

Wright, A.L., Hons, F.M., 2005. Tillage impacts on soil aggregation and carbon and nitrogen sequestration under wheat cropping sequences. Soil Till. Res. 84 (1), 67–75. doi: 10.1016/j.still.2004.09.017.

Zinn, Y.L., Lal, R., Bigham, J.M., Resck, D.V.S., 2007. Edaphic controls on soil organic carbon retention in the Brazilian cerrado: texture and mineralogy. Soil Sci. Soc. Am. J. 71, 1204–1214. doi: 10.2136/sssaj2006.0014.

CHAPTER 2

Soil Organic Matter in Dryland Ecosystems

César Plaza*,,†, Gabriel Gascó‡, Ana M. Méndez‡, Claudio Zaccone§, Fernando T. Maestre****

*Instituto de Ciencias Agrarias, Madrid, Spain
**Universidad Rey Juan Carlos, Móstoles, Spain
†Northern Arizona University, Flagstaff, AZ, United States
‡Universidad Politécnica de Madrid, Madrid, Spain
§University of Foggia, Foggia, Italy

Chapter Outline

INTRODUCTION

Drylands are loosely defined as regions that receive intense solar radiation and relatively low average precipitation, resulting in a water deficit that limits net primary production (Safriel et al., 2005). More precisely, the United Nations Environment Program (UNEP) defines drylands as terrestrial areas where the aridity index (AI), or the ratio of total annual precipitation to potential evapotranspiration (evaporation from soil and transpiration by plants), is less than 0.65 (UNEP, 1997). Based on the AI, drylands are further divided into hyperarid (AI < 0.05), arid (AI within the range from 0.05

The Future of Soil Carbon
http://dx.doi.org/10.1016/B978-0-12-811687-6.00002-X

to 0.2), semiarid (AI from 0.2 to 0.5), and dry subhumid regions (AI from 0.5 to 0.65). The United Nations Convention to Combat Desertification (UNCCD) uses this classification, but addresses specifically only arid, semiarid, and dry subhumid areas as drylands for its purposes, excluding hyperarid areas (UNCCD, 1994). Instead of using the AI, the United Nations Food and Agriculture Organization (FAO) developed an index termed "length of growing period" (LGP) to identify the crop cultivation potential of land. The LGP, referring to the number of days in a year when temperature and soil moisture allow plant growth, is also used to classify general moisture regimes (IIASA/FAO, 2012). Areas with an LGP of 0, less than 60, from 60 to 119, and from 120 to 179 days are classified as hyperarid, arid, dry semiarid, and moist semiarid, respectively, which together correspond closely to the areas defined as drylands in the World Atlas of Desertification (FAO, 1993, 2004; IIASA/FAO, 2012).

Dryland ecosystems are of paramount significance both in terms of their extension and the human population they support. In particular, drylands cover more than 45% of the Earth's land surface, or about 6.7 billion ha (Prăvălie, 2016), more than 70% of which is occupied by developing countries (FAO, 2004; Prăvălie, 2016; Safriel et al., 2005; UN Environmental Management Group, 2011) (Fig. 2.1). Hyperarid, arid, semiarid, and dry subhumid regions account for about 13, 31, 36, and 20% of this area, respectively (Prăvălie, 2016). Most drylands can be ascribed to four biomes of decreasing aridity: deserts, grasslands, Mediterranean scrublands, and woodlands (Safriel et al., 2005). Of the global dryland area, approximately 65% is used as rangelands and 25% as cropland (Safriel et al., 2005). More than 2 billion people, or more than one third of the human population, live in drylands, and about 90% of these live in developing countries (Middleton et al., 2011; Safriel et al., 2005).

Soils make up an essential part of drylands, as in any other terrestrial ecosystem, as they constitute the medium for plant growth, provide a habitat for a wide range of organisms, regulate water fluxes, and store C. The most common soils in drylands are entisols (2.3 billion ha) and aridisols (2.1 billion ha), followed by mollisols, alfisols, and vertisols (FAO, 2004). Compared to humid-region soils, those in drylands are commonly characterized by low organic matter and nutrient contents, frequent water stress, coarse to medium texture, moderate to high base saturation, slightly acid to alkaline surface pH, accumulation of calcium carbonate, shallow to moderate profile development, and low biological activity (Dregne, 1976; Skujins, 1991). These properties result in inherently low fertility (Lal, 2004a).

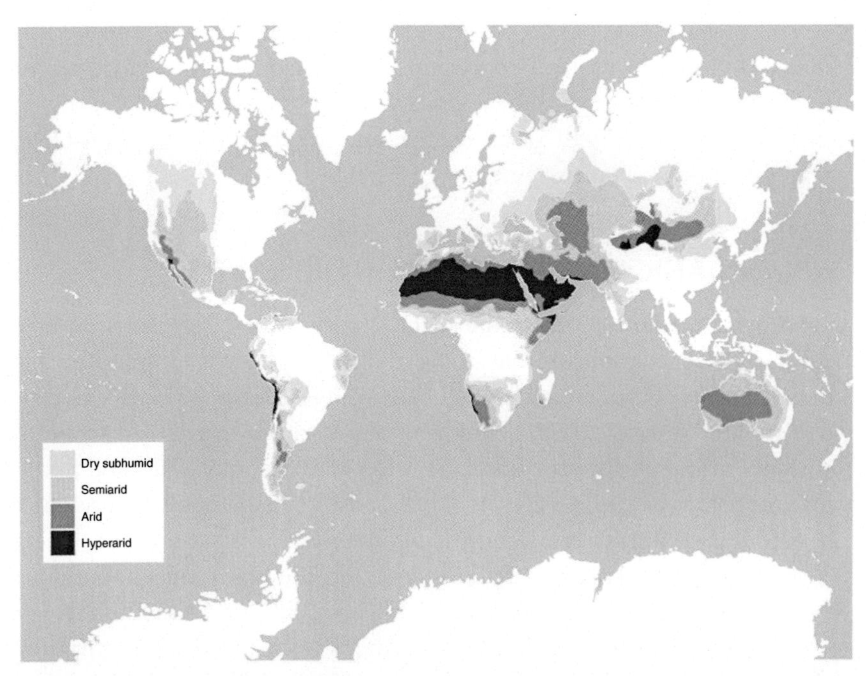

Figure 2.1 ***Map of dryland areas in accordance with the United Nations Convention to Combat Desertification (UNCCD) and Convention on Biological Diversity (CBD) definitions (UNEP-WCMC, 2007).*** Based on the aridity index (AI), or ratio of total annual precipitation to potential evapotranspiration, drylands are divided into hyperarid (AI < 0.05), arid (AI within the range from 0.05 to 0.2), semiarid (AI from 0.2 to 0.5), and dry subhumid regions (AI from 0.5 to 0.65).

Among the many factors that impact the quality and fertility of soils in drylands and their ability to provide essential ecosystems services, soil organic matter deserves special attention. This is because soil organic matter in drylands constitutes a huge global C reservoir and influences almost all physical, chemical, and biological soil properties and processes in these ecosystems (Lal, 2004a, 2009). Among other benefits, soil organic matter promotes the formation of aggregates with mineral particles, which improves soil structure, porosity, and moisture-holding capacity, thus reducing the severity of water scarcity and protecting soils from erosion and compaction. Also, soil organic matter contributes markedly to the soil acid-base buffering capacity and represents an important source of plant nutrients. Furthermore, soil biological diversity and activity are also strongly related to soil organic matter as a major source of food for a variety of organisms (Bot and Benites, 2005; Paul, 2016; Paul et al., 1997).

This chapter provides an overview of the amount of soil organic matter stocks stored in dryland ecosystems. It also explores the processes and factors affecting the balance between accumulation and loss of soil organic matter, its vulnerability to global change, and the main options available for its preservation and enhancement.

SOIL ORGANIC MATTER STOCKS IN DRYLANDS

Soil organic matter represents one of the largest active C reservoirs on Earth and thus plays a significant role in the global C cycle with implications for the mitigation or exacerbation of climate change (Batjes, 1996; IPCC, 2013; Jobbágy and Jackson, 2000; Schmidt et al., 2011). In particular, while the atmosphere and global vegetation are estimated to contain 829 and 520 Pg of C, respectively (IPCC, 2013), global soil organic C mass is thought to be 1301 Pg within the top 1 m, and about 2800 Pg when deeper soil layers are considered. This estimation is based on the Harmonized World Soil Database (HWSD) v. 1.2 (FAO/IIASA/ISRIC/ISSCAS/JRC, 2012; Köchy et al., 2015), corrected for the latest northern permafrost region inventory, which estimates that permafrost soils store 472 Pg of organic C in the top 1 m, 1035 Pg in the top 3 m, and 1330–1580 Pg in the known pool of terrestrial permafrost (Hugelius et al., 2014; Schuur et al., 2015).

Globally, the largest soil organic C concentrations and stocks are in wetlands and peatlands, generally in permafrost and tropical regions (Köchy et al., 2015). Soil organic C concentrations in drylands are markedly smaller, often less than 0.5% of the soil mass, resulting in typical densities of 0–15 kg m^{-2} (Lal, 2004a). However, due to the vast extension of dryland ecosystems, the total amount of organic C stored in dryland soils is significant for the global C cycle (Lal, 2004a). In particular, dryland soils are estimated to contain 431 Pg of organic C (Safriel et al., 2005), which represents about 33% of the global soil organic C reserves in the uppermost 1 m. Hyperarid and arid soils hold 113 Pg, whereas semiarid and dry subhumid soils store 318 Pg. In addition to organic C, dryland soils are estimated to hold 916 Pg of inorganic C, mainly as calcium carbonate, which accounts for 97% of the inorganic C stocks in the world's soils. Hyperarid and arid soils contain 732 Pg while semiarid and dry subhumid soils have 184 Pg (Safriel et al., 2005) (Fig. 2.2).

As in the rest of terrestrial ecosystems, knowledge of the total soil organic C stock and its spatial distribution in drylands is essential to understand

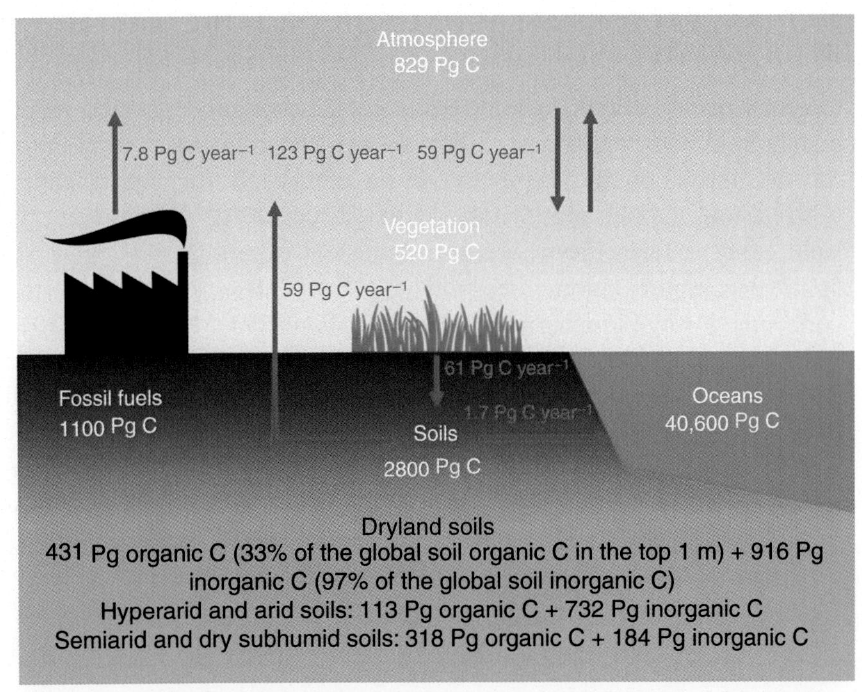

Figure 2.2 ***Simplified global C cycle and soil C stocks in drylands.*** *(Data from Safriel, U., Adeel, Z., Niemeijer, D., Puigdefabregas, J., White, R., Lal, R., et al., 2005. Dryland systems. In: Hassan, R., Scholes, R., Ash, N. (Eds.), The Millennium Ecosystem Assessment Series, Ecosystems and Human Well-being: Current State and Trends, vol. 1. Island Press, Washington, DC, pp. 623–662; IPCC, 2013. Climate change 2013: the physical science basis. In: Stocker, T.F., Qin, D., Plattner, G.-K., Tignor, M., Allen, S.K., Boschung, J., Nauels, A., Xia, Y., Bex, V., Midgley, P.M. (Eds.), Contribution of Working Group I to the Fifth Assessment Report of the Intergovernmental Panel on Climate Change. Cambridge University Press, Cambridge, UK).*

their vulnerability to global change and to develop strategies to recover, preserve, or enhance the services provided by their ecosystems. The current estimates of organic C stocks in dryland soils outlined earlier are still highly uncertain and probably not accurate, being typically derived from soil databases that may underrepresent some regions (Köchy et al., 2015). Moreover, soil profile data are not collected with a consistent methodology and generally do not include soil layers deeper than 1 m (Darrouzet-Nardi, 2013). Nonetheless, most soil C in drylands is believed to reside in topsoil (Ciais et al., 2011), making it highly vulnerable to climate change and erosion process.

NATURAL FACTORS AFFECTING SOIL ORGANIC MATTER IN DRYLANDS

Organic matter content in soils results from the balance between inputs, mainly from plant production, and losses, which occur mainly through in situ decomposition and respiration, but also through leaching, volatilization, erosion, and burning (Schulze and Freibauer, 2005). According to the latest concepts, plant material entering the soil is continuously processed by the decomposer community, producing fragments of smaller size with a concomitant surge in microbial by-products (Courtier-Murias et al., 2013; Lehmann and Kleber, 2015; Plaza et al., 2013) (Fig. 2.3). Soil organic matter decomposition is modulated by a number of interconnected environmental, physical, chemical, and biological controls, such as temperature, moisture, activity, and composition of microbial community, and organomineral interactions (Schmidt et al., 2011; Six et al., 2002; Sollins et al., 1996; Von Lützow et al., 2006). Freezing temperatures and high moisture content (low O_2) help preserve soil organic matter (climatic and environmental protection), as well as its association with aggregates (physical protection) and mineral surfaces (chemical protection) (Plaza et al., 2012, 2013; Torn et al., 1997, 2009;

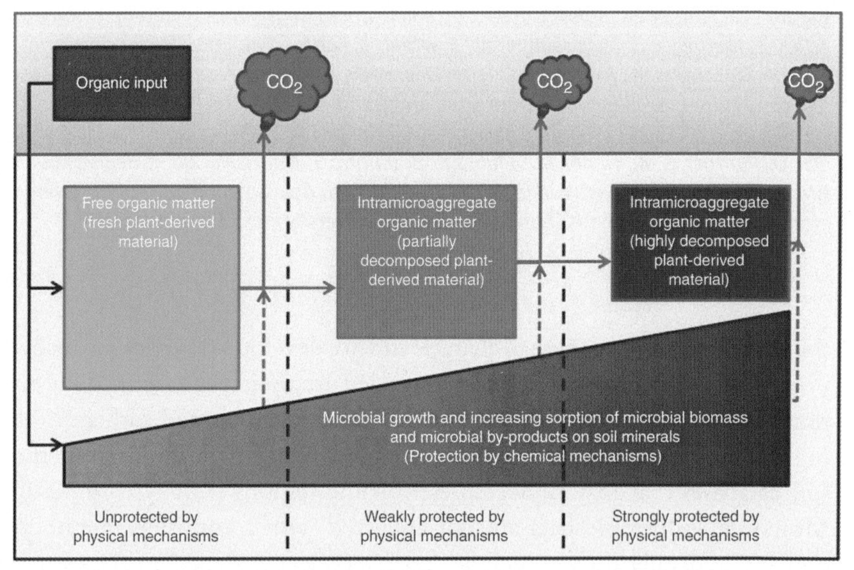

Figure 2.3 ***Organic matter transformation, stabilization, and protection from decomposition in soils.*** *(Redrawn from Courtier-Murias, D., Simpson, A.J., Marzadori, C., Baldoni, G., Ciavatta, C., Fernández, J.M., et al., 2013. Unraveling the long-term stabilization mechanisms of organic materials in soils by physical fractionation and NMR spectroscopy. Agric. Ecosyst. Environ. 171, 9–18, with permission from Elsevier).*

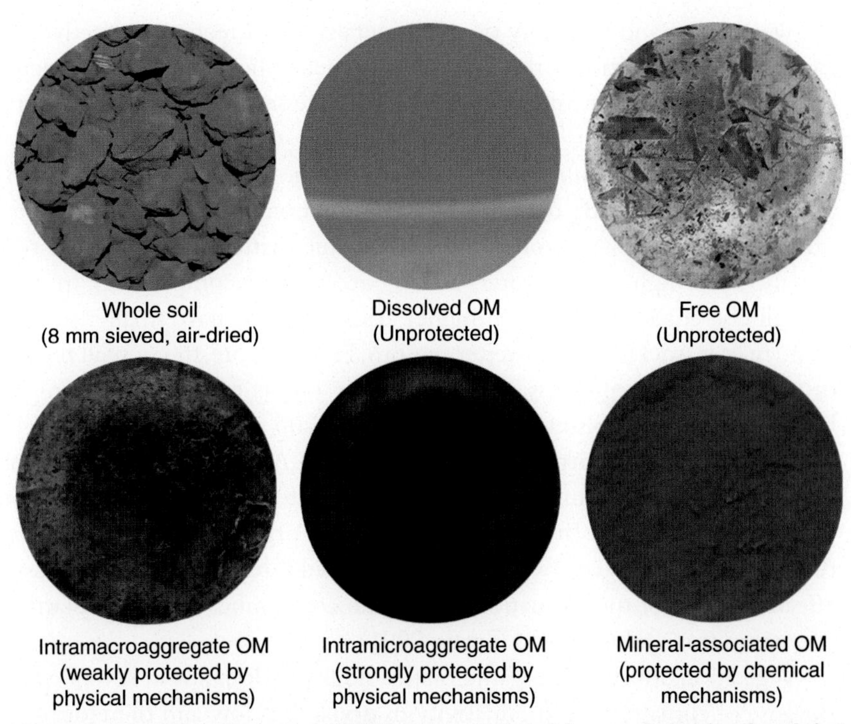

Figure 2.4 ***Soil organic matter (OM) fractions isolated from a dryland soil and their links to conceptual mechanisms of stabilization and protection from decomposition.*** *(Reprinted from Plaza, C., Courtier-Murias, D., Fernández, J.M., Polo, A., Simpson, A.J., 2013. Physical, chemical, and biochemical mechanisms of soil organic matter stabilization under conservation tillage systems: a central role for microbes and microbial by-products in C sequestration. Soil Biol. Biochem. 57, 124–134, with permission from Elsevier).*

Trumbore, 2009) (Fig. 2.4). During the degradation of plant material, organic substances are accumulated and protected from decomposition within aggregates and on mineral surfaces, limiting the access and action of decomposers (Courtier-Murias et al., 2013; Golchin et al., 1994; Lehmann and Kleber, 2015; Plaza et al., 2012, 2013).

More generally, according to the classic work by Jenny (1941), soil organic matter stocks in drylands, like any other soil property in a given ecosystem, is a function of a number of soil forming factors, including climate, organisms (vegetation and soil biota), parent material, relief, and time. The main distinctive soil-forming factors that drive soil organic matter storage in drylands are climate and vegetation. These state factors govern the balance of soil organic matter by influencing both the quantity and quality of plant input and the rate of decomposition. They also strongly influence the weathering and reactivity of soil minerals (thus affecting protective organomineral interactions), the structure and metabolic activity of the microbial community, and erosion as well as fire processes (Lehmann et al., 2014; Morgan, 2005).

Climate is the dominant factor explaining the low organic matter contents in dryland soils. By definition, drylands not only receive low precipitation, which limits plant production and organic matter inputs into the soil, but also have high temperatures, which encourage decomposition. In general, all else equal, regions with higher aridity tend to have smaller soil organic matter stocks than do wetter regions. Furthermore, the rainfall regime of many dryland regions is characterized by events of short duration and high intensity (Nicholson, 2011), which favor soil organic matter losses by runoff and erosion (Martínez-Mena et al., 2002). Also, the low precipitation and high temperature increase the risk of wildfires in drylands (Middleton and Sternberg, 2013). Burning diminishes soil organic matter stocks over the short term; however, over the long term, wildfires might cause opposite effects because of the formation of organic compounds highly resistant to decomposition (Johnson and Curtis, 2001).

Vegetation affects soil organic matter stocks in drylands in several ways. Because of climate, plant productivity in drylands is low and therefore so is the rate of plant C inputs into the soil. However, the low water and nutrient availability generally favor other plant traits that help slow down decomposition and build up soil organic matter, such as poor-quality litter and a high root-to-shoot ratio that favors organic C allocation and protection deep into the soil (De Deyn et al., 2008). The spatial organization of vegetation, which organizes into discrete plant patches surrounded by areas devoid of vascular vegetation (Berdugo et al., 2017), is also relevant to soil organic matter stocks, particularly to their spatial distribution within dryland landscapes (Fig. 2.5). The length of the growing season, which is highly variable in drylands, is also thought to have a significant influence on soil organic matter content (Lal, 2004a).

In addition to vegetation, biological soil crusts are a major biotic factor in organic C fixation in dryland soils. Biological soil crusts, or biocrusts, are communities of organisms dominated by lichens, mosses, bacteria, and fungi, inhabiting the soil surface (Belnap and Lange, 2013) (Fig. 2.6). These communities have been found to be able to take up significant amounts of C and N from the atmosphere by photosynthesis and N fixation (Elbert et al., 2012), thus being an important pool and source of organic inputs into the soil and having a major role in soil respiration in dryland ecosystems (Castillo-Monroy et al., 2011). Elbert et al. (2009) estimated the net

Figure 2.5 ***Typical patchy vegetation of drylands.*** *(Image by Fernando T. Maestre).*

Figure 2.6 ***Biological soil crust.*** *(Image by Fernando T. Maestre).*

uptake of C by biological soil crusts in drylands at 16 g m^{-2} $year^{-1}$, which corresponds to a total of approximately 1.0 Pg $year^{-1}$ for the global dryland area. Biocrusts are also known to protect soil from wind and water erosion, and thus from C loss, by strengthening soil structure through the interaction with mineral particles and formation of aggregates (Belnap, 2013; Bowker, 2007; Eldridge and Leys, 2003).

The abundance and strength of the organo–mineral interactions, and thus the physical and chemical protection of soil organic matter from decomposition, are thought to be lower in dryland soils than in soils of humid regions (Kleber et al., 2015). This is because water largely controls the weathering and formation of reactive sites on minerals needed to create organo–mineral associations. While wet conditions promote plant production, low pH conditions, soil weathering, and the formation of strong inner-sphere bonds between organic matter and mineral surfaces, energetically weaker outer-sphere complexation and H-bonds between organic matter and minerals are thought to dominate in dryland soils because of the lower moisture (Kleber et al., 2015).

MANAGEMENT ACTIVITIES AFFECTING SOIL ORGANIC MATTER IN DRYLANDS

Effects of Grazing on Soil Organic Matter in Drylands

Livestock grazing is the main land-use type in dryland regions and the primary livelihood for many of their inhabitants (Lal, 2004a). Grazing is known to largely affect the structure, composition, and functioning of dryland ecosystems (Eldridge et al., 2016; Gaitán et al., 2017; Maestre et al., 2016), including their capacity to fix and store soil organic matter (Piñeiro et al., 2010). Even though livestock may return a large proportion of nutrients and organic matter in the form of dung and urine (Haynes and Williams, 1999), this effect is likely not strong enough at a large scale to reinforce soil fertility in low-productive ecosystems (Bardgett and Wardle, 2003), such as drylands, and aboveground plant biomass is often reduced significantly by grazing. Using a large database from Australian rangelands, Eldridge et al. (2016) reported reductions of plant biomass by 40% and plant and litter cover by 25% after grazing, with more profound effects in drier environments and noticeable effects even at low grazing intensities. Using the same large set of records, Maestre et al. (2016) found that grazing, even at low intensity, negatively affected the C content and reported average reductions of approximately 8%. In addition to a decrease in aboveground plant inputs into the soil, livestock often increases soil compaction through trampling, thereby reducing water infiltration and availability to plants while encouraging runoff and soil erosion, thus further contributing to soil organic matter losses (Eldridge et al., 2016; Piñeiro et al., 2010). Intensive grazing has also been found to adversely affect soil organic matter stocks by reducing the cover, biomass, and diversity of the biological soil crusts through trampling and burial (Thomas, 2012; Warren and Eldridge, 2003). Increases in C inputs into the soil from greater root biomass due to changes in plant C allocation or species abundance, on the other hand, have been suggested to partially offset soil organic matter loss in drylands (Piñeiro et al., 2010).

Effects of Cultivation on Soil Organic Matter in Drylands

Cultivation is the second most widespread use of drylands (Safriel et al., 2005). The conversion of natural ecosystems to croplands is well known to cause significant losses of soil organic matter (Matson et al., 1997), and has been responsible for a large proportion of the global anthropogenic CO_2 emissions (IPCC, 2013; Lal, 2004b). Meta-analysis results estimate global soil C stock losses at 42% when converting from native forest to crop and at 59% from pasture to crop (Guo and Gifford, 2002). As throughout the rest of the globe, the expansion and intensification of cultivation in drylands have led to a substantial depletion of soil organic matter stocks, concomitantly with more severe nutrient depletion and soil erosion. For example, it has been estimated that cropping of Australian semiarid regions has resulted in long-term losses of soil organic matter often exceeding 60% of the initial content (Dalal and Chan, 2001).

In general, the loss of soil organic matter with conversion to permanent agriculture can be attributed to decreased input rates of plant-derived organic matter and increased decomposition rates. Agricultural crops are harvested and typically leave less organic matter belowground than does native vegetation (Post and Kwon, 2000). The rate and amount of organic matter loss depend not only on natural factors, such as climate and soil properties, but also on the specific cropping system. Farming methods that have particularly exacerbated the loss of soil organic matter in drylands include the use of conventional intensive tillage and removal of crop residues (McLeod et al., 2013).

Conventional intensive soil tillage, used for weed control and seedbed preparation, reduces soil organic matter stocks by increasing soil aeration and facilitating the access of decomposers to organic substrates, thus accelerating decomposition rates (I. Haller, personal communication) (Fig. 2.7). More specifically, soil tillage is thought to increase macroaggregate turnover, thereby inhibiting the formation of microaggregates within macroaggregates

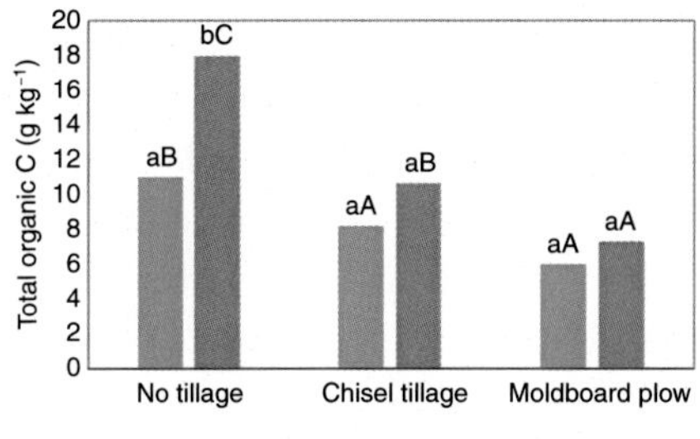

Figure 2.7 ***Total organic C content of semiarid soils under no tillage, chisel plow, and moldboard plow for 23 years (1986–2009) either unamended or organically amended for 2 years (2007–09).*** Within the same tillage system, different lowercase letters indicate significant differences according to the LSD test at the 0.05 level. Within the same amendment system, different uppercase letters indicate significant differences according to the LSD test at the 0.05 level. *(Data from Haller I., 2011. Efectos de la aplicación de lodos E.R.A.R. secados térmicamente sobre la dinámica de la materia orgánica en suelos sometidos a manejos convencionales y de conservación. PhD Thesis. Universidad Autónoma de Madrid, Spain).*

in which particulate organic matter is stabilized over the long term (Six et al., 1998, 1999, 2000). It has also been proposed that the increased macroaggregate turnover may inhibit not only microaggregate formation and thereby the physical protection of crop-derived particulate organic matter, but also the interaction between mineral particles and microbial material that results in the formation of very stable organo–mineral complexes (Plaza et al., 2013). By disturbing the soil structure, tillage also makes soil organic matter more vulnerable to erosion losses.

VULNERABILITY OF SOIL ORGANIC MATTER IN DRYLANDS TO GLOBAL CHANGE

Humanity has a dominant and growing influence on the structure and functioning of all ecosystems on Earth (Vitousek et al., 1997). Because of their inherently low fertility and vegetation cover, dryland ecosystems are particularly vulnerable to human activities and global environmental change drivers, including global warming, invasion of alien species, and increased N deposition, which have profound social and economic implications (Reynolds et al., 2007).

According to the latest data from the population division of the UN Department of Economic and Social Affairs of the United Nation Secretariat (UN-DESA-PD, 2015), the world population continues to grow and is projected to reach 8.5 billion by 2030, 9.7 billion in 2050, and 11.2 billion in 2100. Most of the projected increase will occur in developing countries and in dryland regions. A rapidly increasing population will demand more food and livestock production (Foley et al., 2011; Thornton, 2010; Tilman et al., 2011). This will put more pressure on drylands, as cropland use intensifies and pastures are expanded, negatively affecting soil organic matter stocks as described earlier. In fact, population growth is arguably the most critical challenge facing drylands to avoid further degradation and desertification.

Vulnerability of Soil Organic Matter to Climate Change in Drylands

Continued anthropogenic emissions of greenhouse gases, particularly CO_2, will spur further rises in global mean temperatures over the 21st century (IPCC, 2013). By the end of this century, dryland area is projected to increase by 23% under the current path of CO_2 emissions and global warming (Representative Concentration Pathway RCP8.5) and by 11% under a scenario of intermediate emissions in which total radiative forcing is stabilized shortly after 2100 (RCP4.5), relative to the 1961–90 baseline (Huang et al., 2016). In other words, by 2100, 56% (RPC8.4) or 50% (RPC4.5) of the total land surface area should be covered by drylands (Huang et al., 2016). Dryland expansion and humid region shrinkage will reduce global plant production, which will in turn diminish global soil organic matter stocks and augment CO_2 in the atmosphere.

In drylands, temperatures are expected to rise more than in humid regions, with 4°C increases projected in some areas (IPCC, 2013). Precipitation patterns will be altered, extreme events, such as droughts and heat waves, will become more frequent, and the degree of aridity experienced by drylands will increase (Cook et al., 2015; Fu and Feng, 2014; Huang et al., 2016; IPCC, 2013). All these interwoven factors are expected to reduce the overall plant cover (Delgado-Baquerizo et al., 2013) and the abundance and diversity of soil bacteria and fungi (Maestre et al., 2015), exerting negative consequences on C cycling and storage (Delgado-Baquerizo et al., 2016). Furthermore, the expected increases in aridity with climate change are expected to negatively affect the abundance, composition, and diversity of biocrusts, thereby compromising their organic C sink capacity (Ferrenberg et al., 2015; Maestre et al., 2013).

Vulnerability of Soil Organic Matter to Vegetation Shift in Drylands

Vegetation shifts, and particularly the invasion of native grasslands by woody species (shrub encroachment) or exotic grasses, is a widespread phenomenon long documented in many dryland areas (Archer, 1994; Biedenbender et al., 2004; Brazier et al., 2014; Maestre et al., 2009; Van Auken, 2000). A number of studies suggest that plant invasions of grasslands in dry environments tend to boost the soil organic C content in the absence of human and natural disturbance (Barger et al., 2011; Jackson et al., 2002), while a global meta-analyses reported significant increases in total and organic soil C with shrub encroachment (Eldridge et al., 2011). In North American drylands, trends of soil organic C accumulation with woody plant encroachment were not coupled to aboveground net primary production, which tended to decrease in arid regions and to increase in semiarid and subhumid regions (Barger et al., 2011). Changes in vegetation have often been reported to be accompanied by greater run-off and erosion, which may significantly offset gains in plant C inputs and lead to net soil organic matter losses (Brazier et al., 2014; Puttock et al., 2014; Turnbull et al., 2010). For example, recent studies across a shrub encroachment gradient in the southwestern United States indicated that soil organic C stored beneath the vegetation increased relative to that of bare soil areas as the grass cover became sparser and shrubs became more established in the landscape (Brazier et al., 2014; Puttock et al., 2014). In this case, there was more heterogeneity of soil organic C with encroachment from a more homogeneous distribution under grass to a patchy distribution under shrubs; and this spatial heterogeneity made soil and soil organic C more susceptible to erosion from areas between shrubs.

Vulnerability of Soil Organic Matter to N Deposition in Drylands

Human activities, such as the burning of fossil fuels and the use of N fertilizers, have tremendously accelerated the N cycle and have significantly exacerbated the release of N into the atmosphere (Gruber and Galloway, 2008). This excess of N is not only contributing to global warming but is also resulting in heavier atmospheric deposition of N in terrestrial ecosystems worldwide (Gruber and Galloway, 2008). Plant growth in drylands, as in most ecosystems, is naturally N limited (Maestre et al., 2016; Ochoa-Hueso et al., 2016; Waldrop et al., 2004). The addition of N may thus enhance plant growth and organic matter inputs (Maestre et al., 2016; Ochoa-Hueso et al., 2016). Increased N deposition, however, can also favor fast-growing plant species at the expense of the slow-growing ones, thereby reducing native plant diversity (Bobbink et al., 2010; Lan and Bai, 2012). Other effects of N deposition that may alter the ability of drylands to store soil organic matter include soil acidification, stronger aluminum toxicity, and changes in soil microbial composition, biomass, and activity (Bardgett, 2005; Ochoa-Hueso et al., 2011). In general, N enrichment can stimulate cellulolysis and labile organic matter decomposition, while suppressing the ligninolytic activity of fungi (Fog, 1988). The structure and function of biocrusts are also reportedly affected by N deposition, the most significant responses being reductions in photosynthesis rates and shifts in cyanobacteria community (Ochoa-Hueso et al., 2013). Although the available data are scarce, all these potential operating factors identified in the literature suggest that the effects of N deposition for soil organic matter storage in nonagricultural drylands may be highly context dependent. Soil organic C content declines along a gradient of N deposition in low-productivity semiarid Mediterranean areas (Ochoa-Hueso et al., 2013), while soils of other different ecosystems subject to elevated N additions appear to be soil organic matter sinks (Lloyd, 1999; Zak et al., 2016).

As a whole, ample evidence implies that the ongoing global environmental change will substantially affect the structure and function of dryland ecosystems, including their ability to store soil organic matter. It is important to note that global change phenomena occur simultaneously, and therefore their effects on ecosystems may lead to highly significant interactions between them (Bardgett, 2005). However, little is known about the magnitude of these interactions and how they might affect soil organic matter stocks in drylands, albeit there is evidence of biological feedbacks between climate change and changes in biocrust communities can alter them (Maestre et al., 2013).

PRACTICES TO RESTORE, PRESERVE, AND AUGMENT SOIL ORGANIC MATTER IN DRYLANDS

Soil organic matter is universally recognized to be a central factor enabling ecosystems to provide key services needed for global sustainability, including food and wood production, CO_2 fixation and soil stability (Haygarth and Ritz, 2009; see Chapter 1 by Masciandaro et al. in this book). Yet dryland soils have been a historic source of CO_2 emissions to the atmosphere and have lost massive amounts of their native soil organic matter contents due to land management practices, degradation, and desertification. In the dryland ecosystems of West Asia and North Africa alone, which represent roughly 25% of the global dryland area, the historic organic C loss is thought to have been between 6 and 12 Pg (Lal, 2001a). Desertification has caused the loss of 20 to 30 Pg of C globally (Lal, 2001b, 2004a). The magnitude of this loss shows the potential and the urgent need to develop strategies to restore, preserve, or increase soil organic matter in drylands, to help bolster productivity, prevent erosion, and degradation, and mitigate climate change, with substantial local and global societal, economic, and environmental benefits. Soil organic matter contents can be built up through management practices that increase organic inputs into the soil and decrease organic matter decomposition, erosion, and leaching, or both (Lal, 2004a, 2009). These practices include cropland conversion to grassland and afforestation, improved grazing regime, and improved farming management, such as using organic amendments, mineral N fertilization, cover crops, crop rotations, shifting from conventional tillage to conservation tillage practices, and improving water use efficiency (Lal, 2004a, 2004c) (Fig. 2.8). The potential of increasing soil C in drylands is estimated to be about 60%–70% of the historic C loss in 25–50 years (Lal, 2001a,b, 2004a).

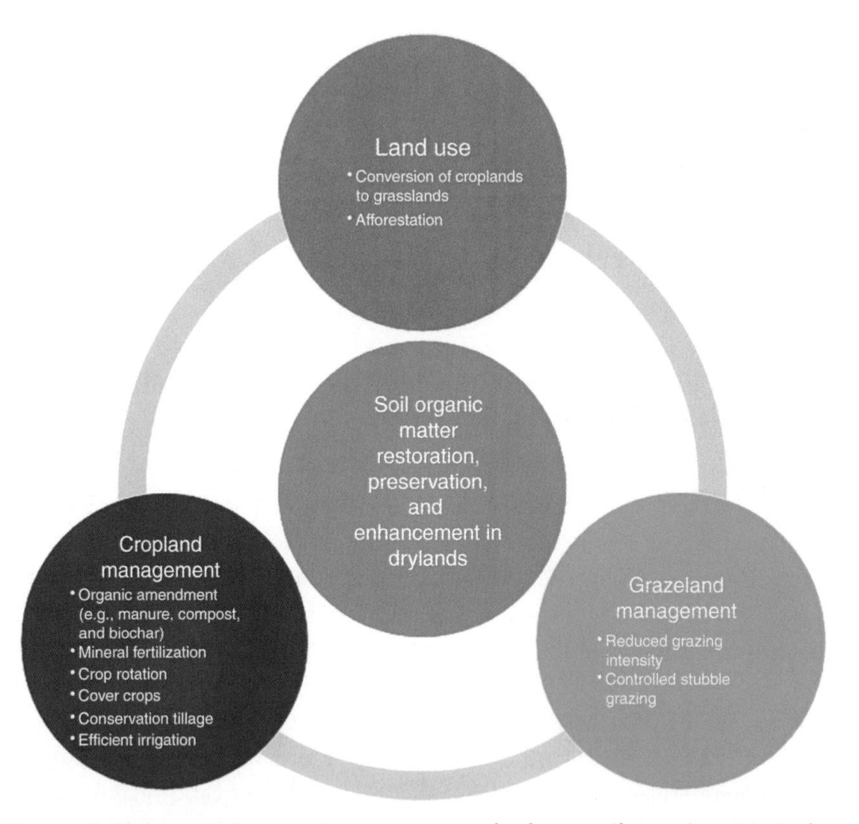

Figure 2.8 ***Main strategies to restore, preserve, and enhance soil organic matter in drylands (Lal, 2004a, 2009).***

Cropland Conversion to Grasslands and Afforestation

Land conversion from natural to agricultural ecosystems has been the main cause for the historic depletion of soil organic matter stocks (Lal, 2004c). There is abundant evidence that, in drylands, shifts from crop cultivation to grass- or tree-based systems, including conversion to pasture and agro-forestry production systems, can help in the recovery, at least partially, of the historic soil organic matter pools (Lal, 2001a,b, 2004a). In a semi-arid watershed in northern Ethiopia, for example, compared to rain-fed crop production, soil organic C stocks to 30 cm increased by 52% with based fruit production, 60% with agroforestry based crop production, 143% with silvopasture, and 227% with open communal pasture after 50 years after land-use conversion (Gelaw et al., 2014). In the Highveld of semiarid South Africa, with the use of a chronosequence and modeling approach, soil organic C stocks to 20 cm deep were estimated to increase after cropland conversion to permanent grassland for use as pasture, reaching maximum values after 10–95 years. These were 30%–94% higher than in adjacent arable soils, but 28%–53% lower than in native grasslands (Preger et al., 2010). These latter results strongly suggest that, in native grassland soil, organic matter loss cannot be entirely recovered after land-use change.

Soil organic matter accrual after afforestation depends on a number of factors, such as previous land use, plantation age, soil properties, tree species, and climate (Berthrong et al., 2012; Laganière et al., 2010; Li et al., 2012). In the long term, C accumulation is likely to be greater in croplands than in grasslands, and in pastures than in natural grasslands, with broadleaf than with coniferous species, in clay-rich than in coarse-textured soils, and in warmer and dryer than in cooler and wetter regions (Berthrong et al., 2012; Laganière et al., 2010; Li et al., 2012). For instance, in the semiarid Loess Plateau of China, cropland afforestation increased soil organic C stocks after an initial depletion period, with higher relative gain as precipitation declined (Chang et al., 2014). In many dryland ecosystems, the impacts of afforestation on soil properties and C sequestration also highly depend on the planting technique employed. In Southeastern Spain, plantations of tree species, such as *Pinus halepensis* frequently show enhanced runoff and soil losses when compared to adjacent natural shrublands, as well as limited improvement in most physicochemical properties, which rarely reach the values shown by natural shrublands even 40 years after planting (Maestre and Cortina, 2004).

Improved Grazing Management

Intensive livestock grazing has been shown to affect the structure, composition, and functions of dryland ecosystems, including their ability to store soil organic matter mainly by reducing plant cover, decreasing plant-derived organic inputs, and destroying biocrusts (Eldridge et al., 2016; Thomas, 2012). In general, improved practices aimed at lowering grazing intensity, alleviating soil disturbance, and lengthening soil and plant recovery time can help reduce soil organic matter loss and restore C sequestration capacity (Thomas, 2012). In semiarid grasslands of South Africa, soil organic C stocks to 20 cm in farms with rotational grazing were higher by a factor of 1.4 relative to areas with continuous grazing, commonly overstocked and with shorter soil and vegetation resting times (Kotzé et al., 2013). Destocking grazed Australian semiarid shrublands for two decades increased soil C stocks by about 6.5 Mg ha^{-1} (Daryanto et al., 2013). Selective grazing can also reduce soil organic matter loss by promoting plant traits connected to slow C turnover (Bardgett and Wardle, 2003; De Deyn et al., 2008). Controlled stubble grazing after crop harvest versus removal of crop residues is another management option to reduce erosion and maintain or build up organic matter in dryland soils (Blanco-Canqui et al., 2016; FAO, 2004; Lal, 2004b; Stavi et al., 2016).

Improved Management Practices in Croplands

Organic amendments. The application of organic amendments to the soil is a long-established management practice proven to be highly effective to replenish and augment soil organic matter in croplands (Ogle et al., 2005; Paustian, 2014; Senesi et al., 2007). Organic amendments can build up soil organic matter both directly by adding organic C and indirectly by enhancing C inputs from crop residues through improving total plant biomass production (Fernández et al., 2009) (Fig. 2.9). Traditional organic products that have been used for thousands of years include crop residues and animal manures. Organic amendments typically originate from waste materials subjected to appropriate treatments, such as composting. These treatments aim to destroy pathogens, eliminate phytotoxic substances, and stabilize the organic matter components, so that once in soil, they decompose relatively slowly (Senesi et al., 2007). Long-term experiments have shown that the application of organic amendments, such as crop residues, manure, sewage sludge, and municipal solid waste compost, can increase soil organic matter content up to 90% (Diacono and Montemurro, 2010)—and the gains in drylands can be long-lasting even after just a single application (Bastida et al., 2008).

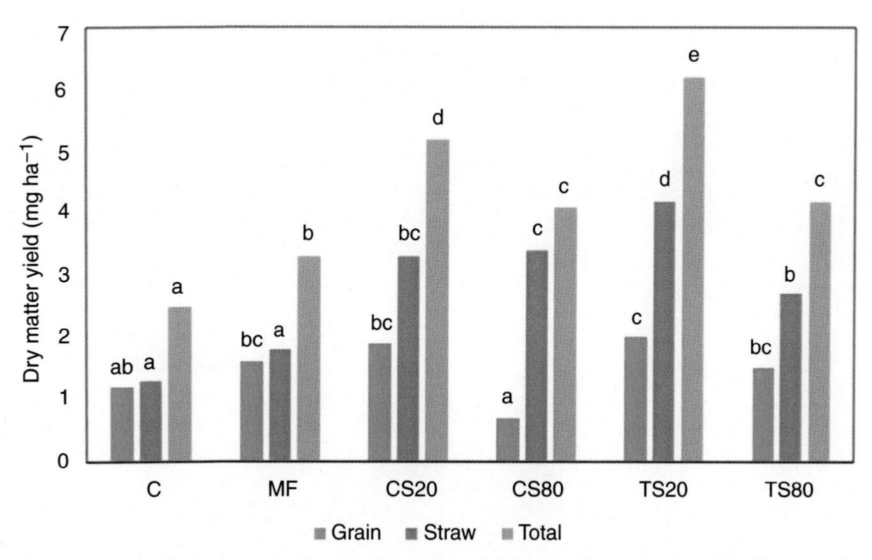

Figure 2.9 ***Grain, straw, and total dry matter yield of winter barley (*****Hordeum vulgare** ***L.) grown on semiarid soils either unamended (C) or amended with mineral fertilizer (MF), 20 and 80 Mg ha^{-1} of sewage sludge compost (CS20 and CS80), and 20 and 80 Mg ha^{-1} of thermally dried sewage sludge (TS20 and TS80).*** Within the same yield component, *different letters* indicate significant differences based on LSD test at the 0.05 level. *(Data From Fernández, J.M., Plaza, C., García-Gil, J.C., Polo, A., 2009. Biochemical properties and barley yield in a semiarid Mediterranean soil amended with two kinds of sewage sludge. Appl. Soil Ecol. 42, 18–24).*

Among the wide range of organic amendments available to farmers, biochar has recently emerged as a highly promising product for climate change mitigation. Biochar is made by the pyrolysis of biomass, and biochar C is believed to form very stable structures, which can be stored in soils for much longer than the biomass C from which it originates (Lehmann, 2007; Lehmann and Joseph, 2015) (Fig. 2.10). The typically high pH values of biochar might be a limiting factor for its application in highly alkaline dryland soils. Nonetheless, under semiarid conditions biochar may also help stabilize organic C from other amendments co-applied to soil by promoting organo–mineral associations (Plaza et al., 2016) (Fig. 2.11).

Mineral fertilization. The correct application of mineral fertilizers, particularly N fertilizers, increases crop production and plant-derived organic inputs into the soil. These inputs, however, may not be substantial enough to offset organic matter losses through mineralization in drylands, and current evidence suggests that it is particularly important to combine mineral fertilization with other recommendable agricultural practices to efficiently

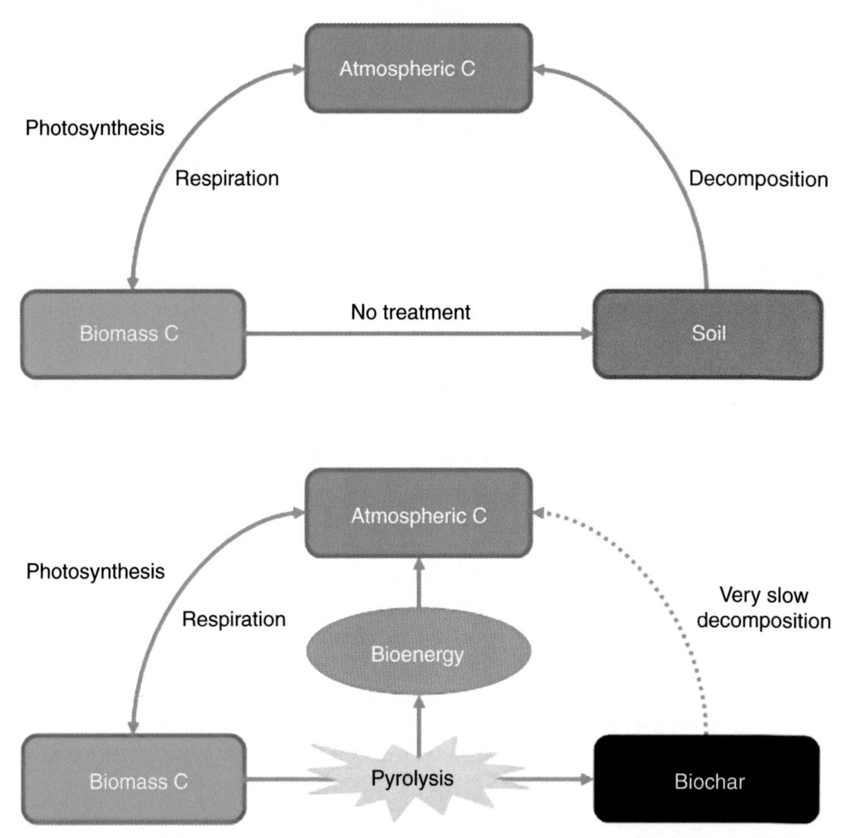

Figure 2.10 ***Carbon biomass cycle and sequestration in biochar.*** **Plants store atmospheric C by photosynthesis in their biomass. Biomass C is released back into the atmosphere by plant respiration or incorporated into the soil. Soil organic C decompose and release C into the atmosphere. The pyrolysis of biomass generates energy and biochar, with very stable C structures that decompose very slowly and can be stored in soils for much longer than the biomass C from which it originates (Lehmann, 2007; Lehmann and Joseph, 2015).**

increase soil organic C stocks. For example, repeated mineral fertilization for 26 years at an arid site in China did not affect the total soil organic matter contents (Hai et al., 2010). Similarly, in a 7-year field experiment in northwestern China under arid conditions, mineral fertilization had little effect on soil organic matter contents, but improved the performance of organic manure (Yang et al., 2016). Practices that retain crop residues in the field have been shown to benefit particularly from mineral fertilization for soil organic matter accrual (Naab et al., 2015).

Crop rotation. Compared to the practice of growing the same crop for many years on the same field (monoculture), crop rotation avoids the

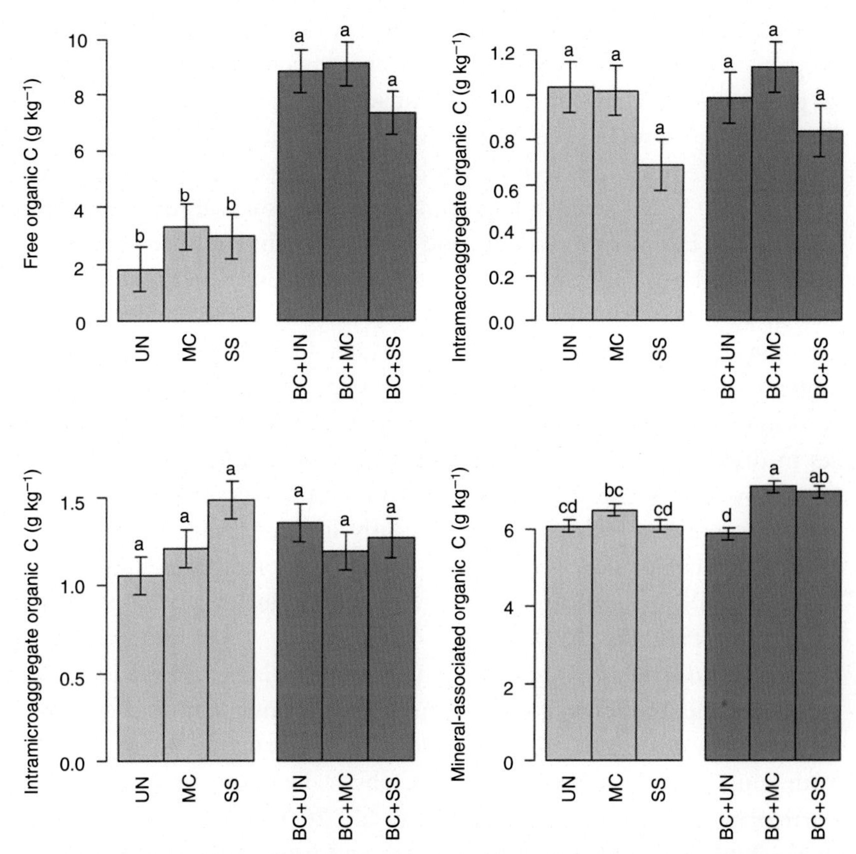

Figure 2.11 ***Free, intramacroaggregate, intramicroaggregate, and mineral-associated organic C content of soils either unamended (UN), amended with municipal solid waste compost (MC), or amended with sewage sludge (SS), without or with biochar (BC).*** *Error bars* indicate pooled standard error. Different *letters* indicate significant differences according to LSD test at the 0.05 level. *(Reprinted from Plaza, C., Giannetta, B., Fernández, J.M., López-de-Sá, E.G., Polo, A., Gascó, G., Méndez, A., Zaccone, C., 2016. Response of different soil organic matter pools to biochar and organic fertilizers. Agric. Ecosyst. Environ. 225, 150–159, with permission from Elsevier).*

disproportional depletion of nutrients and controls weeds as well as pests, thereby improving above- and belowground biomass production. Growing different crops, especially crops that produce large amounts of residue, can augment soil organic matter stocks by increasing organic matter inputs into the soil (Ogle et al., 2005; Palm et al., 2014; West and Post, 2002). For example, an experiment conducted at a semiarid site in the Argentine Pampa showed 3% more soil organic C stocks to 100 cm after 15 years under a soybean–maize alternation than under a soybean monoculture (Alvarez et al., 2014).

Cover crops. A cover crop, consisting of plants planted between cash crops to conserve and enrich the soil can provide multiple benefits, such as preventing soil erosion, storing nutrients, suppressing weeds, controlling pests and diseases, and improving soil structure. Blanco-Canqui et al. (2015) showed that cover crops can boost soil organic C stocks from 0.01 to 0.1 kg m^{-2} $year^{-1}$, depending on factors, such as initial soil organic matter content, biomass production, and climate. In drylands, cover crops can build up soil organic matter by protecting soil from water and wind erosion and by incorporating extra plant residues into the soil. However, cover crops may depress main crop yields in water-limited regions by reducing water availability (Blanco-Canqui et al., 2015), which may counterbalance to some extent the organic matter inputs. Nonetheless, a number of studies in drylands indicate that reducing or eliminating the fallow period by using cover crops, especially if combined with practices that enhance water storage, such as no tillage, can significantly increase plant-derived organic matter inputs and slow erosion and decomposition rates, resulting in net gains in soil organic matter stocks (Blanco-Canqui et al., 2013; Mitchell et al., 2017; Paustian, 2014).

Conservation tillage. In contrast to intensive tillage, conservation practices that entail little or no soil disturbance can maintain and even build up surface soil organic matter by slowing macroaggregate turnover, encouraging microaggregate formation, physically protecting crop-derived particulate organic matter (Paustian et al., 2000; Plaza et al., 2013; Six et al., 2000, 2004; West and Post, 2002), and prompting the formation of primary organo–mineral complexes, particularly with microbial material (Plaza et al., 2013) (Figs. 2.12 and 2.13). A recent meta-analysis using 46 peer-reviewed publications and 174 paired observations showed that tilled soils emitted 21% more CO_2 than untilled soils across different climates, crop types, and soil conditions, and the difference increased to 29% in sandy soils from arid climates with low soil organic C content and low soil moisture (Abdalla et al., 2016). In semiarid central Spain, 25 years of no-tillage, compared to chisel plowing, resulted in 16% more organic C in the uppermost 20 cm of soil, especially because of an enhanced interaction between mineral particles and microbial material that resulted in the formation of highly stable organo–mineral complexes (Plaza et al., 2013). In southern Texas, under subhumid conditions, no-tillage resulted in 64% greater soil organic matter stocks compared with the average pools resulting from different tillage treatments of increasing intensity after 16 years (Salinas-Garcia et al., 1997).

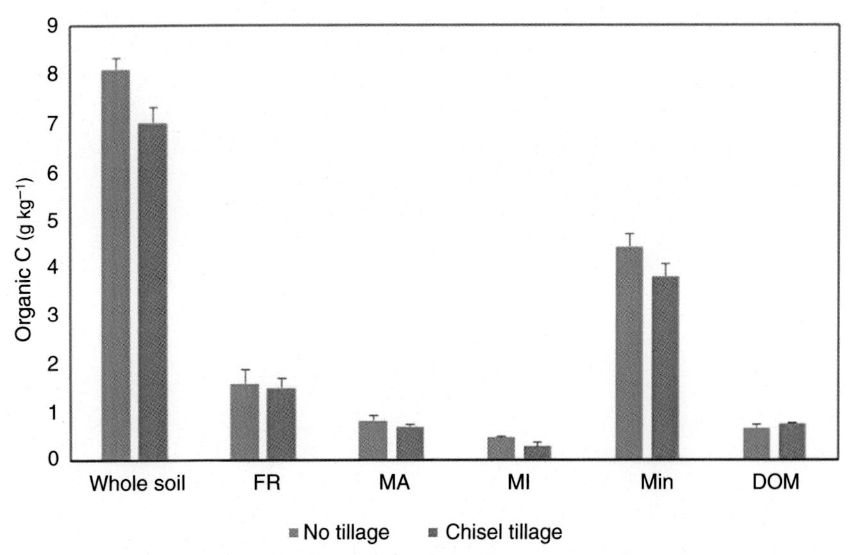

Figure 2.12 ***Organic C content (on whole soil basis) of no-tillage and chisel tilled semi-arid soils and in the corresponding free (FR), intramacroaggregate (MA), intramicroaggregate (MI), mineral-associated (Min), and dissolved (DOM) organic matter fractions.*** Mean ± standard error, $n = 3$. *(Data from Plaza, C., Courtier-Murias, D., Fernández, J.M., Polo, A., Simpson, A.J., 2013. Physical, chemical, and biochemical mechanisms of soil organic matter stabilization under conservation tillage systems: a central role for microbes and microbial by-products in C sequestration. Soil Biol. Biochem. 57, 124–134).*

Efficient irrigation. Matching irrigation to crop water demands can raise yields, crop residue production, and plant-derived inputs into the soil, thus building up soil organic matter stocks (Zhang et al., 2015). Especially in areas where freshwater is scarce, the use of wastewater for agricultural irrigation is steadily becoming more important because of the increasing pressure on water resources (Lado and Ben-Hur, 2009). In addition to being a more sustainable practice of water management, wastewater irrigation has the additional benefit of enriching soil fertility by adding nutrients; nonetheless, the safe and efficient use of effluents requires a careful control of potential risks, such as soil salinization and contamination (Halliwell et al., 2001; Lado and Ben-Hur, 2009). At three sites in southeastern Spain, irrigation with wastewater for 20–40 years was found not to adversely affect soil properties and in fact increased soil organic matter stocks with respect to irrigation with freshwater (Morugan-Coronado et al., 2013). However, some studies have pointed out the risk of greater mineralization of native soil organic matter (positive priming effects) because of the addition of fresh dissolved organic matter (Lado and Ben-Hur, 2009).

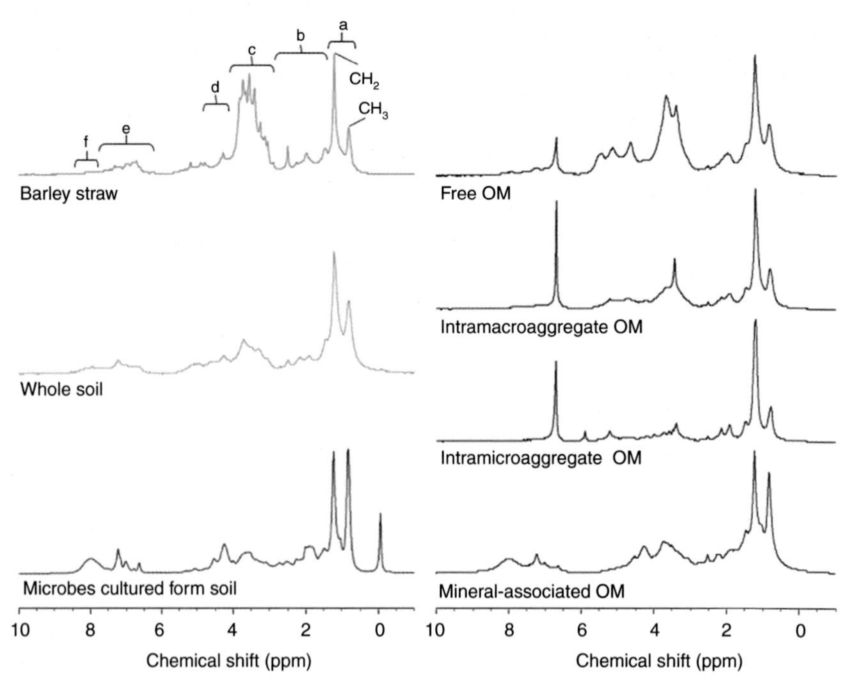

Figure 2.13 ***Diffusion edited ^1H HR-MAS NMR spectra of microbes cultured from soil, barley straw, semiarid soils under chisel tillage for 25 years, and their respective free, intramacroaggregate, intramicroaggregate, and mineral-associated organic matter (OM) pools.*** General assignments of chemical shift ranges are as follows: (a) CH_3 and CH_2 (0.6–1.3 ppm); (b) CH_3 and CH_2 near O and N (1.3–2.9 ppm); (c) O-alkyl, mainly from carbohydrates and lignin (2.9–4.1 ppm); (d) α-H from peptides (4.1–4.8 ppm); (e) aromatic, from lignin and peptides (6.2–7.8 ppm); and (f) amide in peptides (7.8–8.4 ppm). It is noteworthy that the spectrum of the mineral-associated OM pool is very similar to that of soil microbial cultures, whereas the spectra of the other OM fractions more closely resemble that of barley straw. *(Spectra redrawn from Plaza, C., Courtier-Murias, D., Fernández, J.M., Polo, A., Simpson, A.J., 2013. Physical, chemical, and biochemical mechanisms of soil organic matter stabilization under conservation tillage systems: a central role for microbes and microbial by-products in C sequestration. Soil Biol. Biochem. 57, 124–134 and Courtier-Murias, D., Simpson, A.J., Marzadori, C., Baldoni, G., Ciavatta, C., Fernández, J.M., et al., 2013. Unraveling the long-term stabilization mechanisms of organic materials in soils by physical fractionation and NMR spectroscopy. Agric. Ecosyst. Environ. 171, 9–18, with permission from Elsevier).*

CONCLUDING REMARKS

As a whole, soil organic C concentrations in drylands are very low compared to those in other ecosystems. At the same time, drylands cover 45% of Earth's land surface, and thus the total amount of organic C stored in drylands becomes a significant component of the global C cycle. Unfortunately, while other ecosystems traditionally considered as C sinks (e.g., peatlands) have been intensively investigated, estimates of organic C stocks in dryland soils are still very poor and need to be improved strongly. This is extremely important in order to understand their vulnerability to climate change, as well as to develop strategies to recover, preserve, or enhance the ecosystem services provided, which are mostly affected by soil organic matter. Moreover, most drylands are located in poor or developing countries and host about one third of the global human population. As a consequence, further C losses from these soils may have long-term adverse impacts on soil quality and global food security and, in turn, on social unrest and political instability. In conclusion, soil organic matter plays a key role in soil conservation in several countries, and in drylands in particular, where it represents "the link" among global issues, including land degradation, food security, and climate change.

ACKNOWLEDGMENTS

This project has received funding from the European Union's Horizon 2020 research and innovation programme under the Marie Skłodowska-Curie grant agreement No 654132, and from the Spanish Ministry of Economy and Competitiveness under the grants AGL2013-48681 and AGL2016-75762-R. FTM acknowledges support from the European Research Council (ERC Grant Agreement 647038 [BIODESERT]) and the Spanish Ministry of Economy and Competitiveness (BIOMOD project, CGL2013-44661-R).

REFERENCES

Abdalla, K., Chivenge, P., Ciais, P., Chaplot, V., 2016. No-tillage lessens soil CO_2 emissions the most under arid and sandy soil conditions: results from a meta-analysis. Biogeosciences 13, 3619–3633.

Alvarez, C., Alvarez, C.R., Costantini, A., Basanta, M., 2014. Carbon and nitrogen sequestration in soils under different management in the semi-arid Pampa (Argentina). Soil Till. Res. 142, 25–31.

Archer, S., 1994. Woody plant encroachment into southwestern grasslands and savannas: rates, patterns and proximate causes. In: Vavra, M., Laycock, W., Pieper, R. (Eds.), Ecological Implications of Livestock Herbivory in the West. Society for Range Management, Denver, CO, pp. 13–68.

Bardgett, R.D., 2005. The Biology of Soil. Oxford University Press, Oxford, UK.

Bardgett, R.D., Wardle, D.A., 2003. Herbivore-mediated linkages between aboveground and belowground communities. Ecology 84, 2258–2268.

Barger, N.N., Archer, S.R., Campbell, J.L., Huang, C., Morton, J.A., Knapp, A.K., 2011. Woody plant proliferation in North American drylands: a synthesis of impacts on ecosystem carbon balance. J. Geophys. Res. Biogeosci. 116, G00K07.

Bastida, F., Kandeler, E., Hernández, T., García, C., 2008. Long-term effect of municipal solid waste amendment on microbial abundance and humus-associated enzyme activities under semiarid conditions. Microbial. Ecol. 55, 651–661.

Batjes, N.H., 1996. Total carbon and nitrogen in the soils of the world. Eur. J. Soil Sci. 47, 151–163.

Belnap, J., 2013. Biological soil crusts and wind erosion. In: Belnap, J., Lange, O.L. (Eds.), Biological Soil Crusts: Structure, Function, and Management. Springer, Berlin, Germany.

Belnap, J., Lange, O.L., 2013. Biological Soil Crusts: Structure, Function and Management. Springer, Berlin, Germany.

Berdugo, M., Kéfi, S., Soliveres, S., Maestre, F.T., 2017. Plant spatial patterns identify alternative ecosystem multifunctionality states in global drylands. Nature Ecol. Evol. 1, 0003.

Berthrong, S.T., Piñeiro, G., Jobbágy, E.G., Jackson, R.B., 2012. Soil C and N changes with afforestation of grasslands across gradients of precipitation and plantation age. Ecol. Appl. 22, 76–86.

Biedenbender, S.H., McClaren, M.P., Quade, J., Weltz, M.A., 2004. Landscape patterns of vegetation change indicated by soil carbon isotope composition. Geoderma 119, 69–83.

Blanco-Canqui, H., Holman, J.D., Schlegel, A.J., Tatarko, J., Shaver, T.M., 2013. Replacing fallow with cover crops in a semiarid soil: effects on soil properties. Soil Sci. Soc. Am. J. 77, 1026–1034.

Blanco-Canqui, H., Shaver, T.M., Lindquist, J.L., Shapiro, C.A., Elmore, R.W., Francis, C.A., Hergert, G.W., 2015. Cover crops and ecosystem services: insights from studies in temperate soils. Agron. J. 107, 2449–2474.

Blanco-Canqui, H., Tatarko, J., Stalker, A.L., Shaver, T.M., Van Donk, S.J., 2016. Impacts of corn residue grazing and baling on wind erosion potential in a semiarid environment. Soil Sci. Soc. Am. J. 80, 1027–1037.

Bobbink, R., Hicks, K., Galloway, J., Spranger, T., Alkemade, R., Ashmore, M., et al., 2010. Global assessment of nitrogen deposition effects on terrestrial plant diversity: a synthesis. Ecol. Appl. 20, 30–59.

Bot, A., Benites, J., 2005. The Importance of Soil Organic Matter: Key to Drought-Resistant Soil and Sustained Food and Production. FAO Soils Bulletin 80. FAO, Rome, Italy.

Bowker, M.A., 2007. Biological soil crust rehabilitation in theory and practice: an underexploited opportunity. Restor. Ecol. 15, 13–23.

Brazier, R.E., Turnbull, L., Wainwright, J., Bol, R., 2014. Carbon loss by water erosion in drylands: implications from a study of vegetation change in the south-west USA. Hydrol. Process. 28, 2212–2222.

Castillo-Monroy, A.P., Maestre, F.T., Rey, A., Soliveres, S., García-Palacios, P., 2011. Biological soil crust microsites are the main contributor to soil respiration in a semiarid ecosystem. Ecosystems 14, 835–847.

Chang, R., Jin, T., Lü, Y., Liu, G., Fu, B., 2014. Soil carbon and nitrogen changes following afforestation of marginal cropland across a precipitation gradient in Loess Plateau of China. PLoS One 9 (1), e85426.

Ciais, P., Bombelli, A., Williams, M., Piao, S.L., Chave, J., Ryan, C.M., et al., 2011. The carbon balance of Africa: synthesis of recent research studies. Philos. Trans. R. Soc. A 369, 2038–2057.

Cook, B.I., Ault, T.R., Smerdon, J.E., 2015. Unprecedented 21st century drought risk in the American Southwest and Central Plains. Sci. Adv. 1, e1400082.

Courtier-Murias, D., Simpson, A.J., Marzadori, C., Baldoni, G., Ciavatta, C., Fernández, J.M., et al., 2013. Unraveling the long-term stabilization mechanisms of organic materials in soils by physical fractionation and NMR spectroscopy. Agric. Ecosyst. Environ. 171, 9–18.

Dalal, R.C., Chan, K.Y., 2001. Soil organic matter in rainfed cropping systems of the Australian cereal belt. Aust. J. Soil Res. 39, 435–464.
Darrouzet-Nardi, A., 2013. How Much Carbon Is Stored in Dryland Soils? , http://anthony.darrouzet-nardi.net/scienceblog/?p=1664.
Daryanto, S., Eldridge, D.J., Throop, H.L., 2013. Managing semi-arid woodlands for carbon storage: grazing and shrub effects on above- and belowground carbon. Agric. Ecosyst. Environ. 169, 1–11.
De Deyn, G.B., Cornelissen, J.H.C., Bardgett, R.D., 2008. Plant functional traits and soil carbon sequestration in contrasting biomes. Ecol. Lett. 11, 516–531.
Delgado-Baquerizo, M., Maestre, F.T., Gallardo, A., Bowker, M.A., Wallenstein, M.D., Quero, J.L., et al., 2013. Decoupling of soil nutrient cycles as a function of aridity in global drylands. Nature 502, 672–676.
Delgado-Baquerizo, M., Maestre, F.T., Reich, P.B., Jeffries, T.C., Gaitán, J.J., Encinar, D., et al., 2016. Microbial diversity drives multifunctionality in terrestrial ecosystems. Nat. Commun. 7, 10541.
Diacono, M., Montemurro, F., 2010. Long-term effects of organic amendments on soil fertility: a review. Agron. Sustain. Dev. 30, 401–422.
Dregne, H.E., 1976. Soils of arid regions. Developments in Soil Sciencevol. 6Elsevier, Amsterdam, The Netherlands.
Elbert, W., Weber, B., Büdel, B., Andreae, M.O., Pöschl, U., 2009. Microbiotic crusts on soil, rock and plants: neglected major players in the global cycles of carbon and nitrogen? Biogeosci. Discuss. 6, 6983–7015.
Elbert, W., Weber, B., Burrows, S., Steinkamp, J., Büdel, B., Andreae, M.O., et al., 2012. Contribution of cryptogamic covers to the global cycles of carbon and nitrogen. Nat. Geosci. 5, 459–462.
Eldridge, D.J., Bowker, M.A., Maestre, F.T., Reynolds, J.F., Roger, E., Whitford, W.G., 2011. Impacts of shrub encroachment on ecosystem structure and functioning: towards a global synthesis. Ecol. Lett. 14, 709–722.
Eldridge, D.J., Leys, J.F., 2003. Exploring some relations between biological soil crusts, soil aggregation and wind erosion. J. Arid Environ. 53, 453–466.
Eldridge, D.J., Poore, A.G.B., Ruiz-Colmenero, M., Letnic, M., Soliveres, S., 2016. Ecosystem structure, function, and composition in rangelands are negatively affected by livestock grazing. Ecol. Appl. 26, 1273–1283.
FAO, 1993. Land degradation in arid, semi-arid and dry sub-humid areas: rainfed and irrigated lands, rangelands and woodlands. Text for FAO presentation at inter-governmental negotiating committee for the preparation of a convention to combat desertification and drought (INCD). FAO, Rome, Italy.
FAO, 2004. Carbon Sequestration in Dryland Soils. World Soil Resources Reports 102. FAO, Rome, Italy.
FAO/IIASA/ISRIC/ISSCAS/JRC, 2012. Harmonized World Soil Database (Version 1.2). FAO, Rome, Italy and IIASA, Laxenburg, Austria.
Fernández, J.M., Plaza, C., García-Gil, J.C., Polo, A., 2009. Biochemical properties and barley yield in a semiarid Mediterranean soil amended with two kinds of sewage sludge. Appl. Soil Ecol. 42, 18–24.
Ferrenberg, S., Reed, S.C., Belnap, J., 2015. Climate change and physical disturbance cause similar community shifts in biological soil crusts. Proc. Natl. Acad. Sci. U.S.A. 112, 12116–12121.
Fog, K., 1988. The effect of added nitrogen on the rate of decomposition of organic matter. Biol. Rev. 63, 433–462.
Foley, J.A., Ramankutty, N., Brauman, K.A., Cassidy, E.S., Gerber, J.S., Johnston, M., et al., 2011. Nature 478, 337–342.

Fu, Q., Feng, S., 2014. Responses of terrestrial aridity to global warming. J. Geophys. Res. Atmos. 119, 7863–7875.
Gaitán, J.J., Bran, D.E., Oliva, G.E., Aguiar, M.R., Buono, G.G., Ferrante, D., et al., 2017. Aridity and overgrazing have convergent effects on ecosystem structure and functioning in Patagonian rangelands. Land Degrad. Dev.doi: 10.1002/ldr.2694.
Gelaw, A.M., Singh, B.R., Lal, R., 2014. Soil organic carbon and total nitrogen stocks under different land uses in a semi-arid watershed in Tigray, Northern Ethiopia. Agric. Ecosyst. Environ. 188, 256–263.
Golchin, A., Oades, J.M., Skjemstad, J.O., Clarke, P., 1994. Soil structure and carbon cycling. Aust. J. Soil Res. 32, 1043–1068.
Gruber, N., Galloway, J.N., 2008. An Earth-system perspective of the global nitrogen cycle. Nature 451, 293–296.
Guo, L.B., Gifford, R.M., 2002. Soil carbon stocks and land use change: a meta analysis. Global Change Biol. 8, 345–360.
Hai, L., Li, X.G., Li, F.M., Suo, D.R., Guggenberger, G., 2010. Long-term fertilization and manuring effects on physically-separated soil organic matter pools under a wheat-wheat-maize cropping system in an arid region of China. Soil Biol. Biochem. 42, 253–259.
Halliwell, D.J., Barlow, K.M., Nash, D.M., 2001. A review of the effects of wastewater sodium on soil physical properties and their implications for irrigation systems. Aust. J. Soil Res. 39, 1259–1267.
Haygarth, P., Ritz, K., 2009. The future of soils and land use in the UK: soil systems for the provision of land-based ecosystem services. Land Use Policy 26, S187–S197.
Haynes, R.J., Williams, P.H., 1999. The influence of stock camping behaviors on the soil microbial and biochemical properties of grazed pastoral soils. Biol. Fert. Soils 28, 253–258.
Huang, J., Yu, H., Guan, X., Wang, G., Guo, R., 2016. Accelerated dryland expansion under climate change. Nat. Clim. Change 6, 166–171.
Hugelius, G., Strauss, J., Zubrzycki, S., Harden, J.W., Schuur, E.A.G., Ping, C.L., et al., 2014. Estimated stocks of circumpolar permafrost carbon with quantified uncertainty ranges and identified data gaps. Biogeosciences 11, 6573–6593.
IIASA/FAO, 2012. Global Agro-Ecological Zones (GAEZ v3.0). IIASA/FAO, Laxenburg, Austria/Rome, Italy.
IPCC, 2013. Climate change 2013: the physical science basis. In: Stocker, T.F., Qin, D., Plattner, G.-K., Tignor, M., Allen, S.K., Boschung, J., Nauels, A., Xia, Y., Bex, V., Midgley, P.M. (Eds.), Contribution of Working Group I to the Fifth Assessment Report of the Intergovernmental Panel on Climate Change. Cambridge University Press, Cambridge, UK.
Jackson, R.B., Banner, J.L., Jobbágy, E.G., Pockman, W.T., Wall, D.H., 2002. Ecosystem carbon loss with woody plant invasion of grasslands. Nature 418, 623–626.
Jenny, H., 1941. Factors of Soil Formation. McGraw-Hill, New York, NY.
Jobbágy, E.G., Jackson, R.B., 2000. The vertical distribution of soil organic carbon and its relation to climate and vegetation. Ecol. Appl. 10, 423–436.
Johnson, D.W., Curtis, P.S., 2001. Effects of forest management on soil C and N storage: meta analysis. Forest Ecol. Manage. 140, 227–238.
Kleber, M., Eusterhues, K., Keiluweit, M., Mikutta, C., Mikutta, R., Nico, P.S., 2015. Mineral–organic associations: formation, properties, and relevance in soil environments. Adv. Agron. 130, 1–140.
Köchy, M., Hiederer, R., Freibauer, A., 2015. Global distribution of soil organic carbon—Part 1: masses and frequency distributions of SOC stocks for the tropics, permafrost regions, wetlands, and the world. Soil 1, 351–365.
Kotzé, E., Sandhage-Hofmann, A., Meinel, J.-A., du Preeza, C.C., Amelung, W., 2013. Rangeland management impacts on the properties of clayey soils along grazing gradients in the semi-arid grassland biome of South Africa. J. Arid Environ. 97, 220–229.

Lado, M., Ben-Hur, M., 2009. Treated domestic sewage irrigation effects on soil hydraulic properties in arid and semiarid zones: a review. Soil Till. Res. 106, 152–163.
Laganière, J., Angers, D.A., Paré, D., 2010. Carbon accumulation in agricultural soils after afforestation: a meta-analysis. Global Change Biol. 16, 439–453.
Lal, R., 2001a. Carbon sequestration in dryland ecosystems of West Asia and North Africa. Land Degrad. Dev. 13, 45–59.
Lal, R., 2001b. Potential of desertification control to sequester carbon and mitigate the greenhouse effect. Climatic Change 51, 35–72.
Lal, R., 2004a. Carbon sequestration in dryland ecosystems. Environ. Manage. 33, 528–544.
Lal, R., 2004b. Soil carbon sequestration to mitigate climate change. Geoderma 123, 1–22.
Lal, R., 2004c. Soil carbon sequestration impacts on global climate change and food security. Science 304, 1623–1627.
Lal, R., 2009. Challenges and opportunities in soil organic matter research. Eur. J. Soil Sci. 60, 158–169.
Lan, Z., Bai, Y., 2012. Testing mechanisms of N-enrichment induced species loss in a semiarid inner Mongolia grassland: critical thresholds and implications for long-term ecosystem responses. Philos. Trans. R. Soc. B 367, 3125–3134.
Lehmann, J., 2007. A handful of carbon. Nature 447, 143–144.
Lehmann, C.E.R., Anderson, T.M., Sankaran, M., Higgins, S.I., Archibald, S., Hoffmann, W.A., et al., 2014. Savanna vegetation-fire-climate relationships differ among continents. Science 343, 548–552.
Lehmann, J., Joseph, S., 2015. Biochar for Environmental Management: Science and Technology, second ed. Earthscan, London, UK.
Lehmann, J., Kleber, M., 2015. The contentious nature of soil organic matter. Nature 528, 60–68.
Li, D., Niu, S., Luo, Y., 2012. Global patterns of the dynamics of soil carbon and nitrogen stocks following afforestation: a meta-analysis. New Phytol. 195, 172–181.
Lloyd, J., 1999. The CO_2 dependence of photosynthesis, plant growth responses to elevated CO_2 concentrations and their interactions with soil nutrient status. II. Temperature and boreal forest productivity and the combined effects of increasing CO_2 concentrations and increased nitrogen deposition. Funct. Ecol. 13, 439–459.
Maestre, F.T., Bowker, M.A., Puche, M.D., Hinojosa, M.B., Martínez, I., García-Palacios, P., et al., 2009. Shrub encroachment can reverse desertification in Mediterranean semiarid grasslands. Ecol. Lett. 12, 930–941.
Maestre, F.T., Cortina, J., 2004. Are *Pinus halepensis* plantations useful as a restoration tool in semiarid Mediterranean areas? Forest Ecol. Manage. 198, 303–317.
Maestre, F.T., Delgado-Baquerizo, M., Jeffries, T.C., Ochoa, V., Gozalo, B., Quero, J.L., et al., 2015. Increasing aridity reduces soil microbial diversity and abundance in global drylands. Proc. Natl. Acad. Sci. U.S.A. 112, 15684–15689.
Maestre, F.T., Eldridge, D.J., Soliveres, S., Kéfi, S., Delgado-Baquerizo, M., Bowker, M.A., et al., 2016. Structure and functioning of dryland ecosystems in a changing world. Annu. Rev. Ecol. Evol. Syst. 47, 215–237.
Maestre, F.T., Escolar, C., Ladrón de Guevara, M., Quero, J.L., Lázaro, R., Delgado-Baquerizo, M., et al., 2013. Changes in biocrust cover drive carbon cycle responses to climate change in drylands. Global Change Biol. 19, 3835–3847.
Martínez-Mena, M., Álvarez Rogel, J., Castillo, V., Albaladejo, J., 2002. Organic carbon and nitrogen influenced by vegetation removal in a semiarid Mediterranean soil. Biogeochemistry 61, 309–321.
Matson, P.A., Parton, W.J., Power, A.G., Swift, M.J., 1997. Agricultural intensification and ecosystem properties. Science 277, 504–509.
McLeod, M.K., Schwenke, G.D., Cowie, A.L., Harden, S., 2013. Soil carbon is only higher in the surface soil under minimum tillage in Vertosols and Chromosols of New South Wales North-West Slopes and Plains, Australia. Soil Res. 51, 680–694.

Middleton, N.J., Sternberg, T., 2013. Climate hazards in drylands: a review. Earth Sci. Rev. 126, 48–57.
Middleton, N., Stringer, L., Goudie, A., Thomas, D., 2011. The Forgotten Billion. MDG Achievement in the Drylands. UNDP, UNCCD, New York, NY.
Mitchell, J.P., Shrestha, A., Mathesius, K., Scow, K.M., Southard, R.J., Haney, R.L., et al., 2017. Cover cropping and no-tillage improve soil health in an arid irrigated cropping system in California's San Joaquin Valley, USA. Soil Till. Res. 165, 325–335.
Morgan, R.P.C., 2005. Soil Erosion and Conservation, third ed. Blackwell Science, Malden, MA.
Morugan-Coronado, A., Arcenegui, V., García-Orenes, F., Mataix-Solera, J., Mataix-Beneyto, J., 2013. Application of soil quality indices to assess the status of agricultural soils irrigated with treated wastewaters. Solid Earth 4, 119–127.
Naab, J.B., Mahama, G.Y., Koo, J., Jones, J.W., Boote, K.J., 2015. Nitrogen and phosphorus fertilization with crop residue retention enhances crop productivity, soil organic carbon, and total soil nitrogen concentrations in sandy-loam soils in Ghana. Nutr. Cycl. Agroecosyst. 102, 33–43.
Nicholson, S.E., 2011. Dryland Climatology. Cambridge University Press, Cambridge, UK.
Ochoa-Hueso, R., Allen, E.B., Branquinho, C., Cruz, C., Dias, T., Fenn, M.E., et al., 2011. Nitrogen deposition effects on Mediterranean-type ecosystems: an ecological assessment. Environ. Pollut. 159, 2265–2279.
Ochoa-Hueso, R., Delgado-Baquerizo, M., Gallardo, A., Bowker, M.A., Maestre, F.T., 2016. Climatic conditions, soil fertility and atmospheric nitrogen deposition largely determine the structure and functioning of microbial communities in biocrust-dominated Mediterranean drylands. Plant Soil 399, 271–282.
Ochoa-Hueso, R., Maestre, F.T., de los Ríos, A., Valea, S., Theobald, M.R., Vivanco, M.G., et al., 2013. Nitrogen deposition alters nitrogen cycling and reduces soil carbon content in low-productivity semiarid Mediterranean ecosystems. Environ. Pollut. 179, 185–193.
Ogle, S.M., Breidt, F.J., Paustian, K., 2005. Agricultural management impacts on soil organic carbon storage under moist and dry climatic conditions of temperate and tropical regions. Biogeochemistry 72, 87–121.
Palm, C., Blanco-Canqui, H., DeClerck, F., Gatere, L., Grace, P., 2014. Conservation agriculture and ecosystem services: an overview. Agric. Ecosyst. Environ. 187, 87–105.
Paul, E.A., 2016. The nature and dynamics of soil organic matter: plant inputs, microbial transformations, and organic matter stabilization. Soil Biol. Biochem. 98, 109–126.
Paul, E.A., Paustian, K.H., Elliott, E.T., Cole, C.V., 1997. Soil Organic Matter in Temperate Agroecosystems: Long-Term Experiments in North America. CRC Press, Boca Raton, FL.
Paustian, K., 2014. Soil: carbon sequestration in agricultural systems. In: Van Alfen, N. (Ed.), Encyclopedia of Agriculture and Food Systems. Academic Press, London, UK, pp. 140–152.
Paustian, K., Six, J., Elliott, E.T., Hunt, H.W., 2000. Management options for reducing CO_2 emissions from agricultural soils. Biogeochemistry 48, 147–163.
Piñeiro, G., Paruelo, J.M., Oesterheld, M., Jobbágy, E.G., 2010. Pathways of grazing effects on soil organic carbon and nitrogen. Rangeland Ecol. Manage. 63, 109–119.
Plaza, C., Courtier-Murias, D., Fernández, J.M., Polo, A., Simpson, A.J., 2013. Physical, chemical, and biochemical mechanisms of soil organic matter stabilization under conservation tillage systems: a central role for microbes and microbial by-products in C sequestration. Soil Biol. Biochem. 57, 124–134.
Plaza, C., Fernández, J.M., Pereira, E.I.P., Polo, A., 2012. A comprehensive method for fractionating soil organic matter into pools not protected and protected from decomposition by physical and chemical mechanisms. CLEAN – Soil Air Water 40, 134–139.
Plaza, C., Giannetta, B., Fernández, J.M., López-de-Sá, E.G., Polo, A., Gascó, G., Méndez, A., Zaccone, C., 2016. Response of different soil organic matter pools to biochar and organic fertilizers. Agric. Ecosyst. Environ. 225, 150–159.

Post, W.M., Kwon, K.C., 2000. Soil carbon sequestration and land-use change: processes and potential. Global Change Biol. 6, 317–327.
Prăvălie, R., 2016. Drylands extent and environmental issues: a global approach. Earth-Sci. Rev. 161, 259–278.
Preger, A.C., Kösters, R., Du Preez, C.C., Brodowski, S., Amelung, W., 2010. Carbon sequestration in secondary pasture soils: a chronosequence study in the South African Highveld. Eur. J. Soil Sci. 61, 551–562.
Puttock, A., Dungait, J.A.J., Macleod, C.J.A., Bol, R., Brazier, R.E., 2014. Woody plant encroachment into grasslands leads to accelerated erosion of previously stable organic carbon from dryland soils. J. Geophys. Res. G Biogeosci. 119, 2345–2357.
Reynolds, J.F., Stafford Smith, D.M., Lambin, E.F., Turner II, B.L., Mortimore, M., Batterbury, S.P.J., et al., 2007. Global desertification: building a science for dryland development. Science 316, 847–851.
Safriel, U., Adeel, Z., Niemeijer, D., Puigdefabregas, J., White, R., Lal, R., et al., 2005. Dryland systems. Hassan, R., Scholes, R., Ash, N. (Eds.), The Millennium Ecosystem Assessment Series, Ecosystems and Human Well-being: Current State and Trends, vol. 1, Island Press, Washington, DC, pp. 623–662.
Salinas-Garcia, J.R., Matocha, J.E., Hons, F.M., 1997. Long-term tillage and nitrogen fertilization effects on soil properties of an Alfisol under dryland corn/cotton production. Soil Till. Res. 42, 79–93.
Schmidt, M.W.I., Torn, M.S., Abiven, S., Dittmar, T., Guggenberger, G., Janssens, I.A., et al., 2011. Persistence of soil organic matter as an ecosystem property. Nature 478, 49–56.
Schulze, E.D., Freibauer, A., 2005. Carbon unlocked from soils. Nature 437, 205–206.
Schuur, E.A.G., McGuire, A.D., Schädel, C., Grosse, G., Harden, J.W., Hayes, D.J., et al., 2015. Climate change and the permafrost carbon feedback. Nature 520, 171–179.
Senesi, N., Plaza, C., Brunetti, G., Polo, A., 2007. A comparative survey of recent results on humic-like fractions in organic amendments and effects on native soil humic substances. Soil Biol. Biochem. 39, 1244–1262.
Six, J., Conant, R.T., Paul, E.A., Paustian, K., 2002. Stabilization mechanisms of soil organic matter: implications for C-saturation of soils. Plant Soil 241, 155–176.
Six, J., Elliott, E.T., Paustian, K., 1999. Aggregate and soil organic matter dynamics under conventional and no-tillage systems. Soil Sci. Soc. Am. J. 63, 1350–1358.
Six, J., Elliott, E.T., Paustian, K., 2000. Soil macroaggregate turnover and microaggregate formation: a mechanism for C sequestration under no-tillage agriculture. Soil Biol. Biochem. 32, 2099–2103.
Six, J., Elliott, E.T., Paustian, K., Doran, J.W., 1998. Aggregation and soil organic matter accumulation in cultivated and native grassland soils. Soil Sci. Soc. Am. J. 62, 1367–1377.
Six, J., Ogle, S.M., Breidt, F.J., Conant, R.T., Mosier, A.R., Paustian, K., 2004. The potential to mitigate global warming with no-tillage management is only realized when practiced in the long term. Global Change Biol. 10, 155–160.
Skujins, J., 1991. Semiarid Lands and Deserts: Soil Resource and Reclamation. Marcel Dekker, New York, NY.
Sollins, P., Homann, P., Caldwell, B.A., 1996. Stabilization and destabilization of soil organic matter: mechanisms and controls. Geoderma 74, 65–105.
Stavi, I., Argaman, E., Zaady, E., 2016. Positive impact of moderate stuble grazing on soil quality and organic carbon pool in dryland wheat agro-pastoral systems. Catena 146, 94–99.
Thomas, A.D., 2012. Impact of grazing intensity on seasonal variations in soil organic carbon and soil CO_2 efflux in two semiarid grasslands in southern Botswana. Philos. Trans. R. Soc. B 367, 3076–3086.
Thornton, P.K., 2010. Livestock production: recent trends, future prospects. Philos. Trans. R. Soc. B 365, 2853–2867.

Tilman, D., Balzer, C., Hill, J., Befort, B.L., 2011. Global food demand and the sustainable intensification of agriculture. Proc. Natl. Acad. Sci. U.S.A. 108, 20260–20264.

Torn, M.S., Swanston, C.W., Castanha, C., Trumbore, S.E., 2009. Storage and turnover of natural organic matter in soil. In: Huang, P.M., Senesi, N. (Eds.), Biophysico-Chemical Processes Involving Natural Nonliving Organic Matter in Environmental Systems. John Wiley & Sons, Hoboken, NJ.

Torn, M.S., Trumbore, S.E., Chadwick, O.A., Vitousek, P.M., Hendricks, D.M., 1997. Mineral control of soil organic carbon storage and turnover. Nature 389, 170–173.

Trumbore, S., 2009. Radiocarbon and soil carbon dynamics. Annu. Rev. Earth Planet. Sci. 37, 47–66.

Turnbull, L., Wainwright, J., Brazier, R.E., 2010. Changes in hydrology and erosion over a transition from grassland to shrubland. Hydrol. Process. 24, 393–414.

UN Environmental Management Group, 2011. Global Drylands: A UN System-wide Response. UN.

UNCCD, 1994. United Nations Convention to Combat Desertification in Those Countries Experiencing Serious Drought and/or Desertification, Particularly in Africa. A/AC.241/27.

UN-DESA-PD, 2015. World Population Prospects: The 2015 Revision, Data Booklet. ST/ESA/SER.A/377.

UNEP, 1997. World Atlas of Desertification, second ed. Edward Arnold, London, UK.

UNEP-WCMC, 2007. A spatial analysis approach to the global delineation of dryland areas of relevance to the CBD Programme of Work on Dry and Subhumid Lands. Dataset based on spatial analysis between WWF terrestrial ecoregions (WWF-US, 2004) and aridity zones (CRU/UEA; UNEPGRID, 1991). Dataset checked and refined to remove many gaps, overlaps and slivers (July 2014).

Van Auken, O.W., 2000. Shrub invasions of North American semiarid grasslands. Annu. Rev. Ecol. Syst. 31, 197–215.

Vitousek, P.M., Mooney, H.A., Lubchenco, J., Melillo, J.M., 1997. Human domination of earth's ecosystems. Science 277, 494–499.

Von Lützow, M., Kögel-Knabner, I., Ekschmitt, K., Matzner, E., Guggenberger, G., Marschner, B., et al., 2006. Stabilization of organic matter in temperate soils: mechanisms and their relevance under different soil conditions: a review. Eur. J. Soil Sci. 57, 426–445.

Waldrop, M.P., Zak, D.R., Sinsabaugh, R.L., 2004. Microbial community response to nitrogen deposition in northern forest ecosystems. Soil Biol. Biochem. 36, 1443–1451.

Warren, S.D., Eldridge, D.J., 2003. Biological soil crust and livestock in arid ecosystems: are they compatible? In: Belnap, J., Lange, O.L. (Eds.), Biological Soil Crusts: Structure, Function, and Management. Springer, Berlin, Germany, pp. 401–415.

West, T.O., Post, W.M., 2002. Soil organic carbon sequestration rates by tillage and crop rotation: a global data analysis. Soil Sci. Soc. Am. J. 66, 1930–1946.

Yang, R., Su, Y.-Z., Wang, T., Yang, Q., 2016. Effect of chemical and organic fertilization on soil carbon and nitrogen accumulation in a newly cultivated farmland. J. Integr. Agric. 15, 658–666.

Zak, D.R., Freedman, Z.B., Upchurch, R., Steffens, M., Kögel-Knabner, I., 2016. Anthropogenic N deposition increases soil organic matter accumulation without altering its biochemical composition. Global Change Biol. 23, 933–944.

Zhang, F., Li, C., Wang, Z., Glidden, S., Grogan, D.S., Li, X., et al., 2015. Modeling impacts of management on farmland soil carbon dynamics along a climate gradient in Northwest China during 1981–2000. Ecol. Model. 312, 1–10.

CHAPTER 3

Clay Minerals—Organic Matter Interactions in Relation to Carbon Stabilization in Soils

Binoy Sarkar*,, Mandeep Singh*, Sanchita Mandal*, Gordon J. Churchman†, Nanthi S. Bolan‡**

*University of South Australia, Mawson Lakes, SA, Australia
**The University of Sheffield, Sheffield, United Kingdom
†University of Adelaide, Urrbrae, SA, Australia
‡University of Newcastle, Callaghan, NSW, Australia

Chapter Outline

INTRODUCTION

Climate change and resultant global warming is the most debated issue around the globe (Lal, 2004). The temperature of our environment is increasing rapidly due to various natural processes and anthropogenic activities, which are emitting different greenhouse gases in the environment (IPCC, 2006). Among all greenhouse gases, CO_2 plays a vital role in increasing the global temperature. Soil organic C (OC) decomposition as measured by soil respiration is one of the major sources of CO_2 flux to the environment among all other CO_2 generating farm activities (e.g., the practice of residue burning and fossil fuel combustion through farm machinery) (Lal, 2004). Interestingly, global soils have the capacity to provide a sink for CO_2 instead of acting as a source, provided they are properly managed. The appropriate management of agricultural soils in relation to mitigating greenhouse gas (GHG) implies higher biomass input and minimum

The Future of Soil Carbon
http://dx.doi.org/10.1016/B978-0-12-811687-6.00003-1

decomposition of the added organic matter (OM). The net increase in C pools in soils occurs due to the balance between biomass inputs and losses through decomposition or mineralization (Lal, 2004, 2009). It is well known that fertile soils produce more biomass, provided no other factors become a limitation. The stabilization of C in soil is not only important for mitigating greenhouse gases (CO_2, CH_4), but also for improving the soil fertility, and hence sustainable farming.

Soil organic C (SOC) is one of the largest C pools at the global level (Lützow and Kögel-Knabner, 2009). Global soils store approximately 2344 Gigaton (Gt) of OC. Of the total terrestrial OC, 1502 Gt is stored in the first meter of the soil. If the C present in the litter and charcoal is excluded, the total OC in the upper 2-meter soil is about 2157 to 2293 Gt. On an average, soil OC contents in the first 30 cm, 100 cm, and 200 cm are 684–724 Gt, 1462–1584 Gt, 2376–2456 Gt, respectively (Batjes, 1996).

Carbon sequestration means capturing atmospheric CO_2 and locking it in permanent pools so that it does not go back into the atmosphere. The physical, chemical, and biological properties of soil greatly influence the C protection capacity (Lamparter et al., 2009). Both the physical and chemical characteristics of soils are directly or indirectly governed by clay minerals, which are the most reactive particles present in the soil. Apart from the amount of clays, clay types are also very important in protecting carbon. For example, soils rich in smectite protect more OC than kaolinite rich soils (Hassink, 1997; Wattel-Koekkoek et al., 2001). Co-composting of organic amendments with clay materials is effective in the stabilization of C in soil (Fig. 3.1). Similarly, the application of compost to soils of high clay content is likely to achieve a greater C stabilization (Bolan et al., 2012). Clay minerals provide both permanent and variable surface charges and high specific surface areas that are crucial to determine the OC protection in soils. The clay minerals include phyllosilicate minerals (e.g., smectite, kaolinite, illite) and oxide-hydroxides of Fe and Al, which can be either well crystallized or poorly crystallized (e.g., allophane) in nature (Churchman and Lowe, 2012). Due to the adsorption of OC on clay minerals through various mechanisms (e.g., electrostatic attraction, hydrophobic attraction, ligand exchange, π-bonding), OC remains protected against microbial attack (Baldock and Skjemstad, 2000; Singh et al., 2018). This chapter aims to provide a brief overview of the role of various clay minerals in protecting OC in soils and highlights the mechanisms of carbon sequestration by clay minerals.

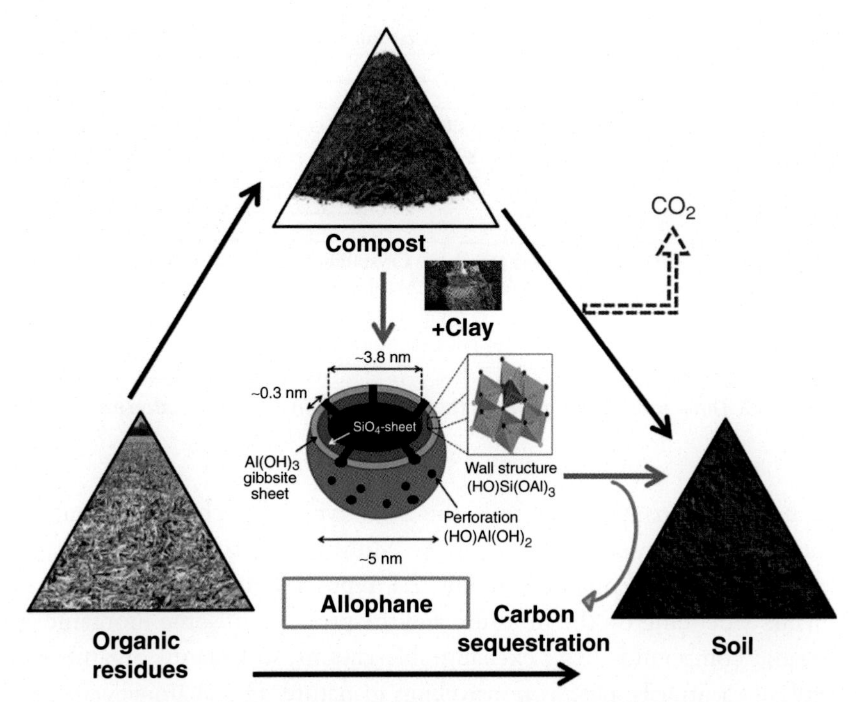

Figure 3.1 ***Schematic diagram illustrating the role of stabilizing agents, such as allophane in enhancing C stabilization in composts.*** *(Modified from Bolan, N., Kunhikrishnan, A., Choppala, G., Thangarajan, R., Chung, J., 2012. Stabilization of carbon in composts and biochars in relation to carbon sequestration and soil fertility. Sci. Total Environ. 424, 264–270).*

CLASSIFICATION OF CLAY MINERALS RELEVANT TO SOIL C PROTECTION

The major groups of clay minerals present in the soil environment include layer and chain silicates, sesquioxides, and other inorganic minerals (Fig. 3.2). The layer silicates can be classified into 1:1, 2:1, and 2:1:1 types. This classification is based on the number of building block sheets (silica tetrahedral and alumina octahedral) that are involved in their structures. The 1:1 type structure is composed of one silica tetrahedral sheet and one alumina octahedral sheet. Contrarily, the 2:1 type structure is composed of one alumina octahedral sheet sandwiched between two silica tetrahedral sheets. The 2:1:1 type clay mineral is formed in a special structure consisting of two silica tetrahedral layers, one alumina octahedral, and one magnesium hydroxide (brucite) octahedral layer (Churchman and Lowe, 2012). While the 1:1 (e.g., kaolinite) and 2:1 (e.g., smectite, illite) clay minerals

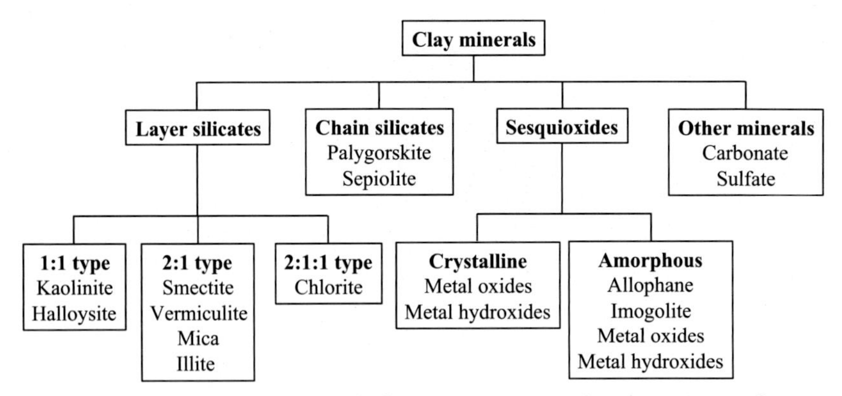

Figure 3.2 ***Different soil clay minerals that participate in soil C protection mechanisms.***

occur predominantly in all soils, the 2:1:1 type (e.g., chlorite) is stable in cool, dry, or temperate climates (e.g., Midwest USA). The smectites, and less commonly, vermiculites in the 2:1 type clay minerals can swell and shrink depending on the moisture and the presence of some inorganic and organic compounds (e.g., exchangeable cations, surfactants, organic molecules). Contrarily, illite is nonswelling in nature. The chain silicates (e.g., palygorskite), which contain ribbons of 2:1 phyllosilicate structures, are also nonswelling in nature, and found in a few soils. The sesquioxides do not have a lamellar structure, and occur universally in all soils with varying degrees of crystallinity. They are mainly the oxides/oxyhydroxides of Fe and Al, and they can account for 50% of the total soil mass (Dixon, 1991; Kampf et al., 2012). The key members of the crystalline sesquioxides include goethite, hematite, and magnetite, whereas ferrihydrite is the most common noncrystalline member. Goethite and ferrihydrite occur very widely under various climatic conditions (Churchman and Lowe, 2012). The amorphous or poorly crystalline aluminosilicates can also occur under special geochemical conditions (e.g., volcanic eruptions). In this subgroup, allophane and imogolite are familiar members (Calabi-Floody et al., 2011; Wada, 1985). Some other metal oxide and hydroxides (e.g., Mn) can also occur in a poorly crystalline form. In addition to the previously mentioned layered silicates and the sesquioxides, some other minerals (e.g., carbonates, sulfates) can also directly or indirectly play a role in the soil C dynamics. The carbonate minerals (e.g., calcite, dolomite) may occur in soils under semi-arid and arid conditions and critically alter the balance between the inorganic and organic C pools (Lal, 2008).

COMPARISON OF C PROTECTION CAPACITY OF CLAY MINERALS

By virtue of their crystalline structure and physicochemical characteristics, clay minerals provide a strong influence on OC stabilization in soils. Both the type and amount of clay minerals are reported to have a direct influence on OC stabilization in soils. In general, soils with higher clay contents are able to protect more C from mineralization (i.e., conversion to CO_2). A soil sample having 23 times higher clay content was found to reduce the respiration or carbon mineralization rate in the soil by about 50% (Wang et al., 2003). Another study reported up to 40% reduction in soil respiration as a result of 12 times increase in soil clay content (Franzluebbers, 1999). It was found that 2:1 type phyllosilicates adsorb a greater quantity of dissolved organic C (DOC) in soil than the 1:1 type clay minerals. This was attributed to some intrinsic physicochemical properties of the respective mineral candidates (Singh et al., 2017b; Stotzky, 1986). For example, the 2:1 type smectitic clay minerals (e.g., montmorillonite) contain a larger amount of specific surface area than 1:1 type clay minerals (e.g., kaolinite). The specific (external) surface area of montmorillonite can be in the range of 15–160 $m^2\ g^{-1}$, whereas kaolinite has a surface area of 6–40 $m^2\ g^{-1}$ (Churchman and Lowe, 2012; Saidy et al., 2013; Singh et al., 2016). The average particle size of montmorillonite is generally much smaller than that of kaolinite, which gives the former a larger surface area per unit mass. The degree of isomorphous substitution (e.g., replacement of Al^{3+} in the octahedral sheet by cations like $Fe^{2+/3+}$, Mg^{2+}) and/or replacement of Si^{4+} in the tetrahedral sheet by cations like Al^{3+}, Fe^{3+}) is negligible in the case of kaolinite, but significant in smectite. This gives smectite a much larger layer charge than kaolinite, subsequently creates more active sites on the former clay mineral's surface. The greater the amount of active sites on the clay mineral surface, the higher is the theoretical tendency to adsorb soil OC. The cation exchange capacity (CEC; which is a measure of layer charge in clay minerals) of kaolinite varies between 0 and 10 cmol (p^+) kg^{-1}, while the value for smectite can be as high as 160 cmol (p^+) kg^{-1}. However, under specific experimental conditions, the specific surface area and pore size distribution of a clay mineral can overshadow the role of CEC of a clay mineral in adsorbing OC. In comparison to kaolinite and smectite, the mica type 2:1 clay minerals (e.g., illite) contain a higher specific surface area (55–195 $m^2\ g^{-1}$) but a medium range CEC (10–40 cmol (p^+) kg^{-1}). The nonexpanding illitic clay minerals are often found interstratified with kaolinite or smectite and play an important role in OC retention in soils. The allophanic materials in the sesquioxide group have unique physicochemical characteristics, such as a very high specific surface area (700–1500 $m^2\ g^{-1}$). The hydroxide/oxide minerals also have a high surface area. As a result, these materials have been reported to retain remarkably greater amount of OM than the layered silicate clay minerals (Alekseeva, 2011; Churchman, 2010; Saidy et al., 2013; Wiseman and Püttmann, 2006). These minerals are also able to take part in specific chemical interactions (e.g., ligand exchange) with soil organic matter (SOM) and prevent microbial mineralization (Kleber et al., 2015).

Along with the surface area, the surface charge of soil minerals imparts a significant influence on their interaction with SOM. The kaolinitic clay minerals have a lower layer charge value, but it can vary according to the pH of the system. The smectitic clay minerals predominantly hold permanent negative charge. On the other hand, the charge of allophanes is highly pH-variable. The surface charge behavior of soil minerals can also arise from the organic components (e.g., humic substances), which display numerous organic functional groups. The humic substances can also make complexes with the soil clay minerals and alter their further interactions with newly added OM in soils. Because of this, the interaction/adsorption of externally added DOC on a Mollisol (which contains more clay–organic complexes) can be quite different from the case of an Ultisol or Alfisol. The native OC content in Mollisols might overshadow the role of sesquioxides in determining the DOC adsorption capacity of the soil (Mayes et al., 2012). The removal of native OM from the clay minerals or sesquioxides can significantly increase the surface area of the mineral materials and consequently increase the material's C adsorption/protection capacity. In the widely used method of specific surface area determination, N_2 gas is adsorbed on the material at a static temperature and variable pressure. When native C remains present in the mineral material, it blocks the pores, prevents migration of the N_2 gas into the material, and thus gives a smaller surface area. However, a different pattern might also be possible when recalcitrant OM in the clay mineral imparts a greater adsorption of external OM through multilayer hydrophobic interactions (Kahle et al., 2003; Singh et al., 2016).

The phyllosilicate minerals are often found in soils intermixed with the sesquioxides of variable crystallinity. The coating of sesquioxides on phyllosilicate minerals not only alters their surface area but also changes their charge behavior. The chemical weathering of native oxides in soils may release Fe and Al ions, which later precipitate as secondary hydroxides/oxides coatings on phyllosilicate clay mineral surfaces (Torn et al., 1997). For example, coatings of illitic and smectitic clay minerals with goethite significantly changed the charge of the clay minerals in a variable pH environment (Saidy et al., 2013). Similarly, phyllosilicate minerals also occur either interstratified or in stacks of a single or multiple layers (Churchman, 2010). This also gives a difference in properties of clay minerals occurring in agricultural soils than those found in clay deposits (Wilson, 1999). Generally, the layered structure clay minerals, when intermingled or coated with sesquioxides, adsorb a greater quantity of DOC and SOM because the oxidic/hydroxidic components provide additional organic interactions through polyvalent cation bridging and ligand exchange reactions. The content of OM in soils is positively correlated with the sesquioxide (Fe and Al oxide/hydroxides) content (Baldock and Skjemstad, 2000; Bolan et al., 2012; Kleber et al., 2015). The removal of sesquioxides from clay minerals or soil primarily reduces the surface area of the mineral material, which directly can reduce the OM adsorption/retention capacity (Kahle et al., 2004; Saidy et al., 2013; Singh et al., 2016). For example, Singh et al. (2016) obtained different surface areas for three soil clay mineral fractions (e.g., kaolinite-illite, smectite, and allophane) by conducting sequential chemical treatments to remove the sesquioxides and native OM. During the adsorption of DOC by these mineral materials from aqueous solutions under various electrolyte types and concentrations, a strong correlation between the surface areas and Langmuir adsorption maxima values ($R^2 = 0.97$; $P < 0.05$) was observed (Fig. 3.3) (Singh et al., 2016).

Allophanic minerals have distinctive effects in enhancing the stabilization of C in soils. By forming stable organo–mineral associations through innersphere complexation together with physical protection mechanisms, allophane plays an important role in increasing the mean residence time of different organic C materials in soils (Bolan et al., 2012; Dahlgren et al., 2004). Due to having higher allophane contents intrinsically, Andisols (volcanic ash soils) contain much higher OM than any other soil types both in terms of equilibrium OM content and mean residence time (Broquen et al., 2005; Parfitt, 2009). Allophanic soils also contain free Fe and Al, which can reduce the availability of DOC to microorganisms by forming Fe/Al-induced precipitation complexes (Scheel et al., 2008; Schwesig et al., 2003). Therefore, the OM content and its preservation in allophanic soils are influenced by multiple processes, such as: (1) protection by complexation with Fe, Al, and allophane; (2) reduced bacterial activity

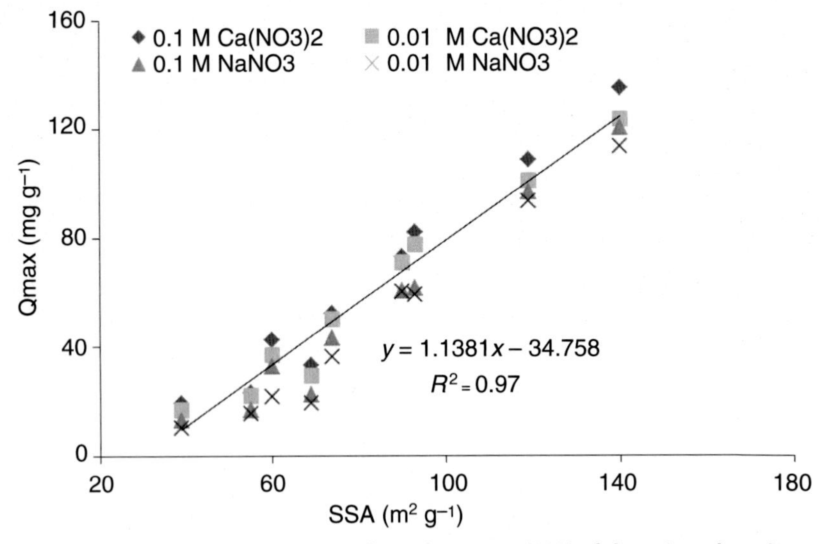

Figure 3.3 ***Relationship between specific surface areas (SSA) of clay minerals and Langmuir maxima (Qmax) values for the adsorption of dissolved organic C under various experimental conditions (values are mean;*** **n = *3).*** *(Adapted from Singh, M., Sarkar, B., Biswas, B., Churchman, J., Bolan, N.S., 2016. Adsorption-desorption behavior of dissolved organic carbon by soil clay fractions of varying mineralogy. Geoderma 280, 47–56).*

due to the presence of free Fe and Al; (3) low pH and poor availability of nutrients, especially phosphorus to soil microorganisms (Bolan et al., 2012; Parfitt, 2009).

MECHANISMS OF OC PROTECTION BY CLAY MINERALS

The strong interactions between clay minerals and SOC can inhibit C mineralization from soils. As a result, soils with different quantities of various clay minerals respond differently to atmospheric CO_2 emissions and hence affect the global warming potential to different extents. The clay minerals prevent or slow down the mineralization of OM in soils through an interrelation among physical, chemical, and biological protection mechanisms (Fig. 3.4). The physical protection mechanism is achieved through stable aggregate formation by the agglomeration of the colloidal particles. The aggregates can be classified as micro- or macroaggregates depending on their sizes. The OM plays the role of a cementing or adhesive agent to hold individual particles together into an aggregate. In the aggregate formation process, some OM may also enter into the pores of the particles and

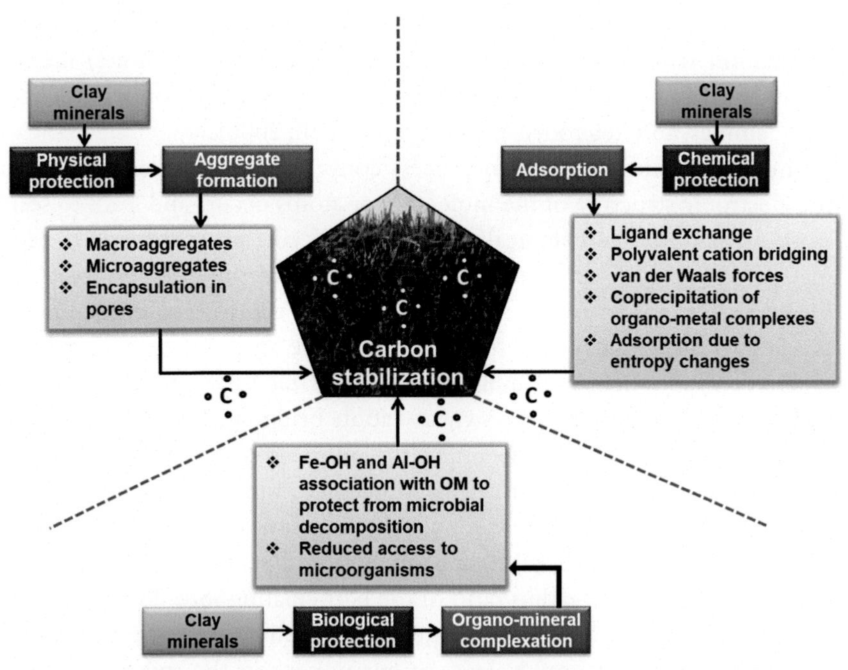

Figure 3.4 ***Mechanisms of soil C protection through the association of clay minerals.***

remain inaccessible to soil microorganisms (Baldock and Skjemstad, 2000; Hassink, 1997; Torn et al., 1997; Wada, 1985).

The chemical protection of SOM by clay minerals occurs mainly through adsorption reactions (Fig. 3.4). Adsorption of OM is achieved through cation bridging, anion exchange, hydrogen bonding, van der Waals interaction, coprecipitation of organo–metal complexes, ligand exchange, and hydrophobic bonding interactions (Stotzky, 1986). The adsorption of SOM on clay minerals can be broadly classified into physical and chemical adsorption processes. The physical adsorption predominantly occurs through van der Waals interaction, where a minimal perturbation of the electronic structure of the molecules or atoms takes place. The van der Waals interaction can be accompanied by a small amount of entropy change in the reaction system. The nonexpanding types of phyllosilicate minerals (e.g., kaolinite, palygorskite) as well as the neutral microsites on expanding smectites adsorb uncharged and nonpolar groups of OM (e.g., aromatic, alkyl-C) through this mechanism. The nonpolar organic substances can also be associated with the clay mineral surfaces through hydrophobic attraction when the latter become coated with hydrophobic persistent OM (i.e., native OM). The formation of H-bonds is another physical adsorption mechanism where minerals with oxygen surfaces (e.g., kaolinite) interact with OM having functional groups, such as carboxyl, carbonyl, phenolic OH^-, amines, and heterocyclic-*N* (Lützow et al., 2006).

In the chemical adsorption process conversely, a significant change in the electronic structure of the molecules or atoms occurs and leads to some bond formation. The chemical adsorption can take place through the formation of outer or inner sphere complexes. In an outer sphere complex, OM remains bound to the hydration shell of the ions without directly binding to the surface. Therefore, it is more like an electrostatic attraction. Different monovalent and multivalent cations take part in such reactions and these cations are also known as cation bridging. Multivalent cations (e.g., Ca^{2+}, Fe^{3+}, Al^{3+}) can facilitate the adsorption of OM in a greater amount and more strongly than monovalent cations (e.g., Na^+, K^+) (Setia et al., 2014). This is because the multivalent cations are able to form polydentate complexes with OM. The strength and stability of the complex depend on the charge-radius ratio of the participating ions. The greater the charge-radius ratio value, the stronger is the binding strength. Among the commonly occurring multivalent cations present in the soil solution, the following order of affinity with OM is generally observed: $Fe^{3+} > Al^{3+} > Pb^{2+} > Ca^{2+} > Mn^{2+} > Mg^{2+}$. The OM, which contains functional groups, such as carboxyl, carbonyl, alcoholic OH^-, and microbial polysaccharides with glucuronic-, galacturonic-, mannuronic-, pyruvic-, succinic-acid groups, becomes associated with mineral surfaces through cation bridging interactions (Kleber et al., 2015; Schwesig et al., 2003; Singh et al., 2016).

The innersphere complex formation involves adsorption of the molecules at specific sites on the mineral surface, which may override the overall electrostatic interaction of the molecules with the bulk surface. This is also known as the ligand exchange reaction, which is strongly exothermic in nature. It takes place through an anion exchange process where the anionic functional groups of OM (e.g., carboxyl and phenolic OH groups) coordinate with the OH groups present on mineral surfaces. Hydroxyl groups of Fe, Al, and Mn sesquioxides often take part in this reaction, which may be associated with an increase of the system pH due to transfer of OH^- ions into the solution phase. The OH^- groups present on the exposed edges of layer silicate clay minerals can also take part in such ligand exchange reactions. In addition, allophane and imogolite types of clay minerals can strongly adsorb OM through this mechanism. The OM rich in aliphatic or phenolic OH^- groups, aliphatic acids (e.g., citric acid, malic acid), amines, ring-NH, and heterocyclic-*N* are mostly adsorbed on mineral surfaces through this mechanism and they form strong organo–mineral associations (Yang et al., 2016).

The clay mineral particles intrinsically have a net negative surface charge. Most of the OM molecules are also negatively charged. Therefore, a repulsion force may come into effect to prevent the adsorption. However, the charge behavior of clay minerals, specially the sesquioxides, may become less negative or even positive due to a variation of the system pH. In such a scenario, electrostatic attraction may take place in addition to the ligand exchange mechanism. Some cationic or zwitterionic organic substances (e.g., enzymes, proteins, fatty acids, or organic acids) are able to migrate into the interlayer space of expandable layer silicate minerals (e.g., smectite), which is also an effective mechanism for protecting these molecules from microbial mineralization (Nannipieri et al., 1996; Stotzky, 1986). However, this mechanism is prevalent only at a pH value (<5) at which these organic molecules are stable and do not undergo a significant dissociation (Violante and Gianfreda, 2000). Additionally, organic cations (e.g., quaternary amines) can be exchanged with the hydrous cations in the interlayer space of the expandable clay minerals, and thus can become stabilized against mineralization (Sarkar et al., 2010, 2013). The metal–organic or clay–organic complexes are formed via one or more of the above mechanisms, which is the key for protecting OM from losses via mineralization, leaching, and runoff. The quantity of layered clay minerals and sesquioxides not only controls the process, but their diverse physicochemical characteristics also play a major role.

ENVIRONMENTAL SENSITIVITY TO C PROTECTION CAPACITY OF CLAY MINERALS

Environmental variables, most dominantly temperature and moisture, can play an important role in SOM protection through the association with clay minerals. A general concept prevails that an increase in temperature can enhance CO_2 evolution from soils due to an increased mineralization rate. For example, Rey et al. (2005) reported temperature sensitivity factors (Q_{10}) of 3.3, 2.7, and 2.2 for C mineralization in the 0–5 cm, 5–10 cm, and 10–20 cm layers, respectively, of a Mediterranean oak forest soil. However, the temperature sensitivity of soil C decomposition is still a matter of debate mainly because of the diverse nature of the OM itself (Davidson and Janssens, 2006; von Lützow and Kögel-Knabner, 2009). It was assumed that the decomposition of resistant components of SOM (e.g., highly aromatic compounds) would not be sensitive to the rise in environmental temperature (Leirós et al., 1999). However, it was experimentally proven that both the resistant and labile pools of SOM in mineral soil might statistically respond to the rise in temperature insignificantly, and therefore contribute similarly to the global warming potential (Fang et al., 2005; Rey et al., 2005). One of the key reasons behind such contradictory reports on temperature sensitivity to SOM mineralization is the adsorption/desorption behavior of OM, which is directly controlled by the mineralogy of soils (Conant et al., 2008; Singh et al., 2017a). As a result of temperature rise, the OM, which is held on clay minerals by weak forces (e.g., van der Waals interaction, hydrogen bonding, dipole–dipole interactions, and ion exchange), may release into the soil solution in labile forms, which become subject to microbial mineralization (Baldock and Skjemstad, 2000; Kalbitz et al., 2000). Conversely, the OM held by strong forces (e.g., ligand exchange, covalent bonds) may desorb extremely slowly, and can thus be protected from mineralization (Marschner and Bredow, 2002). Therefore, while the soil clay mineralogy moderates the mineralization of OM remarkably, the temperature sensitivity of the process largely depends on factors, such as mineralizable pool size and the primary C source utilization pattern of soil microorganisms (Rasmussen et al., 2006).

In comparatively dry soils, a sudden increase in soil moisture content can also raise the C mineralization rate (Rey et al., 2005). A rise in moisture content not only influences the microbial activity in soils, but also affects the adsorption/desorption of OM (Adu and Oades, 1978; McInerney and Bolger, 2000). This effect may be more prominent in the case of swelling–shrinking type of clay minerals (e.g., smectite). It also depends on the pore size distribution of soil clay minerals. Bacteria usually cannot enter into pores, which are generally smaller than 0.5 μm in size (Baldock and Skjemstad, 2000; Sollins et al., 1996). As a result, the OM held in such small pores remain protected from microbial decomposition unless it is released into the soil solution through desorption (Kleber et al., 2015).

CONCLUSIONS

In modern agriculture, we are facing new questions about the capacity of soils to hold C for a long time and how to quantify the processes involved in C storage (Kong et al., 2005). Agriculture is now motivating other industries to look for technological changes so that emission of CO_2 can be decreased (Lal, 2004). Modeling for SOC dynamics is found to be a valuable tool for the prediction of the effect of climate change on C storage and for developing new strategies to mitigate GHG emissions. However, the low availability of authentic and specific data jeopardizes the utility of modeling (Grace et al., 2006). With the use of the right technological and agricultural management practices, we can sequester more amounts of C in soil. Technological practices include application of integrated nutrient management, precision farming, conservation tillage practices, crop rotations, use of farm wastes, and establishment of plant life on slopes and contours (Bindraban et al., 2015; Lal, 2004, 2009).

Furthermore, by establishing a functional link between clay minerals and C stabilization, it will pave the way to develop best management practices and sound climate change policies. However, there are only a few detailed studies on the interaction between clay minerals and stabilization of SOC using naturally occurring soil clay minerals. The climatic control of the mineralization pattern of clay minerals-held SOM is almost unknown, and warrants future research. Similarly, the effect of the introduction of new OM in clayey soils on the carbon mineralization rates and possible priming effect is not properly understood. Along with addressing these questions, soil scientists should also devise methods to determine different SOC pools and fluxes that would make soil GHG emission modeling realistic.

ACKNOWLEDGMENTS

MS and SM are thankful to the University of South Australia and Department of Education and Training, Government of Australia, for awarding them APA PhD Scholarship. This work was partly supported by an Australian Research Council Discovery-Project (DP140100323).

REFERENCES

Adu, J.K., Oades, J.M., 1978. Physical factors influencing decomposition of organic materials in soil aggregates. Soil Biol. Biochem. 10, 109–115.

Alekseeva, T.V., 2011. Clay minerals and organo-mineral associates. In: Gliński, J., Horabik, J., Lipiec, J. (Eds.), Encyclopedia of Agrophysics. Springer Netherlands, Dordrecht, pp. 117–122.

Baldock, J.A., Skjemstad, J., 2000. Role of the soil matrix and minerals in protecting natural organic materials against biological attack. Org. Geochem. 31, 697–710.

Batjes, N.H., 1996. Total carbon and nitrogen in the soils of the world. Eur. J. Soil Sci. 47, 151–163.

Bindraban, P.S., Dimkpa, C., Nagarajan, L., Roy, A., Rabbinge, R., 2015. Revisiting fertilisers and fertilisation strategies for improved nutrient uptake by plants. Biol. Fert. Soils 51, 897–911.

Bolan, N., Kunhikrishnan, A., Choppala, G., Thangarajan, R., Chung, J., 2012. Stabilization of carbon in composts and biochars in relation to carbon sequestration and soil fertility. Sci. Total Environ. 424, 264–270.

Broquen, P., Lobartini, J.C., Candan, F., Falbo, G., 2005. Allophane, aluminum, and organic matter accumulation across a bioclimatic sequence of volcanic ash soils of Argentina. Geoderma 129, 167–177.

Calabi-Floody, M., Bendall, J.S., Jara, A.A., Welland, M.E., Theng, B.K.G., Rumpel, C., et al., 2011. Nanoclays from an andisol: extraction, properties and carbon stabilization. Geoderma 161, 159–167.

Churchman, G.J., 2010. Is the geological concept of clay minerals appropriate for soil science? A literature-based and philosophical analysis. Phys. Chem. Earth 35, 927–940.

Churchman, G.J., Lowe, D.J., 2012. Alteration, formation, and occurence of minerals in soils. In: Huang, P., Li, Y., Sumner, M.E. (Eds.), Handbook of Soil Sciences: Properties and Processes, second ed. CRC Press, Boca Raton, FL.

Conant, R.T., Drijber, R.A., Haddix, M.L., Parton, W.J., Paul, E.A., Plante, A.F., et al., 2008. Sensitivity of organic matter decomposition to warming varies with its quality. Global Change Biol. 14, 868–877.

Dahlgren, R., Saigusa, M., Ugolini, F., 2004. The nature, properties and management of volcanic soils. Adv. Agron. 82, 113–182.

Davidson, E.A., Janssens, I.A., 2006. Temperature sensitivity of soil carbon decomposition and feedbacks to climate change. Nature 440, 165–173.

Dixon, J., 1991. Roles of clays in soils. Appl. Clay Sci. 5, 489–503.

Fang, C., Smith, P., Moncrieff, J.B., Smith, J.U., 2005. Similar response of labile and resistant soil organic matter pools to changes in temperature. Nature 433, 57–59.

Franzluebbers, A.J., 1999. Microbial activity in response to water-filled pore space of variably eroded southern Piedmont soils. Appl. Soil Ecol. 11, 91–101.

Grace, P.R., Ladd, J.N., Robertson, G.P., Gage, S.H., 2006. SOCRATES—a simple model for predicting long-term changes in soil organic carbon in terrestrial ecosystems. Soil Biol. Biochem. 38, 1172–1176.

Hassink, J., 1997. The capacity of soils to preserve organic C and N by their association with clay and silt particles. Plant Soil 191, 77–87.

IPCC, 2006. IPCC guidelines for national greenhouse gas inventories'. In: Eggleston, H.S., Buendia, L., Miwa, K., Ngara, T., Tanabe, K. (Eds.), Prepared by the National Greenhouse Gas Inventories Programme. IGES, Japan.

Kahle, M., Kleber, M., Jahn, R., 2003. Retention of dissolved organic matter by illitic soils and clay fractions: influence of mineral phase properties. J. Plant Nutr. Soil Sci. 166, 737–741.

Kahle, M., Kleber, M., Jahn, R., 2004. Retention of dissolved organic matter by phyllosilicate and soil clay fractions in relation to mineral properties. Org. Geochem. 35, 269–276.

Kalbitz, K., Solinger, S., Park, J.-H., Michalzik, B., Matzner, E., 2000. Controls on the dynamics of dissolved organic matter in soils: a review. Soil Sci. 165, 277–304.

Kampf, N., Scheinost, A.C., Schulze, D.G., 2012. Oxide minerals in soils. In: Huang, P.M., Yuncong, L., Summer, M.E. (Eds.), Handbook of Soil Sciences: Properties and Processes. CRC Press, Boca Raton, FL, pp. 2221–2234.

Kleber, M., Eusterhues, K., Keiluweit, M., Mikutta, C., Mikutta, R., Nico, P.S., 2015. Mineral–organic associations: formation, properties, and relevance in soil environments. Advances in Agronomy, 130, Academic Press, Cambridge, MA, pp. 1–140.

Kong, A.Y., Six, J., Bryant, D.C., Denison, R.F., Van Kessel, C., 2005. The relationship between carbon input, aggregation, and soil organic carbon stabilization in sustainable cropping systems. Soil Sci. Soc. Am. J. 69, 1078–1085.

Lal, R., 2004. Soil carbon sequestration impacts on global climate change and food security. Science 304, 1623–1627.

Lal, R., 2008. Sequestration of atmospheric CO_2 in global carbon pools. Energ. Environ. Sci. 1, 86–100.
Lal, R., 2009. Challenges and opportunities in soil organic matter research. Eur. J. Soil Sci. 60, 158–169.
Lamparter, A., Bachmann, J., Goebel, M.-O., Woche, S., 2009. Carbon mineralization in soil: impact of wetting–drying, aggregation and water repellency. Geoderma 150, 324–333.
Leirós, M.C., Trasar-Cepeda, C., Seoane, S., Gil-Sotres, F., 1999. Dependence of mineralization of soil organic matter on temperature and moisture. Soil Biol. Biochem. 31, 327–335.
Lützow, M., Kögel-Knabner, I., 2009. Temperature sensitivity of soil organic matter decomposition: what do we know? Biol. Fert. Soils 46, 1–15.
Lützow, M.V., Kögel-Knabner, I., Ekschmitt, K., Matzner, E., Guggenberger, G., Marschner, B., et al., 2006. Stabilization of organic matter in temperate soils: mechanisms and their relevance under different soil conditions: a review. Eur. J. Soil Sci. 57, 426–445.
Marschner, B., Bredow, A., 2002. Temperature effects on release and ecologically relevant properties of dissolved organic carbon in sterilised and biologically active soil samples. Soil Biol. Biochem. 34, 459–466.
Mayes, M.A., Heal, K.R., Brandt, C.C., Phillips, J.R., Jardine, P.M., 2012. Relation between soil order and sorption of dissolved organic carbon in temperate subsoils. Soil Sci. Soc. Am. J. 76, 1027–1037.
McInerney, M., Bolger, T., 2000. Temperature, wetting cycles and soil texture effects on carbon and nitrogen dynamics in stabilized earthworm casts. Soil Biol. Biochem. 32, 335–349.
Nannipieri, P., Sequi, P., Fusi, P., 1996. Humus and enzyme activity. In: Piccolo, A. (Ed.), Humic Substances in Terrestrial Ecosystems. Elsevier, Amsterdam, The Netherlands, pp. 293–328.
Parfitt, R.L., 2009. Allophane and imogolite: role in soil biogeochemical processes. Clay Miner. 44, 135–155.
Rasmussen, C., Southard, R.J., Horwath, W.R., 2006. Mineral control of organic carbon mineralization in a range of temperate conifer forest soils. Global Change Biol. 12, 834–847.
Rey, A., Petsikos, C., Jarvis, P.G., Grace, J., 2005. Effect of temperature and moisture on rates of carbon mineralization in a Mediterranean oak forest soil under controlled and field conditions. Eur. J. Soil Sci. 56, 589–599.
Saidy, A., Smernik, R., Baldock, J., Kaiser, K., Sanderman, J., 2013. The sorption of organic carbon onto differing clay minerals in the presence and absence of hydrous iron oxide. Geoderma 209, 15–21.
Sarkar, B., Megharaj, M., Shanmuganathan, D., Naidu, R., 2013. Toxicity of organoclays to microbial processes and earthworm survival in soils. J. Hazard. Mater. 261, 793–800.
Sarkar, B., Megharaj, M., Xi, Y., Krishnamurti, G.S.R., Naidu, R., 2010. Sorption of quaternary ammonium compounds in soils: Implications to the soil microbial activities. J. Hazard. Mater. 184, 448–456.
Scheel, T., Jansen, B., Van Wijk, A.J., Verstraten, J.M., Kalbitz, K., 2008. Stabilization of dissolved organic matter by aluminium: a toxic effect or stabilization through precipitation? Eur. J. Soil Sci. 59, 1122–1132.
Schwesig, D., Kalbitz, K., Matzner, E., 2003. Effects of aluminium on the mineralization of dissolved organic carbon derived from forest floors. Eur. J. Soil Sci. 54, 311–322.
Setia, R., Rengasamy, P., Marschner, P., 2014. Effect of mono- and divalent cations on sorption of water-extractable organic carbon and microbial activity. Biol. Fertil. Soils 50, 727–734.

Singh, M., Sarkar, B., Biswas, B., Bolan, N.S., Churchman, G.J., 2017a. Relationship between soil clay mineralogy and carbon protection capacity as influenced by temperature and moisture. Soil Biol. Biochem. 109, 95–106.

Singh, M., Sarkar, B., Biswas, B., Churchman, J., Bolan, N.S., 2016. Adsorption-desorption behavior of dissolved organic carbon by soil clay fractions of varying mineralogy. Geoderma 280, 47–56.

Singh, M., Sarkar, B., Hussain, S., Ok, Y.S., Bolan, N.S., Churchman, G.J., 2017b. Influence of physico-chemical properties of soil clay fractions on the retention of dissolved organic carbon. Environ. Geochem. Health 39, 1335–1350.

Singh, M., Sarkar, B., Sarkar, S., Churchman, J., Bolan, N., Mandal, S., Menon, M., Purakayastha, T.J., Beerling, D.J., 2018. Stabilization of soil organic carbon as influenced by clay mineralogy. Adv. Agron., https://doi.org/10.1016/bs.agron.2017.11.001.

Sollins, P., Homann, P., Caldwell, B.A., 1996. Stabilization and destabilization of soil organic matter: mechanisms and controls. Geoderma 74, 65–105.

Stotzky, G., 1986. Influence of soil mineral colloids and metabolic processes, growth adhesion, and ecology of microbes and viruses. In: Huang, P.M., Schnitzer, M. (Eds.), Interactions of Soil Minerals with Natural Organics and Microbes, Special Publication 17. Soil Science Society of America, Madison, WI, pp. 305–428.

Torn, M.S., Trumbore, S.E., Chadwick, O.A., Vitousek, P.M., Hendricks, D.M., 1997. Mineral control of soil organic carbon storage and turnover. Nature 389, 170–173.

Violante, A., Gianfreda, L., 2000. Role of biomolecules in the formation and reactivity towards nutrients organics of variable charge minerals and organo-mineral complexes in soil environment. In: Bollag, J.-M., Stotzky, G. (Eds.), Soil Biochemistry, 6, Marcel Dekker, New York, NY, pp. 207–270.

von Lützow, M., Kögel-Knabner, I., 2009. Temperature sensitivity of soil organic matter decomposition: what do we know? Biol. Fert. Soils 46, 1–15.

Wada, K., 1985. Distinctive properties of andosols. In: Stewart, B.S. (Ed.), Advances in Soil Science. Springer, New York, NY, pp. 173–229.

Wang, W.J., Dalal, R.C., Moody, P.W., Smith, C.J., 2003. Relationships of soil respiration to microbial biomass, substrate availability and clay content. Soil Biol. Biochem. 35, 273–284.

Wattel-Koekkoek, E.J.W., van Genuchten, P.P.L., Buurman, P., van Lagen, B., 2001. Amount and composition of clay-associated soil organic matter in a range of kaolinitic and smectitic soils. Geoderma 99, 27–49.

Wilson, M.J., 1999. The origin and formation of clay minerals in soils: past, present and future perspectives. Clay Miner. 34, 7.

Wiseman, C., Püttmann, W., 2006. Interactions between mineral phases in the preservation of soil organic matter. Geoderma 134, 109–118.

Yang, J., Wang, J., Pan, W., Regier, T., Hu, Y., Rumpel, C., et al., 2016. Retention mechanisms of citric acid in ternary kaolinite-Fe(III)-citrate acid systems using Fe K-edge EXAFS and L3,2-edge XANES spectroscopy. Sci. Rep. 6, 26127.

CHAPTER 4

The Molecular Composition of Humus Carbon: Recalcitrance and Reactivity in Soils

Alessandro Piccolo*,, Riccardo Spaccini*,**, Marios Drosos*, Giovanni Vinci*, Vincenza Cozzolino*,****

*Interdepartmental Research Center on Nuclear Magnetic Resonance for the Environment, Agro-Food and New Materials (CERMANU), University of Napoli Federico II, Portici, Italy

**University of Napoli Federico II, Portici, Italy

Chapter Outline

INTRODUCTION

Soil organic C (SOC) is the largest terrestrial reservoir in the biosphere, accounting for 1500–1770 Pg, as compared to C stocks of vegetation (450–650 Pg) (IPCC, 2013). Although humus C represents from 60% to 80% of SOC, its dynamics still remain poorly understood after nearly a century of study, due to the multiplicity of factors that affect stabilization of humic matter. Industrial agricultural practices accelerate the decline of humus content in soil, and, consequently, the reduction of soil fertility, biodiversity, and soil structural stability (Fontaine et al., 2007; Reeves, 1997), while enhancing greenhouse gases (GHG) emissions from soil (Smith et al., 2014). Because it is the specific molecular composition of the soil Humeome that significantly affects SOC storage dynamics (Woo et al., 2014), soil basal respiration (Fang et al., 2005), and humus-plant relationships (Canellas and Olivares, 2014), a rigorous identification of the molecular structure of the components of soil

The Future of Soil Carbon
http://dx.doi.org/10.1016/B978-0-12-811687-6.00004-3

humus C is necessary, if any technological control of its content and dynamics can ever be introduced.

ADVANCES IN THE SUPRAMOLECULAR STRUCTURE OF HUMUS

A novel understanding of the soil Humeome has recently emerged from experimental evidence. Rather than being composed of large molecular weight macropolymers, as traditionally believed, soil humus C is now regarded as a complex noncovalent supramolecular association of the heterogeneous molecules, that survive microbial degradation of plant and animal tissues, and are held together by weak dispersive forces, hydrogen bonds, and metal-bridged intermolecular electrostatic bonds. Piccolo (2001, 2002) extensively reported on the early experimental evidence that brought to delineate this novel paradigm of the chemistry of soil humus C.

The metastable conformation of Humeome extracted from different soils by traditional alkaline solutions was further confirmed when it interacted with dioic acids, such as oxalic, malonic, succinic, and glutaric acids, which are common components of root exudates in soil, and the corresponding changes in molecular size distributions were evaluated by high-performance size-exclusion chromatography (HPSEC) (Piccolo et al., 2003). Lowering the pH from 7 to 3.5 by the addition of dioic acids before HPSEC analysis produced the same dramatic changes in absorbance and HPSEC elution times, which had been observed in earlier works with other organic acids (Piccolo, 2002), and were explained by the disruption of unstable humic superstructures into smaller-sized associations stabilized by the formation of intra- and intermolecular hydrogen bonds among humic molecules stronger than weak dispersive bonds. Dioic acids with an increasing number of C atoms and pKas had progressively larger degrees of protonation at pH 3.5 that enhanced a closer contact with the hydrophobic domains of weakly bound humic superstructures and produced more extensive conformational rearrangement of the Humeome in solutions. Recognition of a process controlling the conformational structure of humic molecules by organic acids exuded in the soil solution by either plant roots or soil microbes was regarded important for the understanding the role of humus in soil–plant interactions.

The supramolecular understanding of the Humeome, despite initial controversies (Swift, 1999), prompted a series of experiments, whose results supported the novel paradigm. Peuravuori and Pihlaja (2004) and Peuravuori (2005) isolated humic matter from strongly colored freshwaters derived from spodosol watersheds and employed different HPSEC experiments to show that humic samples consisted of structurally similar associations of various molecular size ranges, thus showing a nanoscale supramolecular nature. Proton correlation spectroscopy and transmission electron microscopy were applied to study both commercial and river waters humic substances in solution as a function of cation type and concentration, pH, and salinity, and results were consistent with a mutual interconnection of small molecules into more complex assemblies (Baalousha et al., 2005, 2006).

The supramolecular association of humic molecules was invoked to explain the thermal oxidation stability of soil humic extracts saturated with H, Na, Ca, or Al, following the treatment with relatively polar organic compounds, such as methanol, formic acid, and acetic acid (Buurman et al., 2002). While thermal characteristics of H-humates did not change upon addition of the polar molecules, thermal decomposition of Na-humates was shifted to much higher temperatures (750–830°C) than control. Substantially less dramatic was the effect on Ca-humates, whereas hardly any alteration was observed when polar organic compounds were added to Al-humates. These results were explained with the forces holding the Humeome together. Humic molecules are strongly bound to each other by hydrogen bonding in H-humates and by electrostatic bridges in Ca- and Al-humates. These binding forces were not overcome by the simple addition of polar organic molecules, and their stability remained generally unchanged, whereas the treatment altered and strengthened the associations of humic molecules in Na-humates, which are held together only by nonspecific hydrophobic interactions.

Diffusion ordered nuclear magnetic resonance spectroscopy (DOSY-NMR) was applied to a number of different fulvic and humic superstructures to follow their diffusion coefficients (*D*), which are correlated to molecular sizes, as a function of concentration and pH changes upon addition of acetic acid (Šmejkalová and Piccolo, 2008). At increasing concentrations, all solution Humeomes showed invariably lower *D* values in DOSY spectra, thereby indicating an aggregation into apparently larger associations, whose increased hydrodynamic radius was also confirmed by viscosity measurements. When the solutions were brought from alkaline to acidic pH (3.6) by acetic acid addition, the Humeome diffusivity detected by DOSY increased significantly, suggesting a decrease of aggregation and molecular size, inasmuch as it was previously found by other methods (Piccolo, 2002). A larger extent of aggregation and disaggregation was observed for hydrophobic humic acid (HA) than for hydrophilic fulvic acid (FA), whereas no aggregation was detected for a similarly treated true macropolymeric standard.

Again, these results proved the supramolecular nature of the Humeome and its dynamic conformational structure (Šmejkalová and Piccolo, 2008).

While the importance of hydrophobic components in favoring the supramolecular association of the Humeome was further suggested in other works (Chilom et al., 2009; Tarasevich et al., 2013), the disaggregation of a HA adsorbed on graphite by addition of acetic acid into components of smaller size was also indicated by atomic force microscopy (Liu et al., 2011). Moreover, preparative HPSEC was employed to separate the Humeome extracted from a volcanic soil into several fractions of decreasing molecular size and to subject these isolates to advanced physical–chemical analyses (Conte et al., 2006, 2007). These authors showed that variable contact-time pulse sequences of CPMAS-NMR spectra enabled the calculation of cross polarization (TCH) and proton spin–lattice relaxation (T1ρH) times and related these parameters to structural differences among the size fractions. Results indicated that the larger molecular-size fractions contained more hydrophobic molecular domains with slower local molecular motion, while the smaller size fractions were characterized by a larger number of more hydrophilic molecular domains with faster local molecular motion, thereby confirming that the apparently larger molecular size of the Humeome is stabilized by association of hydrophobic components. Furthermore, both UV and fluorescence absorptions were low in the large size-fractions that mainly contained alkyl carbons, increased in the olephinic- and aromatic-rich fractions in intermediate molecular size, and decreased again in the smallest fractions predominantly composed by oxidized carbons. This suggested that the UV, and above all, fluorescence measurements of the bulk Humeome should not be assumed as indexes of environmental or ecological dynamics unless the humic molecular composition is known in details. Similar results for dissolved organic matter (DOM) in freshwater samples were recently explained with the presence of a supramolecular structure and suggested that DOM fluorescence characteristics may be controlled by molecular assemblies with optical properties distributed along a a range of molecular weights (Romera-Castillo et al., 2014).

Following the increasing consensus (Simpson, 2002) on the novel Humeome model of supramolecular association developed by Piccolo (2001, 2002), research attention was devoted in clarifying the average mass of humic molecules. The interest in the electrospray ionisation interphase to mass spectrometry (ESI/MS) increased progressively in recent years due to its advantage in being an extremely soft ionization process and therefore providing unfragmented ions for which not only the absolute molecular weight but also unequivocal structural information (MS/MS) can be obtained (Cole, 2000). These features were proved for ESI of humic molecules in combination with ultra high-resolution instrumentation, such as Fourier transform ion cyclotron resonance mass spectrometry (FT IRC-MS) providing a resolving power up to 120,000 for low-mass molecules (Brown and Rice, 2000; Stenson et al., 2002) and singly charged ions for humic matter in both positive and negative mode (McIntyre et al., 2001). By using ESI-MS, it was determined that the averaged molecular masses of separated size-fractions of a terrestrial HA were generally <1000 Da and not different from the corresponding bulk HA before fractionation by preparative HPSEC (Piccolo and Spiteller, 2003), and that no fragmentation was generated by sample heating to changing the ESI cone voltage (Peuravuori et al., 2007; Piccolo et al., 2010). These findings supported the understanding of the Humeome as a supramolecular association of only apparent large molecular dimension rather than covalently linked macropolymers of large molecular mass. However, the fact that both the bulk HA and their size-fractions showed similar molecular masses and no significant mass differences were observed among size-fractions, was attributed to the limitation of ESI-MS to ionize molecules in complex hydrophobic and hydrophilic systems. In fact, it was proved that a mixture of polar and apolar compounds had a diverse electrospray ionization, depending on their most probable positioning at the surface of the evaporating droplet (Nebbioso et al., 2010). Because the hydrophobic molecules are most favorably positioned at the aqueous-gas interphase, the apolar ions are preferentially transferred in the gas phase, whereas the polar molecules are retained in the droplet interior, and, their ESI-MS detection depressed. These results suggest that the electrospray ionization of different molecules present in complex heterogeneous mixtures of environmental significance such as humic substances is limited by their concentration and reciprocal attracting forces. This limitation was recently acknowledged when a determination of carboxyl groups of humic systems was attempted by FT IRC-MS following deuteromethylation (Zherebker et al., 2017).

The weak dispersive forces at neutral pH and the hydrogen bonds at lower pH are not the only interactions responsible for the conformational stability of the Humeome. Metals also contribute, as bridges among molecules, to stabilize the tridimensional arrangement of humic superstructures with strong intermolecular electrostatic bonds (Aquino et al., 2011; Kalinichev et al., 2011; Orsi, 2014). Siéliéchi et al. (2008) used several techniques to characterize the aggregates formed by humic colloids and hydrolyzed Fe species under various conditions of pH and mixing. They found that at low Fe, the anionic humic network is reorganized into more compact structures, while at increasing Fe, the restructuration of humic colloids is much reduced. They concluded that the interaction between the Humeome and Fe led to a competition between a humic network reconformation and the collision rate of metastable colloids.

Spin-lattice relaxation time in the rotating frame ($T1\rho(H)$) by CP-MAS NMR spectra and diffusion coefficients (D) by 1H-diffusion order spectroscopy (DOSY) NMR spectra were obtained for complexes formed between a HA and either Al^{3+} or Ca^{2+} (Nebbioso and Piccolo, 2009). These measurements showed that the molecular rigidity of humic complexes increased significantly with metal addition throughout the full carbon spectral region, being more pronounced for triple-charged Al than for double charged Ca. However, $T1\rho(H)$ values of spectral intervals suggested that the increase in molecular rigidity was generally in the order: aliphatic C >aromatic/double bonds C >carboxyl C. Concomitantly, DOSY spectra showed that addition of both Al and Ca decreased substantially the diffusivity of humic alkyl components and increased that of aromatic and hydroxyl-alkyl components, thereby indicating that complexation induced a molecular-size increase in the former and a decrease in the latter. These results suggest that saturated and unsaturated long-chain alkanoic acids in humic matter were preferentially involved in metal complexation with Al and Ca, with consequent increase of conformational rigidity and molecular size of humic hydrophobic domains. Conversely, more hydrophilic or mobile humic components appeared relatively less affected by the molecular and intermolecular rearrangements induced in the Humeome by complexation with metals.

These findings were confirmed by following through HPSEC with both UV–vis and refractive index (RI) detectors the conformational changes of a HA and a FA induced by iron complexation (Nuzzo et al., 2013). They observed that the molecular size distribution was reduced for HA and increased for FA with progressive iron complexation (Fig. 4.1).

Due to the electrostatic interactions of Fe with humic components, it is likely that the triple-charged Fe ions formed stronger complexes with the more acidic hydrophilic and hydrated FA than with the less acidic and more hydrophobic HA. The large content of ionized carboxyl groups in FA, thus favored Fe intermolecular bridges among the negatively charged FA molecules, and led to larger size network than for HA. Conversely, the humic conformational arrangements stabilized by only weak hydrophobic bonds were disrupted, by the strong iron complexation, into smaller-size

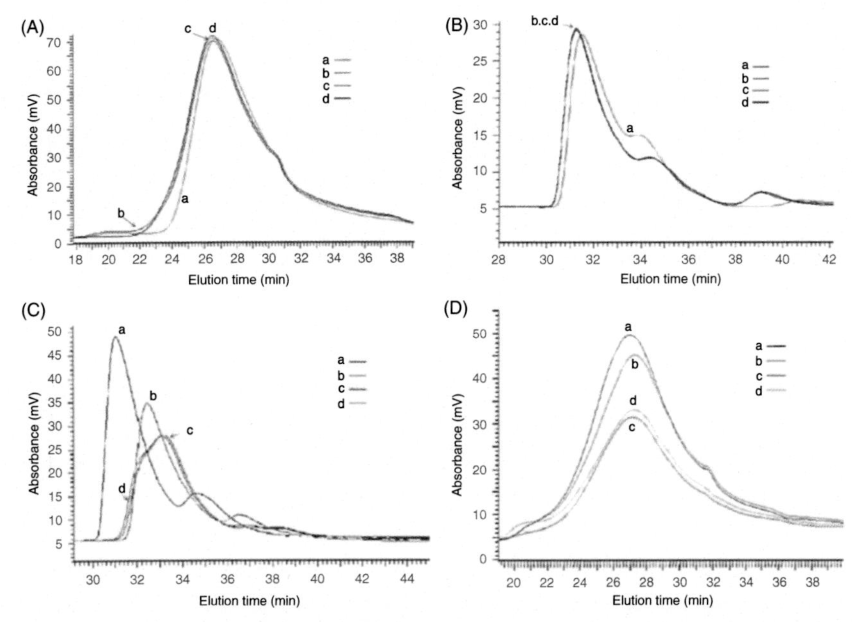

Figure 4.1 (A) UV-detected HPSEC chromatograms of control FA (a) and FA complexed for 80% (b), 90% (c), and 100% (d) with Fe; (B) RI-detected HPSEC chromatograms of control FA (a) and FA complexed for 80% (b), 90% (c), and 100% (d) with Fe; (C) UV-detected HPSEC chromatograms of control HA (a) and HA complexed for 40% (b), 50% (c), and 60% (d) with Fe; (D), RI-detected HPSEC chromatograms of control HA (a) and HA complexed for 40% (b), 50% (c), and 60% (d) with Fe.

aggregates of greater conformational stability. Moreover, the comparison between the HPSEC profiles by either UV- or RI-detector showed that in the case of FA (Fig. 4.1A and B) the Fe complexation resulted only in the shift of the complexed material toward larger molecular dimensions, without significant changes in the absorbance of UV-detected peaks.

Conversely, in the case of HA, while the RI-detector clearly showed a reduction in molecular sizes with progressive Fe complexation, the UV-detector showed a significant reduction in absorbance, but only a slight shift to larger elution volumes (Fig. 4.1C and D). This behavior can be explained by the hypochromic effect (Piccolo, 2002) that occurs when the humic chromophores are separated from each other, because of the disruption of the original metastable conformation into more stabilized smaller conformations, as caused by the formation of strong Fe-bridged intermolecular electrostatic bonds. In fact, the dipole moments of the separated chromophores do not longer overlap, thereby reducing their interaction with the electromagnetic radiation and the peak absorbance in the chromatograms

(Vershin, 1999). These results not only confirmed that humic molecules in solution are organized in supramolecular associations of loosely bound molecules but their conformational stability is greatly increased when in interactions with divalent and trivalent cations, which are abundantly present in soils.

THE SOIL HUMEOME CHARACTERIZED BY HUMEOMICS

The understanding that soil Humeome consists in supramolecular associations of heterogeneous and relatively small (<1000 Da) molecules held together in only apparently large molecular sizes by weak dispersive forces, such as hydrogen and hydrophobic bonds, and metal-bridged intermolecular electrostatic linkages, unveils the possibility to device a chemical procedure for the identification of the molecular components. In fact, it can be envisaged that the molecular complexity of the Humeome may be reduced by a progressive breakdown of those inter- and intramolecular interactions that stabilize the complex suprastructures, thus enabling the isolation of single humic molecules and facilitating their structural characterization. This methodology was named Humeomics, similarly to other omics, and had the aim to separate the single components of the supramolecular Humeome without breaking any C–C covalent bond and identify their molecular structure by applying a combination of analytical techniques. Humeomics may then be described as "a stepwise separation (Fig. 4.2) of molecules from the complex bulk suprastructure by progressively cleaving esters and ether bonds and characterizing the separated molecules by advanced analytical instrumentation."

Nebbioso and Piccolo (2011) applied Humeomics for the first time to a Humic acid isolated from a volcanic soil of the caldera around lake Vico near Rome (Italy) and the sequential fractionation succeeded to identify up to 60% of the Humeome, while the unaccounted 40% was attributed to losses of occluded hydration water and small volatile organic compounds, and to decarboxylation.

The first step of Humeomics consisted in subjecting the bulk humic extracts (RES0) to an organic solvent (dichloromethane and methanol solution) extraction of a fraction (ORG1) that comprises free or unbound humic molecules associated to the humic suprastructure only by weak dispersive interactions. The following step employed a BF_3-MeOH transesterification reaction to cleave weakly bound (or readily accessible) esters present in the Humeome and distribute the products into organosoluble

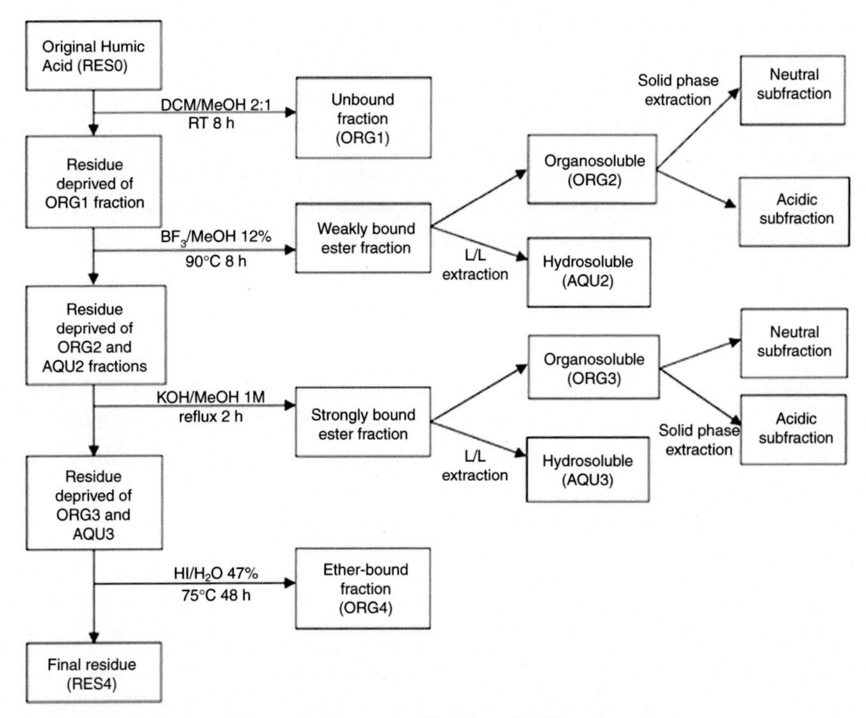

Figure 4.2 ***Scheme of Humeomics sequential chemical fractionation.***

(ORG2) and hydrosoluble (AQU2) fractions. The third step of the Humeomics targeted the esters which became accessible after the removal from Humeome of the moieties solvated by the first two steps. These esters, which are assumed to be strongly bound to the humic matrix, were hydrolyzed by a KOH-MeOH solvolysis reaction and produced components separated into organosoluble (ORG3) and hydrosoluble (AQU3) fractions. The last step applied a 47% HI aqueous solution to cleave the ether bonds in the remaining humic molecules, thus again yielding the freed components into organosoluble (ORG4) and hydrosoluble (AQU4) fractions. The nonextractable humic molecules in the solid residue represented the last fraction (RES4) of the Humeomic fractionation (Fig. 4.2). The molecular composition of the different ORG, AQU, and RES fractions was characterized by chromatography-mass spectrometric techniques (GC–MS and LC–MS) and NMR spectroscopy.

The HPSEC hyphenated with a high-resolution Orbitrap electrospray (ESI) mass spectrometry was applied to characterize the alkaline solution of the bulk HA (RES0). The preliminary separation of the humic superstructure through a HPSEC column before MS analysis provided an easier identification of the RES0 molecular composition that was mainly represented by alkandioic and hydroxydioic acids. Most empirical formulae recognizable by ESI-MS were found in the slower eluting and small-sized HPSEC fraction that comprised alkanoic acids, di- and trihydroxylated C_{18} acids, monounsaturated C_{14}, C_{16}, and C_{18} acids, hydroxy-unsaturated C_6–C_{20} acids, C_4–C_{24} dioic, C_{16}–C_{24} hydroxy-dioic, cyclic acids, and several other unassigned acidic structures. Moreover, a cyclic acid with empirical formula $C_7H_6O_8$ was identified in the diffuse fraction. Due its large unsaturation, it was attributed to a highly substituted aromatic structure, such as a hydroxylated and carboxylated furane ring. Most masses eluting in the earlier HPSEC fraction were not assignable to any definite empirical formula or class structure because of a still great molecular aggregation (Nebbioso and Piccolo, 2011).

Both GG-MS and NMR analyses of the first ORG1 fraction, comprising the unbound organosoluble humic components, showed a greater visibility of alkyl and saturated components than for the bulk Humeome (RES0). In particular, mono- and dicarboxylic acids and some iso- and anteiso-branched alkanoic acids and hydroxyacids, such as ω-C16-18, β-C14, were found to be most abundant components in ORG1 (Nebbioso and Piccolo, 2011).

Though the solution-state ^{1}H NMR spectra and solid-state ^{13}C CPMAS spectra of the organosoluble ORG2 was mostly similar to those of ORG1, except for the intense methoxyl signal due to transesterification in ORG2, the GC–MS characterization revealed a large content of unsaturated $C_{9\text{-}29}$ linear and branched alkanoic acids, α/β $C_{9\text{-}26}$ mono- and dihydroxyacids, the C_{18} unsaturated acid, $C_{8\text{-}30}$ α,ω-diacids, sterols, $C_{12\text{-}28}$ *n*-alkanols, and substituted benzoic and cinnamic acid derivatives. Hence, the removal of unbound alkanoic and saturated acids in ORG1, unveiled the abundance of hydroxyacids and unsaturated compounds in ORG2 (Nebbioso and Piccolo, 2011).

The hydrosoluble AQU2 was characterized by a lesser degree of unsaturated compounds and larger oxygen substitution than in RES0 and in the previous ORG fractions. A large content of *N*-containing molecules in AQU2 implied the greater affinity of this humic nitrogenous components to the aqueous phase. However, HPSEC-ESI-MS showed a still large content of alkyl acidic compounds, such as odd- and even-numbered $C_{10\text{-}18}$ saturated acids, $C_{15\text{-}18}$ unsaturated acids, $C_{15\text{-}18}$ hydroxy-unsaturated acids, and $C_{5\text{-}9}$ diacids. Moreover, $C_{7\text{-}18}$ negatively charged compounds containing one to five *N* atoms were identified in AQU2, together with cyclic acids and $C_xH_{2y}O_y$ saccharide structures. Further unidentified *N*-containing compounds were mostly represented by the $C_8H_7O_3N_3$ empirical formula.

The NMR spectra of the organosoluble fraction ORG3 showed abundance of signals attributed to large aliphatic and CH-X compounds, such as hydroxylated acids. These indications were supported by GC–MS, which revealed the presence in ORG3 of $C_{9\text{-}32}$ saturated and unsaturated *n*-alkanoic acids, $C_{6\text{-}8}$ α,ω-diacids, $C_{16,22,24}$ ω-monohydroxyacids, $C_{22\text{-}26}$ β-monohydroxyacids, C_{18} di- and trihydroxyacids, $C_{12\text{-}28}$ *n*-alkanols, phenolic acids, and steroids. Consistent with ^1H NMR observations, the most abundant saturated acid was C_{16}, whereas $C_{18:1}$ was the largest one among unsaturated acids. Polyhydroxylated acids, diacids, and *n*-alkanols also quantitatively contributed to the molecular description of ORG3. Moreover, two-dimensional NMR spectra (COSY, TOCSY, HSQC) of ORG3 showed many cross-peaks in the aromatic and CH-X regions (Nebbioso and Piccolo, 2011).

While ORG4 and AQU4 were not collected in quantitative amount, the residue remaining after the cleavage of ether bonds (RES4) was depleted of linear structures and completely deprived of alkandioic acids. This was shown by measuring the proton relaxation time (T1ρH) in the solid state, as an indicator of changes in molecular rigidity. A comparison of T1ρH values between RES0 and RES4 revealed changes in molecular flexibility in all spectral regions and especially for sp^2 carbons, thereby indicating a tighter molecular packing for the final RES4 residue. Further dipolar-dephasing NMR experiments showed that RES4 contained a large share of quaternary carbons, thus implying the presence of totally substituted or condensed aromatic carbons, as the remaining type of compounds at the end of the Humeomics fractionation. The application of HPSEC-ESI-MS to the still soluble RES4 in alkaline solution, revealed peaks corresponding to odd- and even-numbered $C_{10\text{-}18}$ saturated acids, $C_{14\text{-}18}$ unsaturated acids, $C_{6,8}$ hydroxy-unsaturated acids, and cyclic acidic structures, being the latters the most abundant compounds (Nebbioso and Piccolo, 2011).

The Humeomics fractionation was later applied to three size-fractions of the same HA after their separation by preparative HPSEC (Nebbioso and Piccolo, 2012), and it was found that it enabled a larger overall analytical response than for the bulk HA. In fact, the number of identified compounds in the Humeome of the three size-fractions was far greater than for the unfractionated HA. This was explained by the weakening of conformational stability of the humic suprastructures during the HPSEC separation, and the consequent separation of less complex humic fractions. In line with the supramolecular theory of Humus stating that hydrophobic interactions play a pivotal role to stabilize the association, it was found that hydrophobic compounds were mainly distributed in the largest size-fraction, while hydrophilic components were eluted in the smallest size-fraction. Compounds with linear chains or stackable aromatic rings were likely associated in regular structures in order to justify their abundance in the Humeome of the largest size fraction. Conversely, irregularly shaped compounds, such as polyhydroxylated compounds and short-chain dialkanoic acids that hindered mutual association, were mostly found in the smallest fraction. Moreover, proteinaceous compounds were mostly found in the middle size fraction, possibly owing to their amphiphilic nature. These findings thus indicated that the structural characteristics of the prevalent humic molecules determined their mutual association in humic suprastructures, as well as their conformational strength and shape. Interestingly, the lack of the residual RES4 fractions in any of the size-fractions, confirmed that the strength of intermolecular association was so reduced by the HPSEC elution that all single molecules became accessible and fully solvated during the Humeomics separation.

Without preliminary HPSEC separation, the residual end product of Humeomics (RES4) still requires to be simplified in order to enhance the identification of its molecular composition. With this aim, an alkaline solution of the RES4 was first subjected to a preparative HPSEC that yielded ten size separates (Nebbioso et al., 2014a). The 10 size-fractions were then injected into an analytical HPSEC hyphenated with a high-resolution Orbitrap ESI-MS. Total ion chromatograms of size-fractions showed two eluting peaks, the molecular masses of which were identified by ESI-MS and the empirical formulae described. Most empirical formulae were easily associated with linear alkanoic, unsaturated, hydroxylated, and hydroxy-unsaturated acids, as well as cyclic acids, although some compound structures could not be identified. Quantitative measurement of components indicated that long-chain saturated acids were more abundant in large-sized fractions than in short-chain homologues, whereas unsaturated, hydroxylated, and most cyclic acids were prevalent in small-sized fractions. This suggests that long, saturated, and unsubstituted linear acids allow formation of large suprastructures, probably as a result of favorable intermolecular packing, compared with the irregularly shaped cyclic, unsaturated or hydroxylated compounds.

Moreover, Orbitrap ESI-MS allowed to run the tandem MS technique and this was essayed on four unknown masses to characterize their structure (Nebbioso et al., 2014a). The masses with *m/z* 129, 141, 155, and 217 were attributed to the empirical formulae of $C_6H_{10}O_3$, $C_6H_6O_4$, $C_6H_4O_5$, and $C_7H_6O_8$, respectively. The mass–mass (MS^2) fragmentation spectra for these four empirical formulae were measured and the analysis of the resulting daughter ions allowed to assign their plausible structure to hydroxy-unsaturated hexanoic acid, two furane rings and a norbornane-like ring, respectively. The latter three compounds were never reported before for terrestrial HA, and may remind the carboxyl-rich alicyclic molecules, which had been proposed in DOM by NMR spectroscopy (Hertkorn et al., 2006).

The size-fractions separated from RES4 of the volcanic soil HA were also analyzed by different NMR techniques (Nebbioso et al., 2014b). 1D ^{1}H NMR spectra did not reveal significant molecular differences among size-fractions, although all of them differed from the spectrum of the bulk RES4 especially in signal intensity for aliphatic materials, which were assigned by 2D NMR to lipidic structures. DOSY-NMR spectra showed that the homogeneity of RES4 was significantly changed by the HPSEC separation. In fact, nominally large size-fractions, rich in lipidic signals, had significantly lower and almost constant diffusivity, due to stable supramolecular associations promoted by hydrophobic interactions among alkyl chains. Conversely, diffusivity gradually increased with the content of aromatic and hydroxyaliphatic signals, which accompanied the reduction of fractions sizes and was related to smaller superstructures. The ^{1}H diffusion coefficients were calculated for signals in the alkyl (2–0.5 ppm), hydroxyalkyl (4.4–2.5 ppm) and aromatic (9–6.5 ppm) spectral regions. In the bulk RES4 material, all signals were substantially aligned to the same diffusivity, with their projections located in a restricted range between -10.2 and -10.4 m^2 s^{-1}, as implied by the homogeneity of molecular domains prior to preparative HPSEC separation. Conversely, diffusion coefficients for the size-fractions differed from those of the bulk RES4, confirming the profound changes in conformational structure induced by HPSEC separation. In particular, the diffusivity for alkyl protons showed a reducing trend with decreasing size of fraction, whereas that for hydroxyalkyl and aromatic protons invariably increased.

Humeomics was also applied to characterize the humin (HU) left in the volcanic soil after extraction of the HAs. The soil HU was characterized by Humeomics before and after a HF treatment used to destroy the soil mineral components (Nebbioso et al. 2015). The molecular characterization of

HU fractions was achieved by GC–MS, thermochemolysis-GC–MS and ^{13}C CPMAS-NMR. Both weight and chromatographic yields were greater for the clay-depleted HU than for bulk HU, thereby increasing the molecular identification in the Humeome of the former one. Saturated and unsaturated alkanoic, ω-alkanedioic, hydroxyalkanoic acids, alkanols, and hydrocarbons were found in both HU samples. Abundant odd numbered *n*-alkanoic acids in ORG1 indicated accumulation of free microbial metabolites, whereas plant derived even-numbered alkanoic acids were preferentially found in ORG2 and ORG3 fractions, thereby implying a tighter interaction with the HU matrix than for microbial metabolites. Unsaturated, *n*-alkanedioic, and hydroxyalkanoic acids were detected only after hydrolysis of complex esters. The aromatic character in HU residues progressively increased with Humeomics sequential steps, while alkyl and hydroxy-alkyl compounds were reduced. In fact, AQU2 extracts prevalently contained alkyl aromatic and carbohydrate-like compounds, whereas aromatic moieties were predominant in the RES fractions. The fact that HU was found to contain similar components as for the HA extracted from the same soil, suggested that traditional humic pools differed more in supramolecular arrangement than in molecular composition.

The "Humeomic" approach has been proved to reach an exhaustive molecular characterization of humic extracts from soil, isolating soil molecules without breaking carbon–carbon bonds, and whose binding to humic matrix is only by weak dispersive forces or ester and ether linkages. The progressive separation of undisturbed single humic components allows their structural identification by modern chromatographic and physical–chemical techniques, thereby advancing knowledge on humus molecular characteristics as related to its origin and formation. Because of its systematic reproducibility, "Humeomics" may lead to develop realistic models of humic conformational architecture, that is required to relate molecular structure to biological activities and functions in environmental ecosystems (Aquino et al., 2011; Orsi, 2014).

The method was recently applied also directly on soil and a last step of alkaline extraction of the final residue was added to the Humeomics fractionation, (Drosos et al., 2017). Fig. 4.3 shows the mass yield (mg) of organic C (OC) obtained by the traditional alkaline extraction protocol adopted by the International Humic Substances Society (IHSS), as compared to the overall yield (mg) of the Humeomics fractionation, when both methods were applied to 100 g of a silt-loamy soil from an agricultural filed near Torino, Italy (Piccolo, 2012). The alkaline extractable soil organic matter

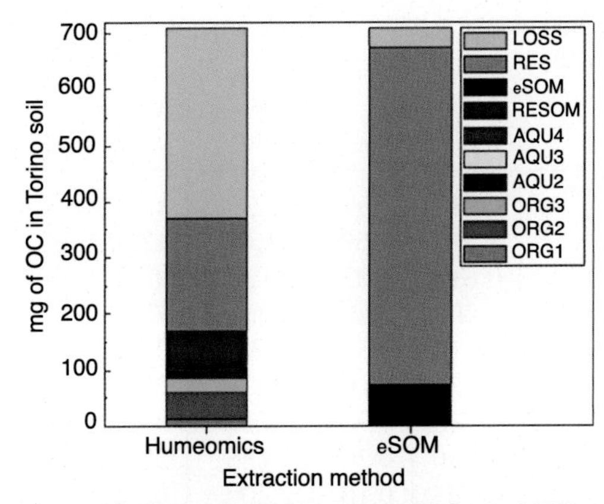

Figure 4.3 ***Organic C yield of Humeomic fractions (ORG1-3, AQU2-4, and RESOM) versus eSOM, including the OC of the unextractable material in both cases (RES) and of the material lost after extraction (LOSS).***

fraction (eSOM) by the IHSS procedure appeared to be greatly less efficient than Humeomics in solvating the Humeome from soil. While the loss of C during eSOM was smaller than for Humeomics, the unextractable fraction (RES) in the case of eSOM was significantly greater. This was possible because Humeomics was capable to progressively increase the accessibility of the Humeome by selectively removing the unbound organosoluble layer, of both the organosoluble (ORG2, ORG3) and hydrosoluble (AQU2, AQU3) compounds after hydrolyses of esters of increasing stability, and of the hydrosoluble AQU4 fraction released after cleavage of ether bonds. As a last step of Humeomics, the final soil residue was again extracted by the alkaline IHSS method to yield a hydrosoluble organo-mineral fraction (RESOM). The amount of the overall organic C extracted from soil by Humeomics (170.3 mg in 100 g of soil) was 2.35 times larger than for the eSOM extraction (72.5 g in 100 g of soil) (Table 4.1). By subtracting the weight of RES after Humeomics (200 mg) from that of RES after eSOM (600 mg), it became clear that Humeomics enabled the solubilization of 66% of the soil Humeome that was otherwise nonextractable and not identified by the traditional IHSS method.

An important aspect of the information gathered by the direct Humeomics application on soil is that they are multidimentional. In fact, the fractionation of the soil Humeome itself is an answer to the question on the extent of association of humic molecules, because they are progressively released from chemically protected domains composed of highly hydrophobic materials (ORG fractions), which are surrounding or contiguous

Table 4.1 Mass yield and elemental analysis of organic matter in eSOM and Humeomic fractions

Sample	C%	N%	C/N ratio	Mass yield	C (mg)	N (mg)
Soil Torino	0.71	0.04	17.75	100 g	710	100
eSOM	**5.80**	**0.50**	**11.60**	**1250 mg**	**72.5**	**6.3**
RES after eSOM	0.61	0.03	20.33	98 g	600	29.2
Loss of material	5.00	0.60	8.33	750 mg	37.5	4.5
ORG1	58.10	0.50	116.20	24 mg	13.9	0.1
ORG2	16.30	0.40	40.75	283 mg	46.1	1.1
AQU2	14.10	0.90	15.67	59 mg	8.3	0.5
ORG3	46.80	0.70	66.85	58 mg	27.1	0.4
AQU3	2.50	0.10	25.00	6 mg	0.2	0.0
AQU4	2.60	0.30	8.67	341 mg	8.9	1.0
RESOM	3.50	0.30	11.67	1880 mg	65.8	5.6
ORGs	23.86	0.44	54.44	365 mg	87.1	1.6
RESOM and AQUs	3.64	0.31	11.72	2286 mg	83.2	7.1
Total Humeomics[a]	**6.42**	**0.33**	**19.57**	**2651 mg**	**170.3**	**8.7**
RES after Humeomics	0.21	0.01	21.00	96 g	200	9.6
Loss of material	25.18	1.61	15.65	1349 mg	339.7	21.7

[a] Total Humeomics refers to the addition of ORG1, ORG2, AQU2, ORG3, AQU3, AQU4, and RESOM fractions. Humeomics and eSOM extractions were conducted in triplicates. The values are referring to the average.

to hydrophilic moieties (AQU fractions). These mixed hydrophilic/hydrophobic humic domains are not only mutually linked by hydrogen, π–π and metal-bridged bonds, but are stabilized in soil by forming organo-mineral assemblies through covalent bonds with either Fe or Al-Si components of oxides, hydroxides, and clay minerals (Drosos et al., 2017).

Also an analysis of the material lost during extractions can provide indirect information on the molecular associations in soil. It is assumed that losses are due to volatile molecules and/or to very small-sized hydrophilic compounds lost during dialysis purification. It was possible to calculate the C/N ratio of the lost humus, by subtracting the C and N content of total Humeomics fractions from that of overall soil. Thus, the C/N ratio for the material lost during Humeomics was 15.65, while that lost during the eSOM extraction was only 8.33 (Table 4.1). This infers that molecules lost while extracting eSOM must have contained more N groups than for the material lost during Humeomics fractionation. Furthemore, the C/N ratio of Humeomics fractions may rapidly indicate their molecular composition.

In fact, the very large C/N ratio of ORG1 (116.20) suggests mainly fatty acids and alkanes components (Drosos et al., 2017), while the progressive increase in *N*-containing molecules is indicated by the drop in C/N values in the subsequent ORG3 and ORG2 (40–67) and AQUs and RESOM (8–16) fractions. These inferences from C/N ratios of fractions well agree with the findings obtained by GC–MS and Orbitrap ESI-MS, which revealed an abundance of fatty acids and sugars in ORG2, of fatty acids and phenolic acids in ORG3, and a large content of amines, amides, and heterocyclic *N*-compounds in AQUs and RESOM (Drosos et al., 2017).

When the H/C and O/C ratios of molecules identified by advanced mass spectrometry are used to build Van Krevelen plots (Fig. 4.4), it becomes evident the greater abundance of molecular information on the soil Humeome provided by Humeomics, in respect to the single alkaline soil extraction represented by eSOM. Moreover, the comparison of plots reveals that the Humeomics fractionation released molecules from the soil Humeome, especially in the lipids-rich ORG fractions, which were not extracted in eSOM. Moreover, the fact that most of the ORG molecules detected are common among the ORG fractions, becomes a supporting evidence of the supramolecular arrangement of the soil Humeome (Piccolo, 2001).

These considerations suggest that Humeomics represents the most advance and useful tool to ultimately unravel not only the molecular composition of the soil Humeome, but also its dynamics due to Humeomics capacity to highlight the changes in SOM composition. For example, the molecular information provided by Humeomics may be used to introduce an index related to C stabilization in soil. In fact, since lipidic compounds are assumed to protect hydrophilic components from mineralization and

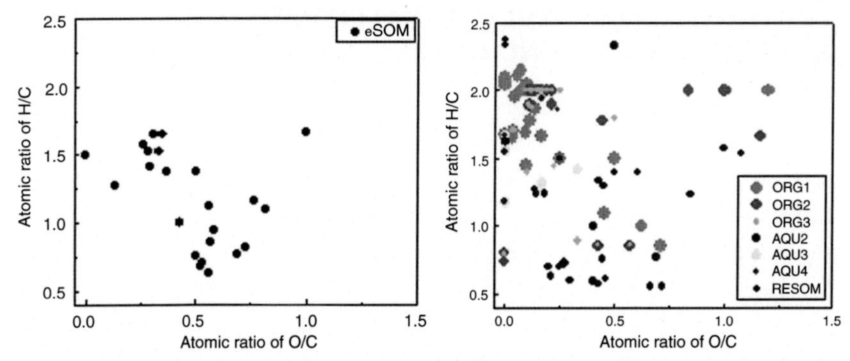

Figure 4.4 ***Van Krevelen plots for eSOM composition (left) and Humeomic fractions (right).***

microbial degradation (Piccolo et al., 2004; Spaccini et al., 2002), the larger the organosoluble/hydrosoluble carbon ratio of the soil Humeome, the greater is the chemical protection of organic matter in a soil. Therefore, in the case of the agricultural soil near Torino, whose Humeomis was reported by Drosos et al. (2017), a possible chemical protection ratio (CPR) can be described as in Eq. (4.1)

$$\text{Extractable OM Chemical Protection Ratio (CPR):}$$
$$\frac{\sum \text{ORG}_i\,\text{OC}_{\text{mg}}}{\text{RESOM OC}_{\text{mg}} + \sum \text{AQU}_i\,\text{OC}_{\text{mg}}} = \frac{87.1}{83.2} = 1.05 \tag{4.1}$$

The CPR ratio may be useful to evaluate a soil management aimed to sequester C in soil. In fact, a particular treatment of the Torino agricultural soil that enhanced the CPR value above 1.05, might indicate that the soil reached a more chemically protected status. Conversely, a decrease of the CPR ratio would suggest that the soil is more prone to mineralization of SOC. It may be envisaged that similar alterations of the soil Humeome may be shown as a function of soil type, and soil cropping and tillage systems. In fact, such management changes may affect the supramolecular structure of the soil Humeome, and consequently alter its dynamics.

Therefore, within the intensification of soil exploitation expected in the next decades with population growth, it becomes essential to rely on an advanced technique, such as Humeomics, that promises to respond to the essential challenge of building up a rigorous relationship between the structure of the complex and heterogeneous soil Humeome and its functional bioactivity in the environment.

THE HUMEOME AS SOIL C: RECALCITRANCE, HYDROPHOBICITY, AND HUMIFICATION

The aftermath of the change from a macroplymeric to a supramolecular paradigm of humus seemed to encourage soil scientists to dissolve the hard-science concepts of humus and humification in a softer ecological vision of soil organic matter (SOC) dynamics in which even the word "humic" should lose significance and use (Lehmann and Kleber, 2015; Piccolo, 2016; Schmidt et al., 2011). Nevertheless, following the indication by Piccolo and coworkers that the recalcitrance of the soil Humeome could not be conferred by the chemical energy stored in the covalent bonds of nonexisting large macropolymers, has also brought others (Kleber 2010; Nuzzo et al.,

2013; Sollins et al., 2007) to conceive that the accumulation of humus in soil should be rather attributed to interactions with mineral particles.

In this perspective, the traditional humification concept based on traditional belief of the coupling of humic molecules into large macropolymers is no longer regarded as a realistic mechanism of SOM accumulation (Dungait et al., 2012; Schmidt et al., 2011). The preservation of soil C is still imagined to be related to the formation of aggregates, which are arranged in a hierarchy whereby SOC pools are protected by the aggregation of soil physical particles (Tiessen et al., 1984; Tisdall and Oades, 1982). By this scenario, the accumulation of soil humus would result from the restricted microbial accessibility of organic molecules due to both adsorption on and/or incorporation in soil mineral components, particle size-fractions and aggregates at variable stages of decomposition (Torn et al., 1997). However, little is still known on the physical–chemical mechanisms by which the humic molecules should interact with the mineral surfaces and create the organo–mineral associations that preserve SOC from rapid mineralization.

The theory of aggregate hierarchy strongly relied on carbohydrates as cementing agents to explain the stabilization of soil structure, although their transient microbial stability did not account for the long-term stability of soil aggregates (Tiessen et al., 1984; Tisdall and Oades, 1982). Conversely, an experiment that followed soil structural stability with time upon addition of either hydrophilic (carbohydratic gum) or hydrophobic (stearic acid, HA) components, suggested that it was rather the hydrophobicity of humic molecules to confer a long-term stabilization to the soil structure (Piccolo and Mbagwu, 1999).

The importance of hydrophobic bonding as a mechanism of chemical protection from microbial decomposition in soil was already highlighted by Piccolo (1996). In fact, the biomolecules reaching the soil may be incorporated into the hydrophobic domains of the existing soil Humeome, and become separated from the soil solution, and, concomitantly, from the microbial decomposing activity. The larger the hydrophobicity of the biomolecules (long-chain alkyl or aromatic compounds) and their mutual association in water, the slower is their degradation in the soil solution, and, the greater is the probability for these compounds to be thermodynamically "squeezed-out" from the aqueous solution due to the "hydrophobic effect" (Israelachvili, 1994; Tanford, 1980), adsorbed onto the soil solid surfaces and the existing hydrophobic humic layers, and, thus, preserved from microbial attack. Conversely, the highly hydrophilic oligo- and macrobiomolecules (carbohydrates and peptides) are more likely to partition into the aqueous soil solution, and although they potentially stabilize soil aggregates, they are rapidly degraded by microorganisms. Soil drying may further contribute to separate hydrophobic compounds from the soil solution and soil microbes, thereby increasing the mean residence time of hydrophobic compounds. The same mechanism may be contributing to the stability of microaggregates by favoring interparticle aggregation by hydrophobic bonding. The buildup of hydrophobic compounds is interrupted by soil ploughing that, by breaking soil macroaggregates, exposes the accumulated organic matter to air and water again, thereby activating the processes of chemical and microbial oxidation, with a concomitant alteration of the interaggregate hydrophobic interactions and an overall decrease of structural stability to water.

This mechanism of accumulation was confirmed by studies on the Humeome characteristics in different African soils of Nigeria (Piccolo et al., 2005a) and Ethiopia (Spaccini et al., 2006) when passing from forested to cultivated conditions. In all locations, HA and FA, HU, and a humic hydrophobic fraction extracted with an acetone–HCl solution (HE) were decreased as a result of deforestation and cultivation, although the magnitude of decrease was site-specific. While the elemental composition of HA, FA, and HU did not vary significantly with land use, that of the hydrophobic HE generally decreased with deforestation. More specifically, the ^{13}C CP-MAS NMR spectra showed that cultivation generally decreased the hydrophobicity of the Humeome in all soils except for the andisol (Spaccini et al., 2006). Although the HE material was the most hydrophobic soil fraction in the forested sites, cultivation also substantially reduced hydrophobicity for HE. When the soils were deforested for cultivation, the larger reduction of apolar as compared to hydrophilic components indicated that the pristine stability of accumulated organic matter was to be ascribed mainly to the selective preservation of hydrophobic compounds.

A sequential treatment of subsoils with NaOCl and HF enabled to estimate the pool size of mineral-protected and recalcitrant OM, as well as the chemical composition of recalcitrant materials (Mikutta et al., 2006). ^{13}C CPMAS NMR spectra showed that alkyl structures made up a large fraction of nonmineral stabilized OC. A study at the nanometer-scale by near edge X-ray absorption fine structure (NEXAFS) spectroscopy on the characteristics of humus in microaggregates have also indicated that the organic matter in coatings on mineral surfaces had more alkyl and carboxylic C (Lehmann et al., 2007). Similar conclusions were reached by a study on three different agricultural soils (Oxisol, Alfisol, Inceptisol) with ^{14}C-measured carbon turnover times of 107, 175, and 680 years, respectively

(Kleber et al., 2011). The density fraction with the youngest ^{14}C age (Oxisol, 107 years) showed the highest relative abundance of aromatic groups and the lowest O-alkyl C/aromatic C ratio as determined by NEXAFS spectroscopy. Conversely, the fraction with the oldest C (680 years in the Inceptisol) had the lowest relative abundance of aromatic groups and highest O-alkyl C/aromatic C ratio. These results suggested that the oldest soil C must have undergone several cycles of oxidation and had lost the hydrophobic compounds that distinguished the younger soil C in the Oxisol.

This experimental evidence provides a new significance for the process of humification in soil. Humification may then be conceived as the progressive segregation from the soil aqueous solution of the poorly soluble hydrophobic compounds derived from the abiotic and abiotic transformation of biomolecules released by dead cells. These highly hydrophobic compounds, such as alkyl and aromatic C, are then adsorbed on the large surface area of fine aluminosilicate minerals by weak dispersive forces. The mutual affinity with other hydrophobic molecules in the soil solution drives the progressive adsorption of OC in multilayer hydrophobic niches in soil aggregates, whose accumulation and long-term persistence is accounted to the chemical protection that excludes water and the accompanying microbes. When the soil is perturbed by plowing or other management practices, the hydrophobic components of SOM are either rapidly lost by mineralization or their oxidized byproducts are stabilized again in complexes with iron oxides and hydroxides (Drosos et al., 2017). A similar but accelerated process of concentration of hydrophobic alkyl and phenolic compounds is that observed with the composting of biomasses (Spaccini and Piccolo, 2007a,b), thereby again suggesting that the humification may be described as the accumulation of hydrophobic molecules during the microbial transformation of biomass.

The difficulty in reaching a distinction between hydrophobic and physical protection of SOC is due to the lack of consistent application of accurate techniques to evaluate the specific characteristics of recalcitrant OM (Kleber, 2010; Marschner et al., 2008). The OM recalcitrant or reactive behaviors are often related to alkaline or acid hydrolyses and oxidation methods, in the attempt to simulate biotic and abiotic degradation processes (Dobbss et al., 2010; Paul, 2016). However, there is no direct evidence of a simple correspondence between classical chemical laboratory treatments and biochemical stability of the soil Humeome (Incerti et al., 2017; Pandey et al., 2014). Moreover, the existing countless, often contrasting, results indicate that the long-term Humeome dynamics rely on the combination of basic physical, chemical, biological variables, climatic conditions, and anthropic perturbations (Bhattacharya et al., 2016; Paul, 2016). Nevertheless, the temporal or spatial record of a set of variables is hardly reproducible but always dependent on the specific experimental conditions and/or specifically designed research program and analytical approaches (Pisani et al., 2016; von Lützow et al., 2007).

The diverse interpretations of recalcitrance and stabilization processes of humus in soil become evident from the poor relationship between the intrinsic complexity of humus and the present poor definition of its molecular composition, as witnessed by the current modeling systems (Incerti et al., 2017). The methodologies developed to simulate SOC dynamics, either for general application (Century, RothC, LPJ, Biome- and Forest-BioGeochemical Cycles) or for specific agroecosystems, (DNDC, DayCent, CANDY), imply the incorporation of defined inputs of OM, including both decaying (above-/belowground litter, crop residues, active, microbial) fractions, and "humic" (stable, passive, recalcitrant) compartments. This assembly of OM components tends to reduce the complexity of SOM in confined pools with predictable turnover and residence time, in order to appraise soil C distribution and transformation within the varying ecological conditions. However, despite the effort to fit an increasing number of variables in flexible and handy mathematical tools, the descriptors of the wide complexity of humus molecular components are limited to the content of few macroindices such as C, N, lignin, and, seldom, cellulose. However, this apriori setting of parameters fails to adequately estimate the liability of features involved in SOC stabilization processes (Castellano et al., 2015; von Lützow et al., 2007). A relatively easy analytical detectability and a wide applicability represent both the strength and weakness of the employed parameters, since the aspecific nature of simple characteristics such as lignin/N or C/N ratios reduces their effectiveness as reliable indicators of either decomposition or accumulation of C (De Marco et al., 2012; Preston et al., 2009). For instance, the use of the C/N ratio to predict decay rates throughout the decomposition process should be adopted with care, because, regardless of its initial value, it progressively decreases as C is lost through respiration, and N is immobilized in the microbial biomass (Bonanomi et al., 2010). Moreover, the role of lignin as a reference fraction for the identification of recalcitrant humic pools has recently been challenged (Preston and Trofymow, 2015; Thevenot et al., 2010), because lignin components display both structural and site-specific rates of decomposition, with a larger bioavailability and faster turnover than the bulk SOM (Bahri et al., 2006). Also in tropical ecosystems, where OM transformation processes are accelerated, it has been questioned the common view that N and lignin control the rate of plant litter decomposition, whereas other nonlignin but stable plant C molecules at smaller initial concentration may play the major role (Hättenschwiler et al., 2011).

The application of modern analytical techniques to determine the molecular complexity of the soil Humeome (such as CPMAS-, DOSY-, and HRMAS-NMR spectroscopies, thermochemolysis-GC–MS, ESI-MS, ^{13}C GC–IRMS) has greatly advanced knowledge of the composition and dynamics of the soil Humeome (Nebbioso and Piccolo, 2013; Paul, 2016). Moreover, results have suggested that it is the molecular composition of the soil Humeome that controls the availability of the humic metabolic C to the microbial communities (Cozzolino et al., 2016; Puglisi et al., 2009), and the consequent processes in soil. In fact, the bioavailability of soil humus fractions affects the differential uptake by microorganisms, which thus mediate the OC accumulation in mineral horizons and the biochemical influence on crop physiology (Canellas et al., 2015; Kemmitt et al., 2008).

The mutual interaction between hydrophobic soil humic components and their association to mineral particles (textural components and aggregates) have been the conceptual framework of a recently developed model of soil C dynamics (Mazzoleni et al., 2012). This model, which is based on information obtained by ^{13}C CPMAS NMR spectroscopy and named Soil Organic Matter DYnamics (SOMDY), accurately anticipates the decomposition/stabilization behavior and the mass losses of a wide range of litter matter across different climates. Prediction capacity emerged from the specific partitioning of humic molecules in physical pools that are affected by organo-mineral associations, humus molecular quality, intermolecular interactions, and mechanisms of physical–chemical protection (Incerti et al., 2017).

The conformation flexibility and complex composition of the supramolecular Humeome determine the environmental diffusion of humic molecules, and, in turn, the functions of SOC and its mineralization/stabilization potential (Chilom et al., 2013; Khalaf et al., 2014; Piccolo, 2012; Šmejkalová et al., 2009). Depending on the dynamic spatial/temporal prevalence of polar and apolar components and the variation of surface tension, the humic compartment may be either preferentially stabilized as suspended colloids in the soil solution (Chilom et al., 2009; Nebbioso and Piccolo, 2013; Piccolo et al., 2001), or undergo a more favored adsorption on soil mineral surfaces, raising its partition into the soil matrix (Drosos et al., 2017; Feng et al., 2005; Kleber et al., 2007; Nebbioso et al., 2015). The coexistence of this apparently opposite equilibria explains the large potential reactivity of soil humus and its variable functionality. In fact, as recalled earlier, the Humeome represents the basic hydrophobic "building block" of the stable and passive SOM pools, both under forest and in agroecosystems (Drosos et al., 2017; Lorenz et al., 2007; Masoom et al., 2016; Piccolo et al., 2005a; Spaccini et al., 2006). Concomitantly, the bioavailable humic molecules may effectively modulate the structural-activity relationships in soil-microbes-plant interactions, as functional components of the active OM pools (Canellas and Olivares, 2014; Cozzolino et al., 2016; Dobbss et al., 2010; Nardi et al., 2007).

TECHNIQUES OF HUMUS STABILIZATION IN SOIL

The investigation on the mechanisms of stabilization of organic matter in soil has a practical importance. In fact, the definition of variables in the accumulation of soil humus is relevant to reach more efficient soil-management practices, whose reference guidelines are related to the increase of C sequestration in soil and the reduction of CO_2 emissions to the atmosphere (European Commission, 2012).

The current conceptual models of SOC accumulation that are mainly limited to physical protection in microaggregates (Six et al., 2004a) has led to the present largely sponsored soil management practices of minimum or no tillage, which are aimed to reduce soil C losses by excluding soil disturbance and consequently exposure of SOC to microbial mineralization (Green Carbon Conference, 2014). However, scientific uncertainties and contrasting outcomes have been reported on the usefulness of reduced tillage in respect to long term SOC accumulation (Du et al., 2017; Powlson et al., 2016). In fact, SOC accumulation by no tillage is not only slow (Powlson et al., 2012; Six et al., 2004b), limited to the soil surface (Luo et al., 2010), and poorly contributing to the stable C pools (Poeplau et al., 2015), but also the little stored C may be rapidly mineralized again should the soil management be periodically reversed to conventional tillage (Jacobs et al., 2010). Moreover, it is also believed the potential protection of the physical mechanism is limited by soil texture as for the maximum fixing capacity (Hassink and Whitmore, 1997).

CARBON SEQUESTRATION BY CHEMICAL PROTECTION IN HUMIFIED DOMAINS

Soil amendment with fresh organic matter is known to increase SOC and N content, soil biological activity, and crop yields, but also GHG emissions (Gattinger et al., 2012; Lehtinen et al., 2014). However, a rising attention is devoted to improve quantity and quality of soil humus by amendments with highly humified OM, such as that reached in composted biomasses (Piccolo, 2012). In fact, while the use of manure as a soil amendment for average EU-27 agricultural lands is assumed to be halved in 2030 (European Commission, 2011), an increasing involvement in long-term SOC sequestration is expected from the soil addition of compost (Mondini et al., 2012). The potential contribution to humified OC content has been estimated to duplicate, for both urban-waste and green composts, attaining in 2030 the average European level of 0.4 Mg ha^{-1} of compost derived OC (European Commission, 2011). Contrary to the short-term effect on crop productivity and the poor humus buildup provided by fresh organic matter, such as sewage sludge, manure, and straw (Poeplau et al., 2017; Powlson et al., 2012), field amendments with mature composts has the potential to effectively promote SOC sequestration and reduce GHG emission (Erhart et al., 2016; Scotti et al., 2016; Spaccini et al., 2009), due to the larger content of stable humified matter added to soil than for fresh biomasses (Spaccini and Piccolo, 2007a, 2009; Spaccini et al., 2016). In fact, the composting process implies an accelerated "humification" driven by the intense initial microbial aerobic decomposition of biomolecules (active or thermophilic phase), followed by a stabilization (ripening) phase. The composting process thus represents an accelerated humification, because it progressively accumulates the biotransformed hydrophobic molecules that escape further microbial transformation, similarly to the new concept of humification of SOM described earlier (Martinez-Balmori et al., 2013; Piccolo, 1996; Spaccini and Piccolo, 2007b; Spaccini et al., 2016). Depending on the starting biomasses the final Humeome in mature compost is composed of a majority of alkyl components, such as lipids, waxes, biopolyesters (cutins and suberins), and of lignin-derived aromatic structures, in whose hydrophobic domains also hydrophilic derivatives (carbohydrates, peptides, cellulose, hemicellulose, pectin, etc.) are protected from microbial mineralization (Pane et al., 2015; Spaccini and Piccolo, 2009).

The addition of mature (humified) compost provides the soil with an external Humeome that is not yet linked to the mineral phase of soil and may then represent a support for the incorporation of labile molecules into the compost hydrophobic domains, thereby exerting a protection from rapid mineralization.

In a short-term incubation experiment, a model biolabile molecule, such as 2-decanol, was ^{13}C labeled and partitioned in two different HAs, isolated from a mature compost and a lignite ore, with different degrees of hydrophobicity (Spaccini et al., 2002). The labeled 2-decanol alone and the two labeled humic solutions were added to a soil that was incubated at water holding capacity for 3 months. Soil samples were periodically collected to measure their ^{13}C content by an isotopic mass spectrometry. It was found that the biolabile ^{13}C labeled 2-decanol was rapidly mineralized when added alone, whereas mineralization was much and significantly reduced when amended to soil in mixture with humic matter, being the retention of ^{13}C carbon in soil a function of the humic hydrophobicity. Moreover, the residual ^{13}C distribution among the soil particle sizes indicated that the hydrophobic protection was most effective in the silt- and clay-sized fractions, where the added humic molecules had been preferentially repartitioned. These results showed that the stable humified compost appeared capable of reducing the biological mineralization of labile molecules due to the protection into the hydrophobic domains of the Humeome.

In another but longer experiment (1 year) both sandy and silty-loamy soils were treated with a bulk compost and its HA extract (HA-C), before or after amendment with a polysaccharidic gum, in order to verify whether the humified compost inhibited mineralization of the biolabile gum (Piccolo et al., 2004). Both bulk compost and HA-C treatments reduced OC losses progressively and significantly at increasing incubation time in both soils, with a final larger OC preservation in the finer textured silty loamy sample. These results further indicated that soil addition with compost may effectively stabilize the biolabile organic matter entering the soil by incorporation into the stable humified domains of compost.

In a similar experiment, soil samples from field plots treated for 20 years with either compost or inorganic fertilizers were added with ^{13}C labeled glucose and incubated for 30 days, in order to follow the dynamics of the labile labeled C in bulk samples, soil aggregates, and textural fractions (Zhang et al., 2015). Isotopic analysis revealed a larger preservation (>53%) of labile ^{13}C OC in compost-amended soils, with an increasing retention of the residual ^{13}C in finer soil fractions. These results confirmed that the exogenous biolabile C is more effectively protected in organic C-rich soil, and is gradually redistributed into the humic matter associated with silt and clay fractions.

The effectiveness of the hydrophobic protection of biolabile molecules by humified organic matter was verified in the field under maize production (Spaccini and Piccolo, 2012). In a 3-year experiment at three different agricultural sites, the OC content in soils under minimum tillage was compared to that of soils under traditional tillage but added with two rates of mature compost (2.7 and 5.4 ton ha^{-1}). After 3 years, minimum tillage showed short-term and even negative effect on OC accumulation, whereas compost treatments fixed in soils from 3 to 22 ton ha^{-1} more OC than for control traditional tillage, thereby indicating that the hydrophobic protection of humified organic matter is well active in field conditions. Moreover, ^{13}C CPMAS-NMR data on humus extracted from treated soils, combined with multivariate analyses, revealed that compost-amended soils progressively incorporated alkyl and aromatic hydrophobic components into the soil Humeome during the experimental time (Spaccini and Piccolo, 2012). These findings further highlight not only that the humification in soil means an increasing accumulation of hydrophobic material that remains progressively unavailable to microbial mineralization, but also that an enhanced hydrophobicity of humus enables to trap and protect from mineralization the biolabile compounds released in cropped soil by root exudates and the processes of crop residues degradation, thereby representing an effective mechanisms of C sequestration in soil.

CARBON SEQUESTRATION BY PHOTOPOLYMERIZATION OF HUMIC MOLECULES UNDER BIOMIMETIC CATALYSIS

An innovative and ecofriendly technology to induce soil C sequestration regards the in-situ chemical modification of soil humus by increasing the intermolecular covalent bonds among humic molecules (Piccolo, 2012). This methodology originates from the consideration that once soil humus is no longer viewed as covalently stabilized macropolymers, but as supramolecular associations of loosely bound heterogeneous small molecules, chemical technologies can then be employed to bind different molecules together and increase the average molecular mass of the soil Humeome with the aim of retarding or even inhibiting its mineralization.

It is known that phenoloxidative enzymes, such as peroxidise and laccase, catalyze a free-radical coupling reaction among humic molecules with a marked increase of the molecular size of the Humeome in solution (Cozzolino and Piccolo, 2002; Piccolo et al., 2000). An alternative to the costly and biological fragile enzymes is represented by a synthetic metal (Fe or Mn) porphyrin ring, that, if adequately substituted to ensure water solubility and the reactive stereochemical structure, is capable of mimicking the bioactivity of the prosthetic group of oxidative enzymes (Fig. 4.5). In fact, the metal-porphyrin biomimetic catalyst was shown to enable a significant increase of the Humeome molecular size in solution when oxidation was provided either by

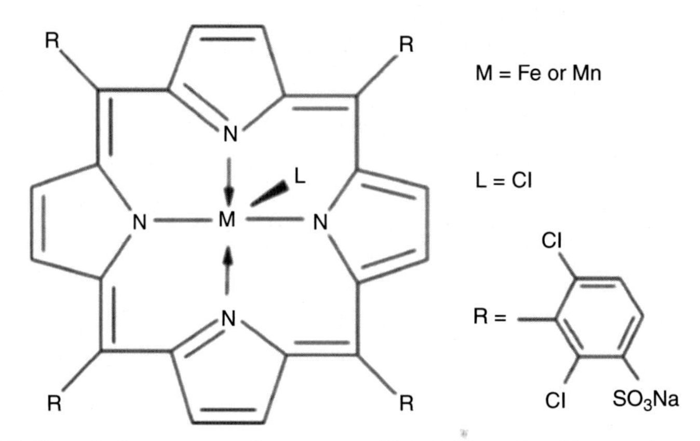

Figure 4.5 ***Chemical structures of meso-tetra-(2,6-dichloro-3-sulfonatophenyl)porphyrinate of Mn(III) [Mn(TDCPPS)Cl] or Fe(III) [Fe(TDCPPS)Cl].***

H_2O_2 (Piccolo et al., 2005b) or by O_2 under UV irradiation (Šmejkalová and Piccolo, 2005). The change in the conformational structure of the Humeome obtained by either oxidative or photo-oxidative biomimetic catalysis, was explained by the enhanced rates of free-radical coupling reactions among humic phenolic monomers (Šmejkalová and Piccolo, 2006), which were turned into oligomers, namely from dimers up to pentamers (Šmejkalová et al., 2006). The photo-oxidative formation of intermolecular C-C and C-O-C bonds in the dissolved Humeome was confirmed and even further enhanced when the biomimetic catalysis was made heterogeneous, by immobilizing the metal-porphyrins to the surface of both kaolinite and montmorillonite through a 3-(1-imidazolyl)propylcarbamoyl-3'-aminopropyl-triethoxysilane (Imi-APTS) chemical spacer (Nuzzo and Piccolo, 2013a,b).

The successful catalytic photopolymerization of dissolved humic molecules induced to apply the same technology in-situ on soil in order to verify whether the controlled formation of coupled oligomers in the soil Humeome may result in a reduced availability to microbial mineralization and an enhanced SOC stabilization and accumulation. Furthermore, the increase in the molecular size of the soil Humeome would also promote the association of soil particles into greater soil aggregates, thus improving the overall soil physical quality. These hypotheses were successfully tested in soil treatments with biomimetic catalysts, in either laboratory or mesocosm experiments, whereby the reached larger chemical stabilization of SOC significantly reduced soil microbial respiration, while improving soil structural stability and the capacity to resist the slaking effect of several wetting and drying cycles (Gelsomino et al., 2010; Nuzzo et al., 2016, 2017; Piccolo et al., 2011).

As in the case of the hydrophobic mechanism of SOC sequestration, the C fixation potential of the technique based on the photopolymerization under biomimetic catalysis was also evaluated in situ in a 3-year field experiment with three different agricultural soils, within the Italian national project MESCOSAGR (sustainable methods for soil organic carbon sequestration in agricultural soils) (Spaccini and Piccolo, 2012). In field plots treated by the water soluble iron–porhyrin catalysis, the average C sequestered in bulk soil for 3 consecutive years, regardless of the heterogeneity of soil matrices and site conditions, ranged from 2.24 to 3.90 Mg ha^{-1} y^{-1} more than control soil. Similar results were obtained by measuring the OC distribution among soil water-stable aggregates, where C preferentially accumulated in macroaggregates (Piccolo et al., 2017). Because such significant SOC sequestration was achieved under a yearly soil perturbation by conventional tillage, the OC incorporation was only accountable to an enhanced chemical and biochemical stabilization of the soil Humeome achieved by the in situ catalyzed photo-oxidative treatment. The soil humic molecules susceptible to undergo the photocatalyzed coupling reactions may be those present in the soil solution when released either by the OM adsorbed on soil particles, or by plant roots during the cropping cycles. All this soluble C may periodically become the substrate of the photopolymerization reactions occurring under the biomimetic catalysis and be subjected to coupling into larger size oligomers. This process is likely to increase the hydrophobicity of the soil Humeome and reduce its aqueous solubility, thereby favoring adsorption on the surfaces of clay-humic aggregates, and consequent protection from microbial mineralization. The result is the observed larger C sequestration in the catalyst-treated soils than for the control. It is important to note that the net carbon sequestration is achieved without reducing crop yields (Grignani et al., 2012), nor altering the soil microbial biodiversity (Puglisi and Trevisan, 2012).

CONCLUSIONS

While the concepts of the new paradigm of soil humus as supramolecular associations of different small molecules, rather than macropolymers, were substantiated by increasing experimental evidence and reached a general consensus, though with diverse phrasing and terminology according to the scientific perspectives of various groups, efforts should be still devoted to the unveiling of the molecular composition of soil humus in order to define the physical–chemical processes of C accumulation in soil and the biological reactivity of humus towards microbes and plants.

An advance to the establishment of accurate and reproducible methods for the identification of the soil C molecules, is represented by the introduction of Humeomics. This procedure combines a simple and reproducible chemical fractionation sequence that does not alter the C-C structure of solvated humic molecules, with the application of modern analytical techniques, such as different pulse techniques of NMR spectroscopy and liquid chromatography separations coupled to high-resolution mass spectrometry. This powerful combination of chemical fractionation and physical–chemical instrumentation was proven to provide an insight in the soil Humeome, both in quantity and quality that it was never attained before by any other approach. The broader application of Humeomics to soils of different nature, covered by different vegetations, with different cropping systems, and managements, promises to advance the molecular knowledge of soil Humeome and its dynamics.

The supramolecular paradigm of soil humus allowed to deepen the understanding of its interaction toward exogenous biolabile molecules and of the processes of C stabilization in soil. In fact, a new concept of humification based on the selective preservation in soil particles of hydrophobic molecules, because of their thermodynamic exclusion from the soil solution, has driven the development of a mechanism of protection of biolabile molecules from rapid microbial mineralization in the hydrophobic domains of humified compost. Moreover, the application of a chemical technology that relies on a newly conceived water-soluble biomimetic catalyst and exploitation of solar energy to promote the oxidative coupling among aromatic humic molecules, was shown to provide an innovative approach to sequester C in agricultural soils.

These recent research results have indicated that if sustainable agriculture is to be pursued to ensure agricultural intensification within environmental protection, it will be essential to reach a higher level of knowledge on the molecular behavior of the soil Humeome. Despite its molecular complexity, an advanced molecular and conformational control of the soil Humeome will enable it to predict not only the activity of the microbial communities in soil but also the mutual microbe–humeome effects toward a more efficient crop growth.

REFERENCES

Aquino, A.J.A., Tunega, D., Schaumann, G.E., Haberhauer, G., Gerzabek, M.H., Lischka, H., 2011. The functionality of cation bridges for binding polar groups in soil aggregates. Int. J. Quantum Chem. 111, 1531–1542.

Baalousha, M., Motelica-Heino, M., Galaup, S., Le Coustumer, P., 2005. Supramolecular structure of humic acids by TEM with improved sample preparation and staining. Microsc. Res. Technol. 66, 299–306.

Baalousha, M., Motelica-Heino, M., Le Coustumer, P., 2006. Conformation and size of humic substances: effects of major cation concentration and type, pH, salinity and residence time. Colloids Surf. A 272, 48–55.

Bahri, H., Dignac, M.F., Rumpel, C., Rasse, D.P., Chenu, C., Mariotti, A., 2006. Lignin turnover kinetics in an agricultural soil is monomer specific. Soil Biol. Biochem. 38, 1977–1988.

Bhattacharya, S.S., Kim, K.-H., Das, S., Uchimiya, M., Jeon, B.H., Kwon, E., et al., 2016. A review on the role of organic inputs in maintaining the soil carbon pool of the terrestrial ecosystem. J. Environ. Manage. 167, 214–227.

Bonanomi, G., Incerti, G., Antignani, V., Capodilupo, M., Mazzoleni, S., 2010. Decomposition and nutrient dynamics in mixed litter of Mediterranean species. Plant Soil 331, 481–496.

Brown, T.L., Rice, J.A., 2000. Effect of experimental parameters on the ESI FT-ICR mass spectrum of fulvic acid. Anal. Chem. 72, 384–390.

Buurman, P., van Lagen, B., Piccolo, A., 2002. Increase in stability against thermal oxidation of soil humic substances as a result of self- association. Org. Geochem. 33, 367–381.

Canellas, L.P., Olivares, F.L., 2014. Physiological responses to humic substances as plant growth promoter. Chem. Biol. Technol. Agric. 1 (3)doi: 10.1186/2196-5641-1-3.

Canellas, L.P., Olivares, F.L., Aguiar, N.O., Jones, D.L., Nebbioso, A., Mazzei, P., 2015. Humic and fulvic acids as biostimulants in horticulture. Sci. Hortic. Amsterdam 196, 15–27.

Castellano, M.J., Mueller, K.E., Olk, D.C., Sawyer, J.E., Six, J., 2015. Integrating plant litter quality, soil organic matter stabilization and the carbon saturation concept. Global Change Biol. 21, 3200–3209.

Chilom, G., Bruns, A.S., Rice, J.A., 2009. Aggregation of humic aicds in solution: contribution of different fractions. Org. Geochem. 40, 455–460.

Chilom, G., Baglieri, A., Johnson-Edler, C.A., Rice, J.A., 2013. Hierarchical self-assembling properties of natural organic matter's components. Org. Geochem. 57, 119–126.

Cole, E.B., 2000. Some tenets pertaining to electrospray ionization mass spectrometry. J. Mass Spectrom. 35, 763–772.

Conte, P., Spaccini, R., Piccolo, A., 2006. Advanced CPMAS-^{13}C NMR techniques for molecular characterization of size-separated fractions from a soil humic acid. Anal. Bioanal. Chem. 386, 382–390.

Conte, P., Spaccini, R., Šmejkalová, D., Nebbioso, A., Piccolo, A., 2007. Spectroscopic and conformational properties of size-fractions separated from a lignite humic acid. Chemosphere 69, 1032–1039.

Cozzolino, A., Piccolo, A., 2002. Polymerization of dissolved humic substances catalyzed by peroxidase: effects of pH and humic composition. Org. Geochem. 33, 281–294.

Cozzolino, V., Di Meo, V., Monda, H., Spaccini, R., Piccolo, A., 2016. The molecular characteristics of compost affect plant growth, arbuscular mycorrhizal fungi, and soil microbial community composition. Biol. Fertil. Soils 52, 15–29.

De Marco, A., Spaccini, R., Vittozzi, P., Esposito, F., Berg, B., Virzo, De Santo, A., 2012. Decomposition of black locust and black pine leaf litter in two coeval forest stands on Mount Vesuvius and dynamics of organic components assessed through proximate analysis and NMR spectroscopy. Soil Biol. Biochem. 51, 1–15.

Dobbss, L.B., Canellas, L.P., Olivares, F.L., Aguiar, N.O., Peres, L.E.P., Azevedo, M., et al., 2010. Bioactivity of chemically transformed humic matter from vermicompost on plant root growth. J. Agric. Food Chem. 58, 3681–3688.

Drosos, M., Nebbioso, A., Mazzei, P., Vinci, G., Spaccini, R., Piccolo, A., 2017. A molecular zoom into soil Humeome by a direct sequential chemical fractionation of soil. Sci. Total Environ. 15, 807–816.

Du, Z., Angers, D.A., Ren, T., Zhang, Q., Li, G., 2017. The effect of no-till on organic C storage in Chinese soils should not be overemphasized: a meta-analysis. Agric. Ecosyst. Environ. 236, 1–11.

Dungait, J., Hopkins, D., Gregory, A., Whitmore, A., 2012. Soil organic matter turnover is governed by accessibility not recalcitrance. Global Change Biol. 18, 1781–1796.

Erhart, E., Schmid, H., Hartl, W., Hülsbergen, K.-J., 2016. Humus, nitrogen and energy balances, and greenhouse gas emissions in a long-term field experiment with compost compared with mineral fertilisation. Soil Res. 54, 254–263.

European Commission, 2011. Soil organic matter management across the EU—best practices, constraints and trade-offs. Final Report for the European Commission's DG Environment.

European Commission, 2012. The state of soil in Europe. Join Research Center European Environmental Agency, Report EUR 25186 EN, Luxembourg Publication Office of European Union.

Fang, C., Smith, P., Moncrieff, J.B., Smith, J.U., 2005. Similar response of labile and resistant soil organic matter pools to changes in temperature. Nature 433, 57–59.

Feng, X.J., Simpson, A.J., Simpson, M.J., 2005. Chemical and mineralogical controls on humic acid sorption to clay mineral surfaces. Org. Geochem. 36, 1553–1566.

Fontaine, S., Barot, S., Barré, P., Bdioui, N., Mary, B., Rumpel, C., 2007. Stability of organic carbon in deep soil layers controlled by fresh carbon supply. Nature 450, 277–280.

Gattinger, A., Muller, A., Haeni, M., Skinner, C., Fliessbach, A., Buchmann, N., et al., 2012. Enhanced top soil carbon stocks under organic farming. Proc. Natl. Acad. Sci. USA 109, 18226–18231.

Gelsomino, A., Tortorella, D., Cianci, V., Petrovicová, B., Sorgonà, A., Piccolo, A., et al., 2010. Effects of a biomimetic iron-porphyrin on soil respiration and maize root morphology as by a microcosm experiment. J. Plant Nutr. Soil Sci. 173, 399–406.

Green Carbon Conference, 2014. Green carbon: making sustainable agriculture real, Conference Proceedings. http://www.greencarbon-ca.eu/post-conference.

Grignani, C., Alluvione, F., Bertora, C., Zavattaro, Fagano, M., Fiorentino, N., et al., 2012. Field plots and crop yields under innovative methods of carbon sequestration in soil. In: Piccolo, A. (Ed.), Carbon Sequestration in Agricultural Soils: A Multidisciplinary Approach to Innovative Methods. Springer-Verlag Berlin, Heidelberg, Berlin, pp. 39–60.

Hassink, J., Whitmore, A.P., 1997. A model of the physical protection of organic matter in soil. Soil Sci. Soc. Am. J. 61, 131–139.

Hättenschwiler, S., Coq, S., Barantal, S., Handa, I.T., 2011. Leaf traits and decomposition in tropical rainforests: revisiting some commonly held views and towards a new hypothesis. New Phytol. 189, 950–965.

Hertkorn, N., Benner, R., Frommberger, M., Schmitt-Kopplin, P., Witt, M., Kaiser, K., et al., 2006. Characterization of a major refractory component of marine dissolved organic matter. Geochim. Cosmochim. Acta 70, 2990–3010.

Incerti, G., Bonanomi, G., Giannino, F., Carteni, F., Spaccini, R., Mazzei, P., et al., 2017. OMDY: a new model of organic matter decomposition based on biomolecular content as assessed by ^{13}C-CPMAS-NMR. Plant Soil. 411 (1–2), 377–394.

IPCC, 2013. Carbon and other biogeochemical cycles. Climate Change 2013: The Physical Science Basis. Contribution of Working Group I to the Fifth Assessment Report of the Intergovernmental Panel on Climate Change. Cambridge University Press, Cambridge, United Kingdom and New York, NY, USA.

Israelachvili, J.N., 1994. Intermolecular and Surface Forces, second ed. Elsevier, Academic Press, London, UK, pp. 282.

Jacobs, A., Helfrich, M., Hanisch, S., Quendt, U., Rauber, R., Ludwig, B., 2010. Effect of conventional and minimum tillage on physical and biochemical stabilization of soil organic matter. Biol. Fertil. Soils 46, 671–680.

Kalinichev, A.G., Iskrenova-Tchoukova, E., Ahn, W.-Y., Clark, M.M., Kirkpatrick, R.J., 2011. Effects of Ca^{2+} on supramolecular aggregation of natural organic matter in aqueous solutions: a comparison of molecular modeling approaches. Geoderma 169, 27–32.
Kemmitt, S.J., Lanyon, C.V., Waite, I.S., Wen, Q., Addiscott, T.M., Birda, N.R.A., et al., 2008. Mineralization of native soil organic matter is not regulated by the size, activity or composition of the soil microbial biomass: a new perspective. Soil Biol. Biochem. 40, 61–73.
Khalaf, M.M.R., Chilom, G., Rice, J.A., 2014. Comparison of the effects of self-assembly and chemical composition on humic acid mineralization. Soil Biol. Biochem. 73, 96–105.
Kleber, M., 2010. What is recalcitrant soil organic matter? Environ. Chem. 7, 320–332.
Kleber, M., Sollins, P., Sutton, R., 2007. A conceptual model of organo-mineral interactions in soils: self-assembly of organic molecular fragments into zonal structures on mineral surfaces. Biogeochemistry 85, 9–24.
Kleber, M., Nico, P.S., Plante, A.F., Filley, T., Kramer, M., Swanston, C., et al., 2011. Old and stable soil organic matter is not necessarily chemically recalcitrant: implications for modeling concepts and temperature sensitivity. Global Change Biol. 17, 1097–1107.
Lehmann, J., Kleber, M., 2015. The contentious nature of soil organic matter. Nature 528, 60–68.
Lehmann, J., Kinyangi, J., Solomon, D., 2007. Organic matter stabilization in soil microaggregates: implications from spatial heterogeneity of organic carbon contents and carbon forms. Biogeochemistry 85, 45–57.
Lehtinen, T., Schlatter, N., Baumgarten, A., Bechini, L., Krüger, J., Grignani, C., et al., 2014. Effect of crop residue incorporation on soil organic carbon and greenhouse gas emissions in European agricultural soils. Soil Use Manage. 30, 524–538.
Liu, Z., Zu, Y., Meng, R., Xing, Z., Tan, S., Zhao, L., et al., 2011. Adsorption of humic acids onto carbonaceous surfaces: atomic force microscopy study. Microsc. Microanal. 17, 1015–1021.
Lorenz, K., Lal, R., Preston, C.M., Nierop, K.G.J., 2007. Strengthening the soil organic carbon pool by increasing contributions from recalcitrant aliphatic bio(macro)molecules. Geoderma 142, 1–10.
Luo, Z., Wang, E., Sun, O.J., 2010. Can no-tillage stimulate carbon sequestration in agricultural soils? A meta-analysis of paired experiments. Agric. Ecosyst. Environ. 139, 224–231.
Marschner, B., Brodowski, S., Dreves, A., Gleixner, G., Gude, A., Grootes, P.M., et al., 2008. How relevant is recalcitrance for the stabilization of organic matter in soils? J. Plant Nutr. Soil Sci. 171, 91–110.
Martinez-Balmori, D., Olivares, F.L., Spaccini, R., Aguiar, K.P., Araújo, M.F., Aguiar, N.O., et al., 2013. Molecular characteristics of vermicompost and their relationship to preservation of inoculated nitrogen-fixing bacteria. J. Anal. Appl. Pyrolysis 104, 540–550.
Masoom, H., Courtier-Murias, D., Farooq, H., Soong, R., Kelleher, B.P., Zhang, C., et al., 2016. Soil organic matter in its native state: unravelling the most complex biomaterial on earth. Environ. Sci. Technol. 50, 1670–1680.
Mazzoleni, S., Bonanomi, G., Giannino, F., Incerti, G., Piermatteo, D., Spaccini, R., et al., 2012. New modeling approach to describe and predict carbon sequestration dynamics in agricultural soils. In: Piccolo, A. (Ed.), Carbon Sequestration in Agricultural Soils. Springer, Berlin, pp. 291–307.
McIntyre, C., Jardine, D.R., McRae, C., 2001. Electrospray mass spectrometry of aquatic fulvic acids. Rapid Commun. Mass Spectrom. 15, 1974–1975.
Mikutta, R., Kleber, M., Torn, M.S., Jahn, R., 2006. Stabilization of soil organic matter: association with minerals or chemical recalcitrance? Biogeochemistry 77, 25–56.
Mondini, C., Coleman, K., Whitmore, A.P., 2012. Spatially explicit modelling of changes in soil organic C in agricultural soils in Italy, 2001–2100: potential for compost amendment. Agric. Ecosyst. Environ. 153, 24–32.

Nardi, S., Muscolo, A., Vaccaro, S., Baiano, S., Spaccini, R., Piccolo, A., 2007. Relationship between molecular characteristics of soil humic fractions and glycolytic pathway and krebs cycle in maize seedlings. Soil Biol. Biochem. 39, 3138–3146.

Nebbioso, A., Piccolo, A., 2009. Molecular rigidity and diffusivity of Al^{3+} and Ca^{2+} humates as revealed by NMR spectroscopy. Environ. Sci. Technol. 43, 2417–2424.

Nebbioso, A., Piccolo, A., 2011. Basis of a humeomics science: chemical fractionation and molecular characterization of humic biosuprastructures. Biomacromolecules 12, 1187–1199.

Nebbioso, A., Piccolo, A., 2012. Advances in humeomics: enhanced structural identification of humic molecules after size fractionation of a soil humic acid. Anal. Chim. Acta 720, 77–90.

Nebbioso, A., Piccolo, A., 2013. Molecular characterization of dissolved organic matter (DOM): a critical review. Anal. Bioanal. Chem. 405, 109–124.

Nebbioso, A., Piccolo, A., Spiteller, M., 2010. Limitations of electrospray ionization in the analysis of a heterogeneous mixture of naturally occurring hydrophilic and hydrophobic compounds. Rapid Commun. Mass Spectrom. 24, 3163–3170.

Nebbioso, A., Piccolo, A., Lamshöft, M., Spiteller, M., 2014a. Molecular characterization of an end-residue of humeomics applied to a soil humic acid. RSC Adv. 4, 23658–23665.

Nebbioso, A., Mazzei, P., Savy, D., 2014b. Reduced complexity of multidimensional and diffusion NMR spectra of soil humic fractions as simplified by humeomics. Chem. Biol. Technol. Agric. 1 (24)doi: 10.1186/540538-014-0024-4.

Nebbioso, A., Vinci, G., Drosos, M., Spaccini, R., Piccolo, A., 2015. Unveiling the molecular composition of the unextractable soil organic fraction (humin) by humeomics. Biol. Fertil. Soils 51, 443–451.

Nuzzo, A., Piccolo, A., 2013a. Enhanced catechol oxidation by heterogeneous biomimetic catalysts immobilized on clay minerals. J. Mol. Catal. A 371, 8–14.

Nuzzo, A., Piccolo, A., 2013b. Oxidative and photo-oxidative polymerization of humic suprastructures by heterogeneous biomimetic catalysis. Biomacromolecules 14, 1645–1652.

Nuzzo, A., Sánchez, A., Fontaine, B., Piccolo, A., 2013. Conformational changes of dissolved humic and fulvic superstructures with progressive iron complexation. J. Geochem. Explor. 129, 1–5.

Nuzzo, A., Madonna, E., Mazzei, P., Spaccini, R., Piccolo, A., 2016. In situ photo-polymerization of soil organic matter by heterogeneous nano-TiO_2 and biomimetic metal-porphyrin catalysts. Biol. Fertil. Soils 52, 585–593.

Nuzzo, A., Spaccini, R., Cozzolino, V., Moschetti, G., Piccolo, A., 2017. In situ polymerization of soil organic matter by oxidative biomimetic catalysis. Chem. Biol. Technol. Agric., in press.

Orsi, M., 2014. Molecular dynamics simulation of humic substances. Chem. Biol. Technol. Agric. 1 (10)doi: 10.1186/S40538-014-0010-4.

Pandey, D., Agrawal, M., Bohra, J.S., Adhya, T.K., Bhattacharyya, P., 2014. Recalcitrant and labile carbon pools in a sub-humid tropical soil under different tillage combinations: a case study of rice–wheat system. Soil Tillage Res. 143, 116–122.

Pane, P., Celano, G., Piccolo, A., Villecco, D., Spaccini, R., Palese, A.M., Zaccardelli, M., 2015. Effects of on-farm composted tomato residues on soil biological activity and yields in a tomato cropping system. Chem. Biol. Technol. Agric. 2 (4)doi: 10.1186/S40538-014-0026-9.

Paul, E.A., 2016. The nature and dynamics of soil organic matter: plant inputs, microbial transformations, and organic matter stabilization. Soil Biol. Biochem. 98, 109–126.

Peuravuori, J., 2005. NMR spectroscopy study of freshwater humic material in light of supramolecular assembly. Environ. Sci. Technol. 39, 5541–5549.

Peuravuori, J., Pihlaja, K., 2004. Preliminary study of lake dissolved organic matter in light of nanoscale supramolecular assembly. Environ. Sci. Technol. 38, 5958–5967.

Peuravuori, J., Bursakova, P., Pihlaja, K., 2007. ESI-MS analyses of dissolved organic matter in light of supramolecular assembly. Anal. Bioanal. Chem. 389, 1559–1568.
Piccolo, A., 1996. Humus and soil conservation. In: Piccolo, A. (Ed.), Humic Substances in Terrestrial Ecosystems. Elsevier, Amsterdam, The Netherlands, pp. 225–264.
Piccolo, A., 2001. The supramolecular structure of humic substances. Soil Sci. 166, 810–832.
Piccolo, A., 2002. The supramolecular structure of humic substances: a novel understanding of humus chemistry and implications in soil science. Adv. Agron. 75, 57–134.
Piccolo, A., 2012. The nature of soil organic matter and innovative soil managements to fight global changes and maintain agricultural productivity. In: Piccolo, A. (Ed.), Carbon Sequestration in Agricultural Soils: A Multidisciplinary Approach to Innovative Methods. Springer-Verlag Berlin Heidelberg, Berlin, pp. 1–20.
Piccolo, A., 2016. In memoriam Prof. F.J. Stevenson and the question of humic substances in soil. Chem. Biol. Technol. Agric. 3 (23)doi: 10.1186//s40538-016-0076-2.
Piccolo, A., Mbagwu, J.S.C., 1999. Role of hydrophobic components of soil organic matter on the stability of soil aggregates. Soil Sci. Soc. Am. J. 63, 1801–1810.
Piccolo, A., Spiteller, M., 2003. Electrospray ionization mass spectrometry of terrestrial humic substances and their size-fractions. Anal. Bioanal. Chem. 377, 1047–1059.
Piccolo, A., Cozzolino, A., Conte, P., Spaccini, R., 2000. Polymerization by enzyme-catalyzed oxidative couplings of supramolecular associations of humic molecules. Naturwissenschaften 87, 391–394.
Piccolo, A., Conte, P., Cozzolino, A., Spaccini, R., 2001. Molecular sizes and association forces of humic substances in solution. In: Clapp, C.E., Hayes, M.H.B., Senesi, N., Bloom, P.R., Jardine, P.M. (Eds.), Humic Substances and Chemical Contaminants. SSSAJ, Madison, USA, pp. 89–118.
Piccolo, A., Conte, P., Spaccini, R., Chiarella, M., 2003. Effects of some dicarboxylic acids on the association of dissolved humic substances. Biol. Fertil. Soils 37, 255–259.
Piccolo, A., Spaccini, R., Nieder, R., Richter, J., 2004. Sequestration of a biologically labile organic carbon in soils by humified organic matter. Clim. Change 67, 329–343.
Piccolo, A., Conte, P., Spaccini, R., Mbagwu, J.S.C., 2005a. Influence of land use on the humic substances of some tropical soils of Nigeria. Eur. J. Soil Sci. 56, 343–352.
Piccolo, A., Conte, P., Tagliatesta, P., 2005b. Increased conformational rigidity of humic substances by oxidative biomimetic catalysis. Biomacromolecules 6, 351–358.
Piccolo, A., Spiteller, M., Nebbioso, A., 2010. Effects of sample properties and mass spectroscopic parameters on electrospray ionization mass spectra of size-fractions from a soil humic acid. Anal. Bioanal. Chem. 397, 3071–3078.
Piccolo, A., Spaccini, R., Nebbioso, A., Mazzei, P., 2011. Carbon sequestration in soil by in situ catalyzed photo-oxidative polymerization of soil organic matter. Environ. Sci. Technol. 45, 6697–6702.
Piccolo, A., Spaccini, R., Cozzolino, V., Nuzzo, A., Drosos, M., Zavattaro, L., 2017. Effective carbon sequestration in Italian agricultural soils by *in situ* polymerization of soil organic matter under biomimetic photo-catalysis. Land Deg. Develop. in press.
Pisani, O., Haddix, M.L., Conant, R.T., Paul, E.A., Simpson, M.J., 2016. Molecular composition of soil organic matter with land-use change along a bi-continental mean annual temperature gradient. Sci. Total Environ. 573, 470–480.
Poeplau, C., Kätterer, T., Bolinder, M.A., Börjesson, G., Berti, A., Lugato, E., 2015. Low stabilization of aboveground crop residue carbon in sandy soils of Swedish long-term experiments. Geoderma 237– 238, 246–255.
Poeplau, C., Reiter, L., Berti, A., Kätterer, T., 2017. Qualitative and quantitative response of soil organic carbon to 40 years of crop residue incorporation under contrasting nitrogen fertilisation regimes. Soil Res. 55, 1–9.
Powlson, D.S., Bhogal, A., Chambers, B.J., Coleman, K., Macdonald, A.J., Goulding, K.W.T., et al., 2012. The potential to increase soil carbon stocks through reduced tillage or organic additions: an England and Wales case study. Agric. Ecosyst. Environ. 146, 23–33.

Powlson, D.S., Stirling, C.M., Thierfelder, C., White, R.P., Jate, M.L., 2016. Does conservation agriculture deliver climate change mitigation through soil carbon sequestration in tropical agro-ecosystems? Agric. Ecosyst. Environ. 220, 164–174.
Preston, C.M., Trofymow, J.A., 2015. The chemistry of some foliar litters and their sequential proximate analysis fractions. Biogeochemistry 126, 197–209.
Preston, C.M., Nault, J.R., Trofymov, J.A., 2009. Chemical changes during 6 years of decomposition of 11 litters in some Canadian forest sites. Part 2. ^{13}C abundance, solid-state ^{13}C NMR spectroscopy and the meaning of "Lignin". Ecosystems 12, 1078–1102.
Puglisi, E., Trevisan, M., 2012. Effects of methods of carbon sequestration in soil on biochemical indicators of soil quality. In: Piccolo, A. (Ed.), Carbon Sequestration in Agricultural Soils: A Multidisciplinary Approach to Innovative Methods. Springer-Verlag Berlin Heidelberg, Berlin, pp. 179–207.
Puglisi, E., Fragoulis, G., Ricciuti, P., Cappa, F., Spaccini, R., Piccolo, A., et al., 2009. Effects of a humic acid and its size-fractions on the bacterial community of soil rhizosphere under maize (*Zea mays* L.). Chemosphere 77, 829–837.
Reeves, D.W., 1997. The role of soil organic matter in maintaining soil quality in continuous cropping systems. Soil Tillage Res. 43, 131–167.
Romera-Castillo, C., Chem, M., Yamashita, Y., Jaffé, R., 2014. Fluorescence characteristics of size-fractionated dissolved organic matter: implications for a molecular assembly based structure. Water Res. 55, 40–51.
Schmidt, M.W.I., Torn, M.S., Abiven, S., Dittmar, T., Guggenberger, G., Janssens, I.A., et al., 2011. Persistence of soil organic matter as an ecosystem property. Nature 478, 49–56.
Scotti, R., Pane, P., Spaccini, R., Palese, A.M., Piccolo, A., Celano, G., et al., 2016. On-farm compost: a useful tool to improve soil quality under intensive farming systems. Appl. Soil Ecol. 107, 13–23.
Siéliéchi, J.M., Lartiges, B.S., Kayem, G.J., et al., 2008. Changes in humic acid conformation during coagulation with ferric chloride: implications for drinking water treatment. Water Res. 42, 2111–2123.
Simpson, A.J., 2002. Determining the molecular weight, aggregation, structures and interactions of natural organic matter using diffusion ordered spectroscopy. Magn. Reson. Chem. 40, S72–S82.
Six, J., Bossuyt, H., Degryze, S., Denef, K., 2004a. A hystory of research on the link between microaggregates, soil biota and soil organic matter dynamics. Soil Tillage Res. 79, 7–31.
Six, J., Ogle, S.M., Breidt, F.J., Conant, R.T., Mosier, A.R., Paustian, K., 2004b. The potential to mitigate global warming with no-tillage management is only realized when practised in the long term. Global Change Biol. 10, 155–160.
Šmejkalová, D., Piccolo, A., 2005. Enhanced molecular dimension of a humic acid induced by photooxidation catalyzed by biomimetic metalporphyrins. Biomacromolecules 6, 2120–2125.
Šmejkalová, D., Piccolo, A., 2006. Rates of oxidative coupling of humic phenolic monomers catalyzed by a biomimetic iron-porphyrin. Environ. Sci. Technol. 40, 1644–1649.
Šmejkalová, D., Piccolo, A., 2008. Aggregation and disaggregation of humic supramolecular assemblies by NMR diffusion ordered spectroscopy (DOSY-NMR). Environ. Sci. Technol. 42, 699–706.
Šmejkalová, D., Piccolo, A., Spiteller, M., 2006. Oligomerization of humic phenolic monomers by oxidative coupling under biomimetic catalysis. Environ. Sci. Technol. 40, 6955–6962.
Šmejkalová, D., Spaccini, R., Fontaine, B., Piccolo, A., 2009. Binding of phenol and differently halogenated phenols to dissolved humic matter as measured by NMR spectroscopy. Environ. Sci. Technol. 43, 5377–5382.
Smith, P.M., Bustamante, H., Ahammad, H., Clark, H., Dong, E.A., Elsiddig, H., et al., 2014. Agriculture, forestry and other land use (AFOLU). In: Edenhofer, O., Pichs-Madruga,

R., Sokona, Y., Farahani, E., Kadner, S., Seyboth, K., Adler, A., Baum, I., Brunner, S., Eickemeier, P., Kriemann, B., Savolainen, J., Schlömer, S., von Stechow, C., Zwickel, T., Minx, J.C. (Eds.), Climate Change 2014: Mitigation of Climate Change. Contribution of Working Group III to the Fifth Assessment Report of the Intergovernmental Panel on Climate Change. Cambridge University Press, Cambridge, United Kingdom and New York, NY, United States, pp. 816–887.

Sollins, P., Swanston, C., Kramer, M., 2007. Stabilization and destabilization of soil organic matter: a new focus. Biogeochemistry 85, 1–7.

Spaccini, R., Piccolo, A., 2007a. Molecular characterisation of compost at increasing stages of maturity: I. Chemical fractionation and infrared spectroscopy. J. Agric. Food Chem. 55, 2293–2302.

Spaccini, R., Piccolo, A., 2007b. Molecular characterization of compost at increasing stages of maturity. 2. Thermochemolysis-GC–MS and ^{13}C-CPMAS-NMR spectroscopy. J. Agric. Food Chem. 55, 2303–2311.

Spaccini, R., Piccolo, A., 2009. Molecular characteristics of humic acids extracted from compost at increasing maturity stages. Soil Biol. Biochem. 41, 1164–1172.

Spaccini, R., Piccolo, A., 2012. Carbon sequestration in soils by hydrophobic protection and in-situ catalyzed photo-polymerization of soil organic matter (SOM). Chemical and physical-chemical aspects of SOM in field plots. In: Piccolo, A. (Ed.), Carbon Sequestration in Agricultural Soils. Springer-Verlag, Heidelberg, pp. 61–105.

Spaccini, R., Piccolo, A., Conte, P., Haberhauer, G., Gerzabek, M.H., 2002. Increased soil organic carbon sequestration through hydrophobic protection by humic substances. Soil Biol. Biochem. 34, 1839–1851.

Spaccini, R., Mbagwu, J.S.C., Conte, P., Piccolo, A., 2006. Changes of humic substances characteristics from forested to cultivated soils in Ethiopia. Geoderma 132, 9–19.

Spaccini, R., Sannino, D., Piccolo, A., Fagnano, M., 2009. Molecular changes in organic matter of a compost-amended soil. Eur. J. Soil Sci. 60 (2), 287–296.

Spaccini, R., Todisco, D., Drosos, M., Nebbioso, A., Piccolo, A., 2016. Decomposition of biodegradable plastic polymer in a real on-farm composting process. Chem. Biol. Technol. Agric. 3 (4)doi: 10.1186/540538-16-0053-9.

Stenson, A.C., Landing, W.M., Marshall, A.G., Cooper, W.T., 2002. Ionization and fragmentation of humic substances in electrospray ionization Fourier transform-ion cyclotron resonance mass spectrometry. Anal. Chem. 74, 4397–4409.

Swift, R.S., 1999. Macromolecular properties of soil humic substances: fact, fiction, and opinion. Soil Sci 164, 790–802.

Tanford, C., 1980. The Hydrophobic Effect: Formation of Micelles and Biological Membranes. Wiley, New York.

Tarasevich, Y.I., Dolenko, S.A., Trifonova, Y.M., Alekseenko, E.Y., 2013. Association and colloid chemical properties of humic acids in aquoeous solutions. Colloid J. 75 (2), 207–213.

Thevenot, M., Dignac, M.F., Rumpel, C., 2010. Fate of lignins in soils: a review. Soil Biol. Biochem. 42, 1200–1211.

Tiessen, H., Stewart, J.W.B., Hunt, H.W., 1984. Concept of soil organic matter transformations in relation to organo-mineral particle size fractions. Plant Soil 76, 287–295.

Tisdall, J.M., Oades, J.M., 1982. Organic matter and water-stable aggregates in soils. J. Soil Sci. 33, 141–163.

Torn, M.S., Trumbore, S.E., Chadwick, O.A., Vitousek, P.M., Hendricks, D.M., 1997. Mineral control of soil organic carbon storage and turnover. Nature 389, 170–173.

Vershin, N.L., 1999. Screening hypochromism in molecular aggregates and biopolymers. J. Biol. Phys. 25, 339–354.

von Lützow, M., Kögel-Knabner, I., Ekschmitt, K., Flessa, H., Guggenberger, G., Matzner, E., et al., 2007. SOM fractionation methods: relevance to functional pools and to stabilization mechanisms. Soil Biol. Biochem. 39, 2183–2207.

Woo, D.K., Quijano, J.C., Kumar, P., Chaoka, S., Bernacchi, C.J., 2014. Threshold dynamics in soil carbon storage for bioenergy crops. Environ. Sci. Technol. 48, 12090–12098.
Zhang, H., Ding, W., Nanthi Bolan, J.L., Yu, H., 2015. The dynamics of glucose-derived ^{13}C incorporation into aggregates of a sandy loam soil following two-decade compost or inorganic fertilizer amendments. Soil Tillage Res. 148, 14–19.
Zherebker, A., Kostyukevich, Y., Kononikhin, A., Kharybin, O., Konstantinov, A.I., Zaitsev, K.V., et al., 2017. Enumeration of carboxyl groups carried on individual components of humic systems using deuteromethylation and Fourier transform mass spectrometry. Anal. Bioanal. Chem. doi: 10.1007/s00216-017-0197-X.

FURTHER READING

Six, J., Elliott, E.T., Paustian, K., Doran, J.W., 1998. Aggregation and soil organic matter accumulation in cultivated and native grassland soils. Soil Sci. Soc. Am. J. 62, 1367–1377.

CHAPTER 5

SOM and Microbes—What Is Left From Microbial Life

Matthias Kästner, Anja Miltner
Helmholtz Centre for Environmental Research—UFZ, Leipzig, Germany

Chapter Outline

INTRODUCTION

The United Nations have defined sustainable development goals (SDG) related to food security, freshwater availability, biodiversity loss, and climate change. Although soils provide a number of ecosystem services, which are essential to human life and for achievement of the SDG, they have not been explicitly considered in the context of SDG (Bouma, 2014). In particular, soils provide the basis for nutrition of a growing global population (Lal, 2007) and need to be managed sustainably in order to keep pace with the growing demand for food. Beside water availability and climate conditions, soil organic matter (SOM) is the most important driver for plant biodiversity and food security and is thus one of the key factors controlling soil fertility and global change mitigation (Bouma, 2014). In addition, soils are considered to be a large potential sink for atmospheric carbon (Lehmann and Kleber, 2015; Schmidt et al., 2011). Maintaining or even increasing the SOM level thus plays a key role in sustainable soil management and global change mitigation. Methods for maintaining SOM levels have been known for a long time; many farmers use organic fertilizers and return crop residues to their fields. Overall, there is growing evidence about the microbial control of terrestrial carbon sequestration (Schulze and Freibauer, 2005).

The Future of Soil Carbon
http://dx.doi.org/10.1016/B978-0-12-811687-6.00005-5

The details of the fate of added carbon and the efficiency in increasing SOM stocks, however, remain to be elucidated. This lack of knowledge hampers directed and efficient optimization of the available soil management options or development of new ones.

As the SOM level will only change slowly in response to management changes, either long observation times or powerful models are needed in order to predict these changes. Development of proper models would therefore substantially reduce the time needed to assess alternatives. Such models need to rely on process understanding in order to cover a wide range of potential scenarios. Although SOM research now has a long tradition, and the high efforts in this field have already provided considerable knowledge on transformation and stabilization of SOM, still too little is known at present: it is not yet possible to precisely predict the changes in SOM levels which will occur under different land-management options or due to the effects of changes in environmental conditions, for example increased temperature or more extreme precipitation distribution, as can be expected in the context of global change.

In the past, recalcitrant plant residues have been regarded as an important source of SOM (von Lützow et al., 2008) although it has been recognized that microbial residues are an additional source of SOM (Kögel-Knabner, 2002). Recently, more sophisticated analytical methods together with theoretical and experimental approaches have made it possible to estimate the contribution of microbial material to SOM. All estimates reported suggest that more than 50% of SOM originates from microbial residues (Fan and Liang, 2015; Liang et al., 2011; Miltner et al., 2012; Simpson et al., 2007a). However, currently the estimates are not precise enough to allow definite conclusions on implications for degradability, stabilization, and availability of the key proximate precursors of SOM. The various origins and dynamics of input will result in different composition, properties, processes, and functions of the SOM formed; therefore, the pathways of SOM formation need to be elucidated in much more detail than in the past (Kallenbach et al., 2016).

Independent of the transformations needed in order to produce SOM, there is a continued input, mainly of plant residues, which are then microbially reworked and mineralized. These processes take place continuously as long as there is active biomass. The SOM level is maintained by the flux through the system. In the steady state, the carbon flux into the system (mainly plant residue deposition) is equal to the flux out of the system (leaching and mineralization). In this situation, a stable SOM level (rather than a pool) may be maintained for a long time, but the SOM is turned over at a certain rate. In order to increase the SOM level, either input of organic material has to increase or mineralization has to decrease. In the opposite case, SOM will be lost slowly from the soil until a new steady state at a lower SOM level is established. In all cases, there is no need to postulate persistent or recalcitrant SOM constituents because of the continued flux of carbon through the system. Here we provide a synopsis of the current state of knowledge on the contribution of microbial biomass to the formation of SOM, based on improved process understanding, and we discuss the resulting consequences as well as open questions.

HUMIC SUBSTANCES AND SOIL ORGANIC MATTER

The benefits of organic matter on soil fertility have been recognized centuries ago by farmers. In the 18th century, scientists started research on humus chemistry. At this time, the interest was mainly in the isolation of humic substances and in the elucidation of the nature of their chemical constituents. Achard (1786) introduced the extraction of peat and soil by dilute alkali in order to obtain humic substances; this approach was the basis for the fractionation of the extracted humic substances based on their solubility in alkali and acid, which has been the standard in humic substances research for many years (Stevenson, 1994; Waksman, 1925). This approach yields operationally defined fractions of humic acids, fulvic acids, and humin; scientists initially aimed at isolating and purifying defined compound classes in order to disentangle the chemical composition of SOM. This approach was not completely successful with respect to the original aims of the research, but the approach did allow a large part of the organic material to be extracted from soil samples and thus made amenable to the analytical methods available (von Lützow et al., 2007). Much of the information we have now on the chemical properties, such as the elemental composition, the contribution of building blocks to SOM and the overwhelming heterogeneity in composition would not have been possible without the humic substances concept. However, the relevance of this fractionation procedure has been questioned increasingly in the last two decades (Lehmann and Kleber, 2015; Schmidt et al., 2011). Nowadays, the research focus has shifted more from chemical characterization of SOM to identification of processes and mechanisms involved in SOM turnover and stabilization (Schmidt et al., 2011). In particular, the complex interactions of SOM with mineral surfaces and the role of water availability and environmental factors, as well as microbes as the actors, have gained importance (Lehmann and Kleber, 2015; Schmidt et al., 2011). Accordingly, fractionation procedures based on particle size or density are currently preferred over the traditional solubility approach (von Lützow et al., 2007). These procedures separate organic matter fractions based on their interaction with mineral particles in soil, with the finer and heavier fractions being more readily processed than the coarse and the light fractions (von Lützow et al., 2007).

New fractionation approaches and novel methods available for analysis of soil samples also contributed to changing the notion of SOM away from macromolecular humic substances toward a complex intramolecular association of smaller molecules (Sutton and Sposito, 2005). In recent years, major advances in analytical approaches and methods have been achieved. Spectroscopic methods, such as infrared and nuclear magnetic resonance (NMR) spectroscopy have been further developed and can now be routinely applied for SOM studies (Ellerbrock et al., 2009; Kelleher and Simpson, 2006; Knicker, 2002; Kögel-Knabner, 2002; Masoom et al., 2016; Simpson et al., 2007b). A general trend is that as highly sensitive methods with high spatial resolution became available, they have been adopted by soil scientists. Examples are atomic force microscopy (AFM) (Cheng et al., 2009; Kunhi Mouvenchery et al., 2016; Schaumann and Kunhi Mouvenchery, 2012), X-ray photoelectron spectroscopy (XPS) (Woche et al., 2017), Raman microscopy (Schmidt et al., 2002), scanning transmission X-ray microscopy and near-edge X-ray adsorption fine structure (STXM/NEXAFS; Solomon et al., 2012), high-resolution mass spectrometry (e.g., Fourier transform-ion cyclotron resonance mass spectrometry [FT-ICR-MS] (Seifert et al., 2016) and nanoscale secondary ion mass spectrometry (NanoSIMS; Mueller et al., 2012; Schurig et al., 2013). Spectroscopic analysis of SOM at small spatial scales became available by these approaches, allowing the localization of SOM components down to a range smaller than clay particles or bacterial cells. This option opened novel insights into small-scale properties of SOM, which is crucial for understanding this heterogeneous material on a relevant scale. The downside of such high resolution, however, is that by studying extremely small sections of a highly complex material, special care has to be taken regarding the representativeness of the results.

In addition, research has moved away from characterizing the composition of SOM to studying processes of SOM formation (Lehmann and Kleber, 2015; Schaeffer et al., 2015); this type of research includes studies of microbial impact on artificial soils, (e.g., Kallenbach et al., 2016; Pronk et al., 2012, 2013, 2017), chronosequence studies (e.g., Schurig et al., 2013), and also laboratory incubation studies, often using isotope-labeled microbial materials, (e.g., Kindler et al., 2006; Schweigert et al., 2015). Using these approaches allowed identification of the processes involved in SOM formation and degradation; this information helped in developing conceptual models for the stabilization of SOM. Nowadays it is accepted that recalcitrance of the organic input is only relevant in the initial stages of litter decomposition (Marschner et al., 2008; von Lützow et al., 2006). Later, physical protection plays the central role. In summary, together with improved analytical possibilities for microbial biomarkers and their isotopic composition, a better estimation of various sources for SOM is now available (Liang et al., 2011; Masoom et al., 2016; Miltner et al., 2012; Simpson et al., 2007a).

BACKGROUND: ROLE OF MICROORGANISMS AS BIOCATALYSTS AND BULK CONTRIBUTORS

The primary origin of the vast majority of soil organic carbon is plant-derived carbon. In contrast to previously considered direct stabilization of plant-derived carbon in soils, current results suggest that the dominant pathway of soil carbon storage proceeds via cycling through microbial biomass. Field experiments have shown that increasing diversity of plant communities results in higher levels of carbon inputs to soil and more favorable microclimatic conditions; denser vegetation in more diverse plant populations reduces evaporation from topsoil and results in higher microbial activity and growth (Lange et al., 2015). This also results in higher diversity of the soil microbial communities. Other experiments in grasslands also showed that high plant diversity is correlated with increased amounts of microbial biomass/necromass biomarkers (lipids and amino sugars), even in deeper horizons (Liang et al., 2016). High plant diversity combined with increased carbon and nitrogen sequestration also appears to be positively correlated with fungal biomarkers related to arbuscular mycorrhizal fungi (Khan et al., 2016; Malik et al., 2016); fungal-based soil food webs were found to be more drought resilient (de Vries et al., 2012; Guhr et al., 2015). Soil microbes generally regulate plant productivity; in nutrient-poor ecosystems, plant symbionts provide nutrients for plant growth. Mycorrhizal fungi and nitrogen-fixing bacteria are responsible for 5%–20% of the nitrogen supply for plant growth in grassland and savanna, and even up to 80% in temperate and boreal forests, as well as for up to 75% of phosphorus supply (van der Heijden et al., 2008). The same authors reported estimates of more than 20,000 plant species being completely dependent on microbial symbionts.

Although the activity of the microbial biomass was considered to be highly important for the majority of soil processes and functions (carbon, nitrogen, phosphorus cycles), active microbial biomass carbon is measured to be relatively low (1%–2% of soil organic carbon; Dalal, 1998). Thus *microbial biomass materials* were considered for a long time to contribute only little to the formation of SOM (Lehmann and Kleber, 2015; Simpson et al., 2007a). Their dominating role was understood as biocatalysts, providing extracellular enzymes for the turnover and transformation of plant-derived macromolecules (Schmidt et al., 2011). This potential for extracellular depolymerization is indeed valid for the majority of microorganisms due to their small size of a few μm and their inability to take up macromolecules, and will result in microbially triggered extracellular biochemistry in soils (Fig. 5.1). However, if macromolecules, such as starch, cellulose, proteins, and lignins are depolymerized to monomers and oligomers of molecular sizes < 600 Dalton in the vicinity of microorganisms (Lehmann and Kleber, 2015), the microorganisms are eager to take up these substrates and use them for cell maintenance and growth: whatever can be taken up by microbes (at least by diffusion through membranes) will be utilized. In addition, whatever can react chemically outside of the cells will react, if not ingested by the microorganisms. Therefore, these low-molecular-weight compounds are only present in very low (steady-state) concentrations in soil, although the flux into the cells may be considerable. These principles apply not only to plant-derived material, but also to dying microbial biomass and necromass materials.

Plant communities drive these processes by the input of organic materials, thus controlling SOM stocks in soils of various land uses (Brant et al., 2006). For example, shifts of croplands to grasslands or forests result in a long-term increase of carbon of around 20 Mg ha^{-1}, whereas similar amounts were released after shifts of grassland to cropland (Poeplau and Don, 2013). In addition, soils under no-tillage practice stored 5%–15% more carbon and 2%–5% more nitrogen in comparison to conventional tillage (Guggenberger et al., 1999; Plaza et al., 2013). However, this is only valid for the upper layers, since less carbon will be transferred to the deeper soil layers.

Complex organic amendments to soil are known to enhance carbon storage in soils; the results of various treatments of forest and arable soils and of land use practices were recently reviewed (Bhattacharya et al., 2016). The comparison of carbon storage on a 2000-year chronosequence of rice cultivation in China showed that long-term paddy rice cultivation results in a strong gain of carbon in the topsoils (~180 kg ha^{-1} a), whereas nonpaddy soils showed only about 1/6 of this gain (Kalbitz et al., 2013). During this period, however, subsoils of the paddies lost up to half of this amount, which alters the total carbon budget heavily.

SOM storage capacity and carbon turnover in terms of mineralization in relation to carbon use by microorganisms is considered to be highly sensitive to increases in temperature; this was found to be particularly relevant for low-quality substrates (Frey et al., 2013). However, long-term studies showed that the observed increases in mineralization in short-term experiments did not persist and that differences diminish over time, finally resulting in turnover rates similar to those observed for lower temperatures (Allison et al., 2010). This suggests that there is a certain limit of carbon storage capacity in various soils.

SOM storage and carbon sequestration is caused by *self-organization* of the *soil-microbe-complex* and is thus highly dependent on the soil microbiome (Young and Crawford, 2004). Recently, the soil continuum model was suggested, whereby SOM is regarded as a continuum of organic compounds at various stages of decay (Lehmann and Kleber, 2015). In fact, it was found that soil bacterial community structures and decomposer functions converged during plant succession at two deglaciated chronosequences of glacier forefields (Castle et al., 2016).

For the processes controlling the *steady-state concentrations of SOM* in soils (and thus the carbon sequestration), the rates of microbial decay of plant-derived organic matter and the influxes into the soil are important. However, the most relevant flux- and concentration-determining processes, which are still only partly understood, are the processes and conditions causing "persistence" and "protection" of organic materials, which are biodegradable in principle but persist in soil for extended periods (Liang and Balser, 2011; Schimel and Schaeffer, 2012). Soil science thus must come to life (East, 2013): the relevant research fields of microbiome-based services in the triangle of "soil structure—microbiome—plant roots" are plant-growth promotion (increasing stress tolerance, nutrient availability, mycorrhiza effects) together with SOM formation and carbon storage.

However, there is a critical assumption in many terrestrial ecosystem models that *soil microbial communities* in common environments are highly redundant and will function in an identical way independently of the community composition (Nielsen et al., 2011; Strickland et al., 2009). This assumption has led to the development of various approaches for experiments using community mixtures from sites with different conditions and land uses. However, mixing experiments using different litter layers showed that the assumption of functionally equivalent communities, analyzed in terms of mineralization rates, is actually not correct (Strickland et al., 2009). Rates of CO_2 formation from litter decomposition were found to depend on the microbial inoculum: differences in the microbial community alone accounted for substantial (>20%) variation in total mineralization. These results were also supported by experiments using mixtures of different soils and inoculum communities (Don et al., 2013, 2017). The expected average mineralization of the mixtures in comparison to the original soil samples was not observed; instead, high-mineralization samples showed retarded mineralization when mixed with other soils of lower mineralization. However, these experiments left many important questions unanswered, since the observed effects could not be explicitly related to the impact of microbial communities because they were often impaired by other soil-matrix-related effects.

Mixing and drying/rewetting always result in the *mobilization of easily degradable substrates* from SOM. Mixing of soil samples also affects SOM turnover rates, which thus may result in positive or negative priming effects (Kuzyakov, 2010; Kuzyakov and Blagodatskaya, 2015). The mobilization of easily degradable substrates may be not only due to changes in SOM availability but also to the provision of recently died microbial biomass; for example, mixing may particularly damage fungal mycelia from forest or grassland samples, thus providing degradable biomass residues leaking from the hyphae. Mobilization may have severe impacts on the mineralization rates and on the interpretation of the results, in particular if the analytical methods for community analysis are exclusively focusing on bacterial communities. Fungi are of particular relevance, since fungi primarily start the degradation of macromolecular plant-derived materials, whereas bacteria dominate the turnover in the long run (Amelung et al., 2001; Lemanski and Scheu, 2014; Malik et al., 2016).

Recent results showed that *spatial isolation* of SOM from degrading microbes, in pores of various sizes, is also a mechanism by which easily degradable organic compounds can be protected in soils (Bailey et al., 2017). This is particularly relevant during changes of water contents with drying and rewetting, which may lead to spatial separation or remobilization and thus to microbially colonized locations more favorable to organic matter decomposition, also providing a potential explanation for the Birch effect, i.e. the pulse of SOM mineralization after fast rewetting of a soil. In addition, pore-water-related dissolved organic matter (DOM) samples showed different molecular compositions (as analyzed by means of FT-ICR-MS) as well as microbial degradability by various microorganisms with different metabolic properties, all in relation to the water potential used for mobilization: −15 kPa leads to pore water containing DOM with a more complex composition and higher mineralization in comparison to −1.5 kPa, mainly because pores of different sizes are drained at the two water potentials (Bailey et al., 2017).

In addition, altered *mineralization rates* may be related not only to shifts of the microbial community compositions but also to changes in their activity, for example, substrate-use efficiency and growth. Therefore, kinetic effects must also be considered: if high amounts of biomass are present in relation to the amount of substrate, *first-order degradation kinetics* with higher mineralization can be observed. In contrast, if microbial biomass is low compared to the amount of substrate, the microbes will grow and thus more carbon will be converted into microbial biomass, resulting in *sigmoidal kinetics* (Simkins and Alexander, 1984). These effects need to be considered, particularly for experiments with inoculation of different communities to various sterilized soils.

Microorganisms are well accepted as the dominating drivers of turnover of carbon fixed by the photosynthetic primary production in *marine systems*; this fact has been the subject of research for decades (Jiao et al., 2010). The research in marine biogeochemistry resulted in the development of two conceptual approaches for carbon sequestration: (1) the general *biological pump*, leading to the formation of sedimenting particulate organic matter (POM) derived from modified biological detritus and necromass, and (2) the *microbial carbon pump* (MP), a side loop of the biological pump, converting easily degradable DOM into microbial biomass and later into necromass/POM, resulting in hardly biodegradable so-called refractory DOM (RDOM), which can persist for years. Three processes of DOM formation by the MP were identified: (1) microbial growth and exudation (anabolism), (2) death and decay, mostly enhanced by viral lysis of the cells, and (3) general degradation of POM. These processes obviously lead to persisting DOM, in which molecular species derived from bacteria such as porins, D-amino acids, muramic acid, and lipopolysaccharides can be identified. This material resembles fragments of microbial cell wall material, which may contribute to the DOM fraction due to their very small size. Both types of pump eventually pump organic carbon from low concentrations of reactive carbon to high concentrations of more refractory carbon, building up a huge reservoir of DOM and sedimentary carbon (Jiao et al., 2010). The current opinion is that they are connected by export of carbon from POM via the microbial loop to the MP.

There are good reasons to assume that similar processes are also valid for soil systems, since the microbial drivers are highly similar. In particular, carbon-turnover research based on microbial ecology combined with biogeochemistry has a much longer tradition in marine systems, whereas in soil systems this type of research started flowering only recently. However, there are differences between marine and soil systems that may shape the processes in different ways, for example, varying chemical activities of water as well as much higher amounts of sorptive mineral surfaces and much slower mixing and diffusion processes in soil. Presumably that is why active soil layers generally have a much smaller extension in comparison to marine sediments. Indeed, microbial necromass residues in soils have already been identified and visualized in terms of cell-envelope fragments (Miltner et al., 2012) and necromass nanoparticles of small size (<0.45 μm) that fall into the DOM fraction. Therefore, for soil systems, the most relevant questions are highly similar to those for marine systems (Jiao et al., 2010): (1) why do heterotrophic microorganisms not degrade the refractory DOM (and also POM), and (2) what are the structural and biochemical constraints to degradability? However, in soils *substrate accessibility* is of much greater importance than in marine systems.

Recently, an analogous approach for a *terrestrial microbial pump* dominated by bacteria and fungi was suggested (Fan and Liang, 2015). Here, the processes leading to refractory or persisting compounds are caused by "burying" carbon in huge amounts of large macromolecular aggregates, which are derived from plant residues and microbial necromass (which is sometimes misleadingly termed as microbial "tissues"). This stabilization of intra-aggregate SOM may simply be caused by the low accessibility of these materials to microbial degradation and even to exoenzymes, as well as by the reduced solubilization into the free-water phase. Modeling approaches based on this concept resulted in estimated microbial biomass contributions to SOM at equilibrium stage in amounts ranging from 47% to 80%, depending on the substrate quality and the amount of remaining plant-derived biomass.

In aqueous systems without sorptive surfaces, it is well known from *substrate turnover and growth* models (Michaelis and Menten, 1913, cited in Don et al., 2013; Monod, 1949) that increasing substrate concentrations result in increasing turnover and microbial growth rates. Vice versa, decreasing concentrations will thus result in decreasing turnover rates that depend on the likelihood of an enzyme or an organism encountering substrate molecules. This general relation was also proven for soil systems (Don et al., 2013);

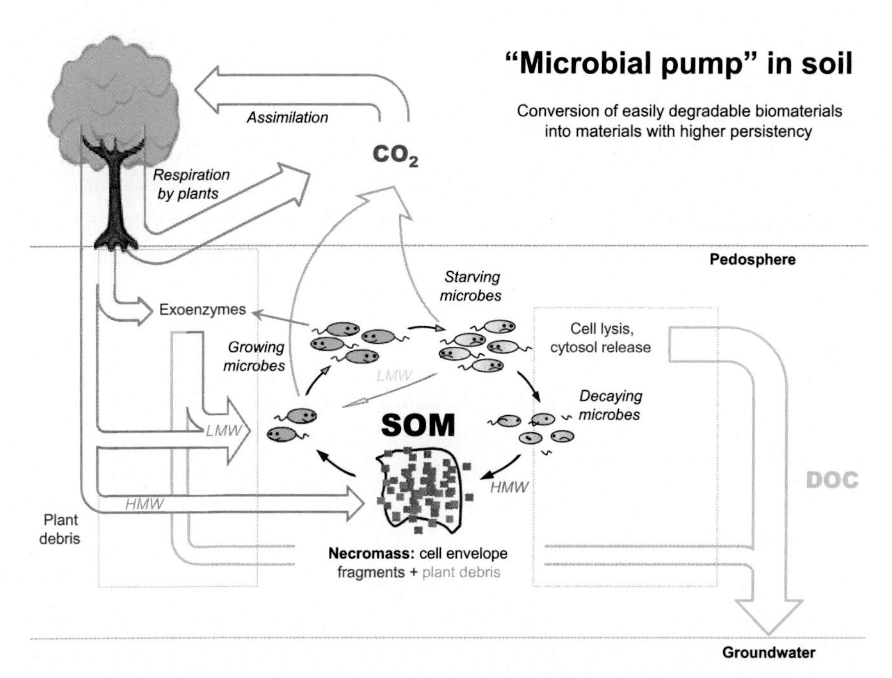

Figure 5.1 ***Processes fueling the microbial pump in soil.*** *hmw*, High molecular weight compounds; *lmw*, low molecular weight compounds; *SOM*, soil organic matter. *Dotted line* represents nonordered extracellular bio(geo)chemistry. *(Modified from Gleixner, G., 2013. Soil organic matter dynamics: a biological perspective derived from the use of compound-specific isotopes studies. Ecol. Res. 28, 683–695 and Miltner, A., Bombach, P., Schmidt-Brücken, B., Kästner, M., 2012. SOM genesis: microbial biomass as a significant source. Biogeochemistry 111, 41–55).*

this is a function not only of concentration but also of accessibility (Bailey et al., 2017; Dungait et al., 2012; Moinet et al., 2016).

However, the very general important open question on SOM turnover and storage is: are there maximum (steady state) levels of SOM for different soil types or can the contents increase without general limits? Functional diversity assessments in primary successions across glacier forefields in the European Central Alps suggest temporary steady-state levels of SOM in the range of 7 to 20 mg g^{-1} (Tscherko et al., 2003).

FATE AND TURNOVER ANALYSES OF MICROBIAL BIOMASS

Besides being responsible for the turnover of primary plant-produced carbon by catabolic activity, microorganisms are also material contributors to the formation of SOM. They can increase SOM via synthesis of biomass (anabolism) and subsequent necromass accumulation and stabilization, as suggested by several authors (Benner, 2011; Bradford et al., 2013; Cotrufo et al., 2013; Gougoulias et al., 2014; Liang and Balser, 2011; Miltner et al., 2009, 2012; Schaeffer et al., 2015; Schimel and Schaeffer, 2012; Schimel and Weintraub, 2003). This concept was suggested more than a decade ago, considering *cell wall fragments*, *exoenzymes*, and particularly *osmolytes* as major contributions of microorganisms to SOM (Schimel et al., 2007). Stimulating conceptual frameworks based on the general microbial necromass approach are currently being developed, but the details of the conversion of easily degradable organic compounds into more-or-less persistent SOM still remain to be clarified. The current understanding of SOM formation is such that all labile plant constituents are consumed by microorganisms with varying efficiencies and that microbial products are subsequently the main precursors of SOM, which are stabilized by *aggregation and bonding to the mineral matrix* of the soil (Cotrufo et al., 2013). These materials were in fact found to be mostly associated to the mineral phase of soils (Kunhi Mouvenchery et al., 2016; Ludwig et al., 2015; Schurig et al., 2013; Woche et al., 2017). In particular, bacterial residues were found to be predominantly associated to mineral surfaces (Plaza et al., 2013).

This universal genesis of SOM from *microbial substrate use* combined with *matrix stabilization* is consistent with the observed similarity in the chemical composition of SOM in various soil ecosystems. Modern analytical tools and many results on biomolecules in soils and SOM turnover suggest a direct and very rapid contribution of microbial biomass residues (necromass) to SOM (Lehmann and Kleber, 2015; Miltner et al., 2012). The concept is supported by recent results from comprehensive multiphase NMR, showing the origin of SOM to be dominated by microbial and plant-derived macromolecular aggregates (Masoom et al., 2016). In addition, a very important result of these data is that around 75% of the necromass-derived SOM material is actually *not in contact with water*. In addition to the mineral association, the lack of water access immediately provides the simplest explanation for stabilization: without access to the free water phase with sufficient chemical activity of water, the biodegradation of these molecules is practically impossible (Potts, 1994), as is known for example from the conservation mechanism of jam. Such stabilization may be caused by embedding organic matter in hydrophobic domains and by interactions with negatively charged functional groups in soil, in particular with clay, causing exclusion of water, or with other organic compounds (Sollins et al., 1996). Proteins are a particular class of molecules that can be protected against degradation by this mechanism (Fan et al., 2004), for example, on mineral surfaces, where they can form clusters of hydrophobic layers around a particle (Kleber et al., 2007).

In the last two decades the general evidence has been growing that microbial residues may be more important in the genesis of SOM than previously estimated. A significant quantitative contribution of microbial necromass to SOM was deduced from the *distribution pattern of organic nitrogen* in soil. Previous studies showed contents of *amino sugars* significantly exceeding the total amount of microbial biomass nitrogen (Amelung, 2001; Appuhn and Joergensen, 2006; Guggenberger et al., 1999; Joergensen et al., 2010). Some amino sugar polymers and biomarkers are partly indicative for fungal (glucosamine) and bacterial (muramic acid) cell envelope structures, and thus reflect the remnants after cell death of these general groups of microbial biomass (Appuhn and Joergensen, 2006). Therefore, amino sugars were considered to provide information on "soil legacy" related to total carbon, clay content, and culture age rather than current processes and drivers, such as plant communities or microbial communities (Liang et al., 2013).

Specific evidence was provided using improved 2 D NMR studies: Kelleher and Simpson (2006) found that nearly all of the NMR signals in traditional humic acid fractions could be assigned to intact and degrading biopolymers of plants and microbes. In other experiments, the spectra were directly associated to peptides, lipids, carbohydrates, and lignins (Simpson et al., 2007b). Finally, the authors found that in some soils microbial biomass residues, in particular peptidoglycans from microbial cell envelope materials, were one order of magnitude higher than reported by other researchers and may account for up to 50% of the signals (Simpson et al., 2007a,b), which is in good agreement to the results obtained by analyses of amino sugars in soils (Liang et al., 2013).

In addition, it is well known from other NMR studies that almost the *entire organic nitrogen* in soils is found in *amide bonds*, which is indicative of proteins (Knicker, 2011). Proteins may be derived from plants or from microorganisms. If, however, biodegradable plant material is biodegraded, proteins are normally also degraded and converted to microbial biomass. Therefore, proteins in SOM need to be analyzed for their molecular origin. Currently, protein extractions from soils have shown a high contribution of bacterial proteins, but the extraction efficiency is still much too low for a final representative assessment (Bastida et al., 2014). In addition, protein databases need to be expanded for reliable assessment of the phylogeny of soil microorganisms.

Direct *quantitative measurement* of microbial necromass contribution to SOM in soil is extremely difficult because it is hard to differentiate microbial from nonmicrobial carbon in soil organic matter (Bradford et al., 2013). Therefore, approaches enabling the carbon flow in soil to be traced using stable and radio isotope tracer compounds have been applied. However, real evidence in the form of turnover mass balances has rarely been provided, although supporting research and data sets are currently emerging.

In order to develop a conceptual framework of the microbial contribution to SOM formation and to balance the incorporation of the carbon derived from microbial biomass, we incubated *^{13}C-labeled microbial biomass* in reference soils and analyzed the fate of the labeled carbon. In general, there are four dominating construction forms of microbial cells that are mainly related to the architecture of the cell envelope (Madigan et al., 2017): (1) *Gram-negative bacteria* containing phospholipid membranes with peptidoglycan (amino sugar polymer: *N*-acetyl-glucosamine and *N*-acetyl-muraminic acid) and outer layers coated with lipopolysaccharides, (2) *Gram-positive bacteria* containing phospholipid membranes with dominantly thick peptidoglycan layers in which teichoic and sometimes mycolic acids are embedded, (3) *archaea* with ether lipid membranes and protein outer layers (S-layer), and (4) *fungi* with mycelial growth of many species and phospholipid membranes plus chitin-based envelopes (amino sugar polymer, *N*-acetyl-glucosamine). For more details of the cell organization, see "Factors and mechanisms causing persistence of microbial necromass."

We incubated an agricultural reference soil with a ^{13}C-labeled Gram-negative model bacterium (*Escherichia coli*), which is not known to colonize soils and thus served as a defined source of dying biomass (Kindler et al., 2006). The cells died rapidly during incubation in soil; after 225 d of incubation, 56% of the introduced bulk biomass ^{13}C was mineralized and 44% remained in SOM. Analyses of fatty acids and protein-derived amino acids in the total versus living microbial biomass amounts showed that the majority was bound to nonliving SOM. Less than 15 % of the label was found in living organisms other than *E. coli* (Kindler et al., 2009; Lueders et al., 2006; Miltner et al., 2009). More than 80% of the ^{13}C-label in lipids was lost, whereas the loss from the fraction of protein amino acids only accounted for less than 10%. Therefore, it is very important to analyze not only the living organisms but also the total remainder in nonliving SOM. If only the biomolecules of the living organisms are analyzed, for example, polar lipid fatty acids (PLFA), the real carbon conversion of the substrate and the contribution of microbes to SOM will be considerably underestimated.

Later, similar experiments were performed with the same soil but with ^{13}C-labeled Gram-positive *Bacillus subtilis* cells. Gram-positive cells also died rapidly; 34% of the label was mineralized while 66% remained in SOM. Again, loss of the label in proteins was found to be in the range of 10% (unpublished data). Recent experiments with the ^{13}C-labeled ectomycorrhizal fungus *Laccaria bicolor* in a forest reference soil incubated over a similar period of time showed an initial rapid decay of the mycelia, as found with the other classes of organisms: 64% of the bulk carbon label remained in SOM and 37% was mineralized (Schweigert et al., 2015). About 90% of the fungal label in lipids was lost; much less than 10 % was found in phospholipids of other microorganisms. For these experiments, the assessment of protein turnover is in progress, but first data are indicating significant persistence of fungal proteins (unpublished data).

The results of the turnover mass-balance experiments with labeled microorganisms of various classes showed a significantly *slower turnover of ectomycorrhizal fungal* than of bacterial biomass, with the Gram-positive bacteria being mineralized slightly faster than Gram-negative bacteria. At the end of the experiments, however, between 34% and 66% of the initial biomass carbon remained in the soil. Less than 15% of the label is present in other living microbial biomass; thus about 20%–50% of the label from the various classes of microorganisms remains in nonliving SOM. *Proteins showed very high meta-persistence* irrespective of the type of biomass studied, whereas lipids were turned over to a much higher extent than the bulk biomass carbon. These results show that the analysis of the bulk carbon flux does not allow any conclusions to be made about the turnover of specific biogenic macromolecules. However, around 20%–50% of the microbial biomass carbon can be generally considered to be stabilized as SOM; this carbon is obviously dominated by proteins and presumably also by cell wall residues, including amino sugars as shown previously (Miltner et al., 2012). Cell wall residues contain considerable amounts of amino sugars; these compounds can thus be considered as the second dominating class of biomolecules derived from microbial biomass (Liang et al., 2013).

Summarizing, more than 50% of the biomolecular mass from photosynthetic primary production is more or less easily biodegradable; it will be degraded and utilized, but also partly mineralized again and again by microorganisms leaving up to 50% of their biomass as remnants in SOM. Therefore, we can state that the consuming microorganisms convert the easily degradable primary plant-derived biomass carbon partly to secondary microbial biomass, which is presumably cycled repeatedly through the microbial food web. However, it is currently not clear whether these materials remain as original materials or whether they are repeatedly recycled, resulting in a steady-state amount. The presumably high persistence of proteins is only indicated by the high amount of ^{13}C label remaining in proteins, but this may also be a result of continuous recycling of peptides and amino acids in the microbial food web. This can only be clarified with detailed proteomic analyses of the remaining proteins in soil. In conclusion, we can expect that a minimum of 25%–50% of SOM is derived from microbial biomass residues, as suggested by Miltner et al. (2009). Reliable measured data for various soil types are still needed in order to understand fully the carbon storage capacities of soils.

Supporting results also come from recent experiments with *artificial soils*. Direct evidence for SOM formation derived from microbial biomass growth was provided by experiments with various mixtures of mineral soil components inoculated by soil bacteria which were fed weekly with glucose, cellobiose, syringol (a lignin monomer), or with plant-derived DOM (Kallenbach et al., 2016). After 18 months of incubation, surprisingly, all mixtures produced stable and highly similar microbe-derived SOM-like materials, irrespective of the substrate. The various substrates used had carbon use efficiencies (CUE) of 0.18–0.32, the highest values were found for mixtures of DOM with kaolinite. These values are somewhat lower than the CUE observed with the isotope-labeled experiments discussed earlier. Increased CUE with higher formation of SOM in soils was observed with higher fungal contribution (Kallenbach et al., 2016). The data provide clear evidence that *stable SOM is formed during microbial growth on low-molecular-weight substrates*.

Other experiments with eight different artificial soil mixtures containing quartz, illite, montmorillionite, ferrihydrite, boehmite, and charcoal, together with a common inoculum from an arable topsoil and supply of a uniform complex organic substrate (sterilized farmyard manure), showed similar kinetics of respiration as well as similar aggregate formation and partitioning of organic carbon and nitrogen into the fractions $<20\ \mu m$. About 50% of the substrate (POM fraction) was lost as CO_2 during the 18-month incubation. The remaining organic fraction was analyzed by means of ^{13}C-NMR, which revealed an accumulation of protein-rich microbially produced substances (Pronk et al., 2012, 2013). The ferrihydrite-amended mixtures did not contain higher amounts of SOM, indicating that the silicate clay minerals were more important for initial SOM accumulation. However, metal oxides, clay minerals, and to a lesser extent charcoal influenced the microbial community composition (Ding et al., 2013). Nevertheless, similar respiration rates were found, again supporting the hypothesis of highly redundant microbial communities, at least with respect to general microbial activities, such as organic matter mineralization.

In the experiments studying the fate of labeled microbial biomass, the turnover and incorporation into SOM data were used in *simulation models*. The results suggest that three generations of microbes cycling in the microbial food web were sufficient to establish a stable level of 50% contribution of microbial biomass residues to SOM (Miltner et al., 2009, 2012). The flow of carbon through microbial pathways in soils has been modeled by a theoretical Absorbing Markov Chain approach (Liang et al., 2011), providing an approach to estimate the relative proportions of living microbial biomass and necromass as two different compartments of the soil carbon pool. The simulations suggest that the size of the microbial necromass carbon pool could be 40 times as high as the living microbial biomass carbon pool in soil, resulting in a total of up to 40%–80% based on a realistic estimate for microbial biomass carbon in soil of 1%–2% of the soil organic carbon (Liang et al., 2011). Recently, another conceptual model has been developed for the assessment of the relative contributions of plant-derived carbon and microbial necromass to SOM (Fan and Liang, 2015). This model is based on the microbial response to plant-derived substrate supply (*anabolism and growth*) and on parameters of bacterial and fungal biomass and necromass turnover. The simulations again showed a very high contribution of microbial-derived carbon to soil organic carbon at "equilibrium stage," ranging from 47% to 80%. Microbe-derived carbon is thus important for SOM sequestration, but the variation of microbial-derived versus plant-derived carbon is presumably ecosystem-specific and may change due to shifts in environmental conditions and nutrient availability.

It was already stated five years ago that little progress has been made in reconciling biochemical properties and processes with the kinetically defined pools of current carbon modeling approaches (Dungait et al., 2012). Because only data from model experiments and modeling approaches are available at present, we need reliable experimental data about the actual contribution of microbial necromass to SOM in various soils, and in particular about the variability of the contribution in specific soil types.

Accordingly, there is an urgent demand for quantitative data about the share of microbial necromass in SOM in order to allow the assessment of the contribution of microbial carbon in natural ecosystems and the parameterization of process-based global and local carbon turnover models

(Liang and Balser, 2011). This is of growing importance as microbial processes are increasingly being integrated into modeling approaches (Wieder et al., 2013, 2014). It is also important for *reliable assessment and simulation of alterations of carbon stocks triggered by climate change*, because microbially explicit models, such as MIMICS and the currently applied carbon turnover models for general Earth–System carbon balance predictions have predicted contradictive responses to climate warming and increased CO_2 (Wieder et al., 2014). Wieder et al. (2015) have thus provided a *roadmap for the incorporation of microbial theory* and explicit models into Earth–System models in order to improve their performance. However, details of the microbial processes are still not fully understood and thus cannot yet be parametrized. In addition, yields or CUE as well as decay rates of living microbes are not considered. In this context, it may be advantageous to combine these approaches with models developed in other areas for the degradation of organic contaminants and pesticides (Adam et al., 2014; Rein et al., 2016; Trapp et al., 2017).

IMPLICATIONS OF THE MICROBIAL NECROMASS—MATRIX STABILIZATION CONCEPT

Previous general reviews on carbon cycling in soils report a positive effect of microbial diversity on decomposition of plant input material and the amount of microbial biomass (Nielsen et al., 2011). So far, however, these results were not specifically related to soil carbon storage. The increase in microbial activity and growth, with shifts to higher CUE or anabolic carbon yields, results in increased amounts of necromass in SOM, as shown earlier. In conclusion, the combination of experimental results and theoretical considerations finally lead to the paradigm shift from "humic substances" toward the concept of "necromass–matrix stabilization." This shift not only results in a different view but for the first time also potentially provides *in-depth mechanistic understanding* of soil processes. Based on this concept, we have to consider that remnants of biomass (necromass) will be found as fragments of cell envelopes and as a general coating of cytosolic materials (proteins, ribosomes, nucleic acids, lipids, and various kinds of mono- and oligomers) on mineral surfaces in soil. Upcoming research on these processes should thus focus on providing sufficient understanding of general microbial physiology, behaviors, and microbial ecology in soils (Schimel and Schaeffer, 2012).

The concept also has an impact on understanding of the *relation between carbon and nitrogen* (and also phosphorus). Previous considerations suggested that plants consume predominantly inorganic nitrogen species, whereas bacteria and fungi also use organic nitrogen (Zang et al., 2016). However, microbes do compete with plants for inorganic nitrogen species. Nitrogen fertilization was shown not only to increase plant productivity and aboveground biomass production but also to reduce SOM decay within the rhizosphere (Zang et al., 2016). This is presumably due to the higher potential for *microbial growth and higher CUE* after fertilization (Manzoni et al., 2012). Microbial biomass and related necromass is thus not only part of SOM: it is (1) a *combined carbon, nitrogen, and phosphorus storage material*, since necromass contains on average 10%–15% organic nitrogen and 1%–4% phosphorus (Vrede et al., 2002), (2) the *mobilizing driver* (depolymerization of macromolecules) in recycling of plant and microbial debris, and (3) a *nutrient distributor* for plants, for example, by mycorrhiza, and for other microbes in terms of the "fungal pipeline" concept (Worrich et al., 2017). Therefore, it is not surprising that fungal-based food webs have higher carbon and nitrogen storage potentials and are also more drought-resistant than bacterial-based food webs. Excess of carbon with nitrogen limitation always leads to carbon losses due to changes in microbial anabolism and CUE; vice versa, sufficient nitrogen supply leads to lower mineralization and higher anabolism with carbon and nitrogen storage.

It is well known that high *fungal-bacteria ratios* are commonly related to high carbon storage and SOM contents as well as microbial communities shifted toward more copiotrophic rather than oligotrophic organisms (Fierer et al., 2007; Trivedi et al., 2013). This is also consistent with the fact that the turnover of fungal necromass is lower than that of bacterial residues. However, we are currently far away from being able to assign soil turnover processes to specific organisms. Therefore, more *trait-based approaches in soil microbial ecology linked to turnover* are needed, in particular for real process-based turnover modeling.

However, restricting research to fungi and bacteria may overlook significant contributions by other groups of organisms, since the relatively high amounts of *microalgae* found in soils (Otto et al., 2010) may also be of relevance. Algal biomass is considered to be relatively stable, and in particular the extracellular polymeric substances (EPS) produced by these organisms provide an appropriate glue for aggregate formation from POM (Rossi and De Philippis, 2016; Rossi et al., 2012). This clearly reduces diffusion and transport processes, as observed for microbial crusts (Belnap et al., 2005). In turn, these EPS compounds will also affect the *water storage properties* of soils since they can be considered as water-absorbing polymers.

The factors controlling the *anabolism of microorganisms* were reviewed a few years ago by Manzoni et al. (2012). A major driver for SOM formation is the *carbon use efficiency* (*CUE*) (ratio of growth = biomass formed to substrate consumed, also known as yield coefficient in microbiology), showing an expected high sensitivity and negative correlation to temperature; C:N ratios of plant inputs are also negatively correlated to CUE. The highest CUE in terrestrial ecosystems was found with C:N ratios around 10:1, which is in the range of microbial biomass itself (Manzoni et al., 2012) and thus favors microbial anabolism. The *temperature sensitivity* may be explained by the consideration that microbial turnover kinetics is related to the temperature by the Arrhenius equation (doubling of the turnover by a factor of two per 10°C increase); thus, reactions requiring higher activation energy must be more sensitive to changes in temperature (Schmidt et al., 2011). However, soil microbial processes have been shown not to follow the Arrhenius equation in the long run; longer periods of warming leveled-out the initial increase and the communities returned to mineralization rates similar to those found before warming as they adapt to the new conditions (Rousk et al., 2012). Increase of temperature above the temperature optimum of the community will result in a shift of the community toward higher optimum temperatures (Rousk and Bååth, 2011). Similar effects can be expected for the CUE.

Microbes usually strive for C, nutrients, and energy. Therefore, substrates are typically considered as *sources of carbon and energy* allowing cell maintenance and growth; however, if multiple substrates are available, the organisms can use substrates more for energy provision or for carbon supply, depending on the molecular composition, nutrient and electron acceptor availability (Madigan et al., 2017). The substrate use usually depends on the molecular size and the energy content of the substrates, considered as Gibbs free energy of the coupled redox reactions: the oxidation reaction of the substrate (electron donor) combined with the reduction reactions of the terminal electron acceptor for respiration (i.e., O_2, NO_3^-, SO_4^{2-}, CO_2). Depending on the amounts of the available reaction partners, the net energy potential is then partly transformed into microbially usable energy, feeding maintenance and growth; thus it is the dominating driver of the ratio of CUE and mineralization (Brant et al., 2006; Frey et al., 2013; LaRowe and van Cappellen, 2011; McCarty, 2007; Trapp et al., 2017). Some substrates, such as many sugars and some amino acids, support high CUE, whereas others, such as hydrocarbons and aromatic compounds, provide much lower CUE. CUE appears to peak around a nominal zero oxidation state of carbon in the compound (LaRowe and van Cappellen, 2011). However, the cell's need for energy and carbon may vary depending on its status and the overall environmental conditions.

In addition, microbial activity, and thus also the extent of the use of the chemical energy potential and the CUE on the cell level, is mainly triggered by the *chemical activity of water* (resulting from osmotic and matric potential) in the vicinity of the metabolizing cells and the substrate. Similarly, the availability of nutrients, such as nitrogen, phosphorus, and macroelements as well as electron acceptors also depends on diffusion and thus on the matric potential. An important fact, which is often forgotten, is that when microbes degrade organic chemicals they always produce water via these reactions, which in turn increases the water activity. Moreover, high formation of microbial biomass and a contribution of the necromass to SOM of up to 48% were found in neutral arable and slightly acidic grassland soils; this is presumably related to *near-neutral pH values* and high clay contents (Khan et al., 2016). In addition, arbuscular mycorrhizal fungi appear to contribute more to carbon sequestration than saprophytic fungi.

Microbial necromass and hardly degradable plant debris are biomaterials that share many properties: some of the materials can store high amounts of water. These materials thus contribute to the *water-holding capacity* and modify the water retention curve, that is, the relation between gravimetric water content and *matric potential.* The great water storage capacity is also related to other properties, in particular the shrinking and swelling dependent on the water content. Because high amounts of water are stored in these materials, the frequently used model concept of free-water menisci and phases between mineral particles in soils (Or et al., 2007) is perhaps an oversimplification and only valid for very high water contents near the maximum water holding capacity. Microbial motility by swimming and swarming is only possible if free water films and menisci are large enough (Tecon and Or, 2016); however, if water films are too thin or water is only present in gel-like structures, as is the case if soil moisture is lower than water-holding capacity, perhaps only gliding motility and retarded diffusion of substrates are possible. Such conditions may be of advantage for fungi, which are able to transport water (and also C substrates and nutrients; Worrich et al., 2017) over longer distances even through dry pores ("fungal pipeline concept"; Banitz et al., 2013; Guhr et al., 2015). Because these organisms are able to transport water to substrates that do not contain sufficient water, they benefit strongly from the ecological advantage of being able to exploit niches and substrates in soils that are otherwise not bioavailable due to limited water activity. The best example of such properties is the dry rot of timber wood in roofs by the fungus *Serpula lacrimans*, which transports water from the basement of a house to the roof where the wood decay is performed (Schmidt, 2007).

Dependent on the gas exchange (O_2, CO_2, and partly CH_4) with the atmosphere as well as the availability of nutrients and electron acceptors, *microorganisms drive the redox conditions* in soils (Bondici et al., 2016). They produce and consume the redox-active nitrogen and sulfur molecules (such as NO_3^-, $NH4^+$, S^{2-}, S^0, SO_4^{2-}). In addition, the dominating redox-active metals Fe and Mn are oxidized and reduced due to microbial metabolic activity (Madigan et al., 2017). Oxidation (mostly) results in precipitation and reduction in dissolution processes, which lead to permanent mobilization or immobilization of these metals, depending on changes in environmental conditions, such as water saturation or substrate and litter input. In the complex environments of soils, precipitation of Fe, Al, and Mn occurs not only by microcrystal formation in solution but also by precipitation on surfaces and within biomass or necromass macromolecular aggregates (see "Factors and mechanisms causing persistence of microbial necromass").

Other results provide evidence that the *accessibility of organic matter* to microbes is the regulating process for the decay of SOM in soil, rather than chemical structure and the changes in soil water content (Dungait et al., 2012; Moinet et al., 2016). In turn, decreasing accessibility will lead to increased "persistence" of degradable organic materials. Persistence may thus also be caused by the formation of *hydrophobic domains* with decreased chemical activity of water (Masoom et al., 2016) by drying in general, by loss of pore water connectivity or simply by the absence of degrading microorganisms due to their distance or limited motility and diffusion. In addition, microbial cell envelope residues, in particular if the microbes died after adaption to water stress, may leave their hydrophobic surface properties as an impact on mineral surfaces (Achtenhagen et al., 2015). Therefore, inaccessibility may be a result of low water contents, and thus it is not a real contradiction as suggested by the authors mentioned earlier. In addition, inaccessibility can be caused by occlusion or incrustation of organic molecules or necromass, not only in aggregates but also in metal precipitates or other particles, such as clay or other mineral associations (see "Factors and mechanisms causing persistence of microbial necromass"). Other litter materials, such as wood or lignocellulose particles may enter soil as dry materials and remain, only partly hydrated, as dry spots in SOM. Moreover, some bacteria appear to protect themselves against predation and other environmental stress by forming "clay hutches" in their surroundings (Lünsdorf et al., 2000).

In addition, *carbon age determinations* using ^{13}C *and* ^{14}C *isotopic signatures* assume that fresh carbon derived from plant material is used for microbial turnover in soil (Ahrens et al., 2015; Amelung et al., 2008). However, according to the concept of "you are what you eat" in relation to isotopic signatures of substrates and the consuming biomass (Boschker et al., 1998), the assumption of a substrate containing fresh carbon is not generally valid in soil and sediment systems, because carbon atoms are recycled again and again through the microbial food web. It has to be kept in mind that SOM is not inert. All organic compounds in soil are turned over constantly, although the rates differ depending on environmental conditions, microorganisms involved, and related substrates. Carbon turnover has been quantified in the past by ^{14}C-based age determinations (Rethemeyer et al., 2005) or by changes in ^{13}C abundance after change from C_4 to C_3 vegetation or vice versa (e.g., Lobe et al., 2005). Such age determinations can be made for bulk soil C, but also compound-specific ages have been determined for biomarkers (Eglinton et al., 1996; Kramer and Gleixner, 2006). When interpreting these results, it has to be considered that the age determinations give the age of the carbon atoms, not of the molecules in which it is bound. Actively growing microorganisms may have very old ^{14}C signatures, even for recently assimilated carbon, if the carbon in their growth substrate was removed from the atmosphere by autotrophic processes a long time ago. The isotope analyses give only limited information on the metabolic path the carbon atoms took until they were bound in the biomarker molecule, and on how long each of the steps took. In other words, this means in particular that *microbes eating old carbon appear old* in terms of isotopic signatures, even if they used this substrate recently. In addition, heterotrophic CO_2 fixation by microorganisms (Miltner et al., 2005; Santruckova et al., 2005) may also impair the isotopic signatures in soil, depending on the source of CO_2 and its relative contribution to microbial biomass carbon. Nevertheless, these analyses provide valuable information on how long the carbon remained in soil and on the age of the carbon in various compound classes. This information could help to identify potential levers for sustainable carbon management.

Age determinations are also tightly connected with considerations of *mean residence times* (Amelung et al., 2008; Schmidt et al., 2011) of compounds. They are generally considered to be compound-specific; however, according to the macromolecular "architecture" and *complexity of microbial necromass materials*, in particular complex cell envelope materials, this may not be an adequate perception if tested with pure compounds. For mixed aggregates and polymers derived from proteins, lipopolysaccharides, peptidoglycans, and sometimes mycolic acids, the "lumped" mean residence time needs to be considered, because the specific compounds cannot be degraded independently of the decay of the overall structures. This is also reflected by the degradation data from decaying microorganisms, which showed a high variability of mineralization of compound groups, for example, 90% for PLFA and around 10% for proteins, whereas the typical bulk carbon turnover is in the range of 50% (Miltner et al., 2012; Schweigert et al., 2015).

Surface layer proteins of many bacteria and of most archaea, which do not have rigid outer layers of the cell envelope, are considered to have specific properties for complexing ions (Albers and Meyer, 2011; Sara and Sleytr, 2000). These proteins exhibit the general ability to *self-aggregate on surfaces*, thereby changing the surface properties. In particular, S-layer proteins were found to have specific properties for binding and enrichment of heavy metals (Lederer et al., 2014). Protein layers on mineral surfaces may also alter the wettability of minerals (Achtenhagen et al., 2015).

The amount of SOM and microbial necromass depends on the flux from the plant input via microbes because SOM presumably represents a *non-equilibrium steady state* of matter with very high internal turnover, which is generally overlooked. Therefore, there is an urgent need for quantification of the microbial contribution and processes involved in SOM turnover (Kögel-Knabner, 2017). The resulting important related research questions are:

- To which extent are structures of plant input materials modified and controlled by environmental conditions and exoenzymes? This is of particular importance for providing the substrates actually taken up and metabolized by microorganisms.
- How are soil properties affected by the structures of SOM and its precursors? This is relevant for understanding feedback loops on microbial activity. The self-organization in soils (soil-microbe-complex), in particular the interactions of the lithospheric components with the microbial players and their leftovers, was already considered as an upcoming research field a decade ago (Young and Crawford, 2004).
- What do all these processes mean for the global carbon turnover on the landscape scale? What is the impact of land use in general on these processes, and how can we optimize land management in terms of carbon sequestration based on modeling of the global carbon turnover?

FACTORS AND MECHANISMS CAUSING PERSISTENCE OF MICROBIAL NECROMASS AND SOM

A significant accumulation in soil and the observed high contribution of microbial residues to SOM (Lehmann and Kleber, 2015; Liang and Balser, 2011; Miltner et al., 2012; Simpson et al., 2007a) requires that this material be either recalcitrant to microbial degradation or strongly stabilized in soil. We cannot expect *biological materials* to be very recalcitrant to degradation, due to their chemical structures. Persistence of the material is therefore not a result of chemical recalcitrance, but must be caused by stabilization processes. For sustainable SOM management, we need to know how this material is stabilized in soil and how to use these mechanisms efficiently.

First, at least part of the *microbial biomass residue is intrinsically meta-stable*, although the isolated components are considered to be biodegradable. This is mainly due to physical aggregation by the cell organization. The cell envelope has evolved to resist attack by other microbes or extracellular enzymes (Madigan et al., 2017). This is achieved by the bio-organization of the cell wall components in a stable agglomeration. In this complex, access to the individual components is highly limited because each component is embedded into a surrounding structure, which has to be removed before degradation is possible. Thus, a large number of different enzymes are needed at the same place for full degradation; this is relatively unlikely to occur in soil, in particular if extracellular enzymes are involved in depolymerization of macromolecular compounds. Stabilization by *physical arrangement of the components* is further enhanced if cell envelope residues attach to each other and form piles of fragments. Formation of such piles has been observed in soil (Miltner et al., 2012). This results in considerable protection against microbial degradation, as the piles can only be attacked from the outside.

In addition, lipids are a key component of the cell envelope organization of the four general types of microbial biomass (Gram-positive and Gram-negative bacteria, archaea, and fungi; see "Fate and Turnover Analyses of Microbial Biomass"). In particular, the cell membrane, present in all organisms, consists of lipid bilayers (Madigan et al., 2017). In addition, Gram-negative bacterial cells have an outer membrane, which also contains lipids. Gram-positive bacterial cells do not have an outer membrane, but have other hydrophobic compounds, for example mycolic acids, at the surface of their cell envelope (Madigan et al., 2017). The *cell envelope fragments* would tend to attach to mineral surfaces *with their hydrophilic part*, exposing the hydrophobic part to the soil solution. Interaction of microbial cells and cell fragments has recently been shown to significantly reduce wettability of soil minerals (Achtenhagen et al., 2015). This effect may create domains with hydrophobic exteriors, which form efficient barriers for water. Water exchange with these domains will therefore be very slow, and microbial activity in these domains will be limited by low availability of water. Degradation of the cell envelope residues will therefore be suppressed, resulting in stabilization of the material. Perhaps the materials shown to lack contact to the water phase by NMR studies (Masoom et al., 2016) may be located in these domains.

Furthermore, cell envelope residues, like any other organic material in soil, may be stabilized by interaction with minerals (Amelung et al., 2001; von Lützow et al., 2008). In particular, Fe and Al oxides may play an important role (Eusterhues et al., 2003). These minerals are redox-sensitive; they are frequently oxidized and reduced due to microbial metabolic activity (Madigan et al., 2017). Oxidation generally results in precipitation and reduction in dissolution of Fe oxides (Schwertmann, 1988). These minerals are therefore either mobilized or immobilized at every *redox potential change* of sufficient extent in the soil. In the complex environments of soils, precipitation of Fe occurs not only by *microcrystal formation* in solution but also by *precipitation on surfaces and within biomass* or necromass macromolecular aggregates, including cell envelope residues. This *incrustation* leads to significant changes in the activity, structures, and properties of extracellular enzymes and to changes in *bioavailability and biodegradability* of organic compounds. These processes will contribute significantly to the "persistence" of biodegradable molecules in soils. It is well known that organic molecules are stabilized when associated to Fe and Mn mineral surfaces (Mikutta et al., 2009); however, the incorporation of oxidized metal species into macromolecular aggregates and polymers of biotic origin was seldom considered in the past. In addition, incrustation of necromass fragments and biomacromolecular aggregates may occur not only with metal oxides (including Al oxides) but presumably also by incrustations with Si or carbonates, which will result in similar inaccessibility effects.

The stabilizing effect of Fe minerals can be considered to be particularly strong. First of all, Fe oxides are abundant minerals in almost all soils. Second, a whole suite of *microbial biomass components has a high affinity to Fe oxides.* This includes teichoic acids in cell envelopes of Gram-positive bacteria as well as nucleic acids, which contain phosphate groups and thus are able to form innersphere complexes with Fe oxides (Atkinson et al., 1972). Proteins also interact strongly with Fe oxides and with other minerals (Kleber et al., 2007) and may go undergo changes in conformation that decrease their accessibility for proteases. These reactions reduce the availability of the involved compounds considerably, resulting in stabilization against microbial decay, as found with ^{14}C-labeled bacterial cell envelopes (Achtenhagen, unpublished data). The pronounced effect of organo-mineral interactions on mineralization of organic matter has been reported by Keil et al. (1994) for marine organic matter, which persisted in organo-mineral associations, but could be degraded within very short times once it was separated from the minerals.

Finally, as for all other organic compounds in soil, *unfavorable environmental conditions* will contribute to the persistence of necromass. However, it is not enough to consider large-scale environmental conditions. Soils are very heterogeneous, so the availability of water, nutrients, substrate, and electron acceptors may also differ at very small scales. Therefore, many different habitats and niches for microorganisms may exist with different environmental conditions, and thus different microbial activities, in a small volume of soil (Or et al., 2007; Young and Crawford, 2004). Degradation activity may differ in these microhabitats, depending on the presence of specific microorganisms and the conditions prevailing.

PERSPECTIVE—IMPACT OF THE NECROMASS CONCEPT UPON PROCESS UNDERSTANDING

The necromass concept presented in this contribution has a pronounced impact on our understanding of the processes involved in *SOM turnover and persistence*. According to this concept, plant litter and root exudates are the major carbon input into soils and will be reworked by the microbial pump (Fig. 5.1). The low-molecular fraction of this input will be rapidly taken up by soil microorganisms, whereas high-molecular-weight compounds (i.e., plant polymers and aggregates of microbial necromass) have to be hydrolyzed to oligomers or monomers by extracellular enzymes (or by leached cytosolic enzymes after cell lysis) before they can enter microbial cells and be utilized by the organisms. Microorganisms grow on these low-molecular-weight substrates whenever they can obtain them. However, when the supply ceases or environmental conditions become unfavorable, they will die and eventually disintegrate. This decay leaves DOM and cell envelope fragments behind and also produces CO_2 from degradation of these compounds (Kästner et al., 2014). Given sufficient water supply and transport to deeper layers, the more persistent part of the DOM will leach to the groundwater. The particulate cell envelope fragments, in contrast, will be preserved and contribute to SOM, presumably at the sites of their formation. It has to be kept in mind, however, that this material may serve as a substrate for other bacteria and thus be recycled several times. Each rotation of the microbial pump will result in carbon loss as CO_2 and DOM, but part of the carbon is always preserved in SOM.

Assuming that this conceptual model describes the processes correctly, the amount of carbon stabilized in soil depends on the following factors and parameters:

- The amount of plant input. Plant input fuels the whole system and drives the microbial pump.
- The speed of the microbial pump. This determines how much of the plant residue carbon is routed through the microbial pump and thus transformed to microbial residues (or CO_2 and DOM) and finally to SOM. If the microbial pump is slower than plant material input, unaltered plant residues will accumulate, whereas they will be completely transformed if the microbial biomass removes that material as fast as it enters the soil. The decisive factor here is the activity of the microbial community, which is determined by both the composition of the microbial community and by external factors, such as substrate quality and environmental conditions for the microbes.
- The CUE of the microorganisms. Each "rotation" of the microbial pump will produce SOM, CO_2, and DOM. The higher the share of plant-derived carbon which is utilized by the microbes to synthesize their biomass (instead of producing CO_2 or DOM), the higher will be the amount of carbon which potentially can accumulate in the soil. The ratio of CO_2 versus biomass/necromass formation is reflected by the CUE, which appears to be higher with increased fungal contribution to the microbial food web. CUE of soil microorganisms can vary greatly, depending on substrate, nutrient supply (nitrogen and phosphorus), and environmental conditions.
- The stabilization of the cell envelope fragments and DOM. The microbial pump as such already keeps organic carbon in the soil system because it is recycled in the microbial food web. In addition, both the dissolved and the particulate fractions interact with the solid minerals; furthermore, the particulate matter interacts not only with precipitating metals during redox changes, but also with Al oxides, Si, and carbonates. These processes result in necromass-matrix-stabilization. This includes a group of processes related to molecule properties, physicochemical conditions, and soil system conditions.

Stabilization of microbial necromass relies on the same mechanisms as for any other organic material in soil. Inaccessibility of organic compounds (or physical stabilization) has been identified as an important mechanism. In the case of cell envelope fragments, there are several factors contributing to inaccessibility and thus to stabilization of the material. The nanoscale architecture of the cell envelopes is very complex: the material is composed of lipid bilayers, proteins embedded into these bilayers, peptidoglycan, and lipopolysaccharides or teichoic acids. All these materials tend to self-assemble in relatively stable structures. The individual compounds are therefore protected from degradation as long as other compounds are located around them. It has also been observed that cell envelope fragments frequently occur in piles of several layers, sometimes coating preferential flow channels in soils, further decreasing the accessibility for degrading enzymes, in particular for the inner layers. Finally, the cell envelope fragments also interact with mineral surfaces, providing further stabilization.

Cell membrane lipids and a number of other cell envelope compounds are amphiphilic molecules. When these compounds sorb to hydrophilic mineral surfaces, they tend to interact with the hydrophilic part, exposing the hydrophobic part to the surface. If the surface coverage of the minerals with this type of material is sufficient, the particles may become hydrophobic. This has been demonstrated in model experiments with pure minerals and pure bacterial cultures. As a consequence, hydrophobic domains may form, where water cannot enter, as observed by NMR. Microbial activity in such domains will be very low, simply because of insufficient water availability, resulting in reduced degradation of the cell envelope material. Another type of interaction between microbial residues and minerals is incrustation by Fe, Mn, Al, or Si oxides. Incrustation may be related to mineral dissolution and precipitation, for example due to redox oscillations. It reduces the accessibility of the material considerably and thus contributes to stabilization.

As a result of all the described processes, the consideration of defined pools or fractions of organic material may be misleading. Instead, we can assume a continuum from plant residues to microbial biomass and necromass in various stages of decay with a certain steady-state level of bulk carbon. The stability of the material is predominantly controlled not only by system properties, such as mineralogy, environmental conditions, nutrient and water availability, but also by litter quantity and quality together with microbial community composition and activity.

Some of these aspects have recently been addressed in SOM models. For example, the COMISSION model explicitly includes sorption, microbial interactions, and DOM transport (Ahrens et al., 2015). Including microbial processes, it has explained more of the observed variance and rendered the model predictions more reliable. Old ^{14}C ages could be explained by this model without the need for a carbon pool with millennial turnover times, even though the fact that microorganisms may grow on ^{14}C-depleted, "old" substrates was not considered. This underlines how important the microbial pump may be for carbon stabilization. Also Gougoulias et al. (2014) described the importance of the microbial communities and their activities as drivers for soil processes. Their approach, however, lacks the possibility of high-throughput analyses. Thus it is highly difficult to track the quantitative portion of carbon use by different phylogenetic groups using this approach. In the future, therefore, it should be further developed to include this option. A potential way to improve predictions of soil processes is to include environmental factors, microbial community composition, and microbial biomass parameters (Graham et al., 2016). More than half of the models investigated by the authors were improved by incorporating information on both microbial community composition and microbial biomass in addition to environmental variables. This again underlines the importance of understanding microbial communities and processes for prediction of ecosystem processes, including the carbon cycle.

Earlier, we described how plant material is used by soil microbes and how soil microbial biomass may cause "dynamic persistence" (Fig. 5.1). This modifies the plant molecular fingerprint and generates a microbial molecular fingerprint. Microorganisms use the plant material as a carbon and energy source, which results in a steady flux from plant material through microbial biomass to necromass. This flux is modified depending on the energy provided by the substrate and on environmental conditions. Microorganisms have a high potential to adapt to changes in environmental conditions, either by physiological adaptation or by modifications of the microbial community composition. However, if environmental conditions become too unfavorable, for example the input does not meet the energetic requirements of the microbial community, the cells die. This necromass, in particular the cell envelope, is stabilized mainly by its macromolecular architecture, which provides resistance to microbial attack not only while the cell lives but also after its death, and by interaction with the mineral matrix. Finally, there is a steady flux of carbon from plant material via the microbial food web toward microbial necromass, even if no net changes in the carbon pools can be detected because the system is in a steady state. This results in apparent carbon persistence, as there is always a certain amount of carbon cycling through the microbial food web.

Thus, in the long run, recalcitrance of organic input does not play a significant role (von Lützow et al., 2008). Instead, various physical stabilization processes seem to be the crucial drivers, mainly by controlling bioavailability of the organic compounds. We also described how microbial necromass can be stabilized in soil. This can induce water repellency and contributes to nitrogen and phosphorus storage in soil. Preservation of biomolecules may also have implications for the assessment of analytical data obtained in soil research. For example, it is expected that nucleic acids are preserved in soil to a certain extent, because they may strongly interact with oxides via their phosphate groups. If these preserved nucleic acids are extracted for metagenomic analyses, they may contribute to the metagenome found, although the organisms are no longer alive. The contribution of preserved nucleic acids to the amount extracted in such studies thus needs to be quantified in order to assess the metagenome of the living microbial community, as opposed to the phylogenetic footprint. Such considerations also apply to protein stabilization in soils (which is presumably much higher) and proteomic studies in general. Similarly, age determinations using ^{14}C dating of bulk SOM or of biomarkers may be partly misleading, because the age determined corresponds to the time when the carbon was removed from the atmosphere by the plants. Turnover in soil may change the chemistry of the carbon, but not its ^{14}C signature. Old ^{14}C dates thus do not necessarily relate to old molecules. They still indicate preservation of the carbon in the soil; however, the mechanisms and pathways remain to be elucidated.

Here we focused on the contribution of microbial biomass and necromass to SOM formation. However, soil microorganisms also have other functions, all of which are important to plant growth. The soil microbiome has been shown to affect plant growth, nutrient availability, and drought resistance (East, 2013). Enhanced plant growth will in turn provide more carbon and energy to the soil microorganisms, thus enhancing their activity. This will also have implications for SOM accumulation and turnover. Therefore, we must always consider the triangle of "soil structure—microbiome—plants" (East, 2013) if we want to understand and control soil processes.

Recent research results on soil microbial processes and the fate of microbial necromass in soil should significantly improve the predictive power of soil carbon models, as they provide important process understanding.

A key issue may be quantifying the CUE of soil microbes because this is a crucial parameter which changes depending on the availability of carbon, nutrients, energy, and water in relation to the microbes' requirements. Another important issue is to characterize the microbial community in soils, in terms of both community composition and the activities performed by the microorganisms. Soil proteomics may play an important role in this context in the next years. If CUE as well as microbial community structure and functions can properly be incorporated into carbon turnover models, and energy fluxes are considered in detail, then we will be able to predict how efficient SOM management measures will be. It will also allow benefits or risks of land management practices to be identified without the need for long-term experiments. Finally, this will be an important step toward a better and more directed manipulation of soil carbon storage and CO_2 sequestration.

ACKNOWLEDGMENT

This contribution was financed by the general funding of the Helmholtz-Association.

REFERENCES

Achard, F.K., 1786. Chemische untersuchungen des torfs. Crell's Chem. Annal. 2, 391–403.

Achtenhagen, J., Goebel, M.-O., Miltner, A., Woche, S.K., Kaestner, M., 2015. Bacterial impact on the wetting properties of soil minerals. Biogeochemistry 122, 269–280.

Adam, I.K.U., Rein, A., Miltner, A., da Costa Fulgêncio, A.C., Trapp, S., Kästner, M., 2014. Experimental results and integrated modeling of bacterial growth on an insoluble hydrophobic substrate (phenanthrene). Environ. Sci. Technol. 48, 8717–8726.

Ahrens, B., Braakhekke, M.C., Guggenberger, G., Schrumpf, M., Reichstein, M., 2015. Contribution of sorption, DOC transport and microbial interactions to the ^{14}C age of a soil organic carbon profile: insights from a calibrated process model. Soil Biol. Biochem. 88, 390–402.

Albers, S.-V., Meyer, B.H., 2011. The archaeal cell envelope. Nat. Rev. Microbiol. 9, 415–426.

Allison, S.D., Wallenstein, M.D., Bradford, M.A., 2010. Soil-carbon response to warming dependent on microbial physiology. Nat. Geosci. 3, 336–340.

Amelung, W., 2001. Methods using amino sugars as markers for microbial residues in soil. In: Lal, R., Kimble, J.M., Follett, R.F., Stewart, B.A. (Eds.), Assessment Methods for Soil Carbon. CRC/Lewis Publishers, Boca Raton, FL, USA.

Amelung, W., Brodowski, S., Sandhage-Hofmann, A., Bol, R., 2008. Combining biomarker with stable isotope analyses for assessing the transformation and turnover of soil organic matter. In: Sparks, D.L. (Ed.), Advances in Agronomy. Academic Press, Burlington, NC.

Amelung, W., Miltner, A., Zhang, X., Zech, W., 2001. Fate of microbial residues during litter decomposition as affected by minerals. Soil Sci. 166, 598–606.

Appuhn, A., Joergensen, R., 2006. Microbial colonisation of roots as a function of plant species. Soil Biol. Biochem. 38, 1040–1051.

Atkinson, R.J., Posner, A.M., Quirk, J.P., 1972. Kinetics of isotopic exchange of phosphate at the a-FeOOH-aqueous solution interface. J. Inorg. Nucl. Chem. 34, 2201–2211.

Bailey, V.L., Smith, A.P., Tfaily, M., Fansler, S.J., Bond-Laberty, B., 2017. Differences in soluble organic carbon chemistry in pore waters sampled from different pore size domains. Soil Biol. Biochem. 107, 133–143.
Banitz, T., Johst, K., Wick, L.Y., Schamfuß, S., Harms, H., Frank, K., 2013. Highways versus pipelines: contributions of two fungal transport mechanisms to efficient bioremediation. Environ. Microbiol. Rep. 5, 211–218.
Bastida, F., Hernández, T., Garciá, C., 2014. Metaproteomics of soils from semiarid environment: functional and phylogenetic information obtained with different protein extraction methods. J. Proteom. 101, 31–42.
Belnap, J., Welter, J.R., Grimm, N.B., Barger, N., Ludwig, J.A., 2005. Linkages between microbial and hydorlogic processes in arid and semiarid watersheds. Ecology 86, 298–307.
Benner, R., 2011. Biosequestration of carbon by heterotrophic microorganisms. Nat. Rev. Microbiol. 9, 75.
Bhattacharya, S.S., Kim, K.-H., Das, S., Uchimiya, M., Jeon, B.H., Kwon, E., et al., 2016. A review on the role of organic inputs in maintaining the soil carbon pool of the terrestrial ecosystem. J. Environ. Manage. 167, 214–227.
Bondici, V.F., Swerhone, J.J.D., Lawrence, J.R., Wolfaardt, G.M., Warner, J., Korber, D.R., 2016. Biogeochemical importance of the bacterial community in uranium waste deposited at Key Lake Northern Saskatchewan. Geomicrobiol. J. 33, 807–821.
Boschker, H.T.S., Nold, S.C., Wellsbury, P., Bos, D., de Graaf, W., Pel, R., et al., 1998. Direct linking of microbial populations to specific biogeochemical processes by ^{13}C-labelling of biomarkers. Nature 392, 801–805.
Bouma, J., 2014. Soil science contributions towards sustainable development goals and their implementation: linking soil functions with ecosystem services. J. Plant Nutr. Soil Sci. 177, 111–120.
Bradford, M.A., Keiser, A.D., Davies, C.A., Mersmann, C.A., Strickland, M.S., 2013. Empirical evidence that soil carbon formation from plant inputs is positively related to microbial growth. Biogeochemistry 113, 271–281.
Brant, J.B., Sulzman, E.W., Myrold, D.D., 2006. Microbial community utilization of added carbon substrates in response to long-term carbon input manipulation. Soil Biol. Biochem. 38, 2219–2232.
Castle, S.C., Nemergut, D.R., Grandy, A.S., Leff, J.W., Graham, E.B., Hood, E., et al., 2016. Biogeochemical drivers of microbial community convergence across actively retreating glaciers. Soil Biol. Biochem. 101, 74–84.
Cheng, S., Bryant, R., Doerr, S.H., Wright, C.J., Williams, P.R., 2009. Investigation of surface properties of soil particles and model materials with contrasting hydrophobicity using atomic force microscopy. Environ. Sci. Technol. 43, 6500–6506.
Cotrufo, M.F., Wallenstein, M.D., Boot, C.M., Denef, K., Paul, E., 2013. The microbial efficiency-matrix stabilization (MEMS) framework integrates plant litter decomposition with soil organic matter stabilization: do labile plant inputs form stable soil organic matter? Glob. Change Biol. 19, 988–995.
Dalal, R.C., 1998. Soil microbial biomass: what do the numbers really mean? Aust. J. Exp. Agr. 38, 649–665.
de Vries, F.T., Liiri, M.E., Bjørnlund, L., Bowker, M.A., Christensen, S., Setälä, H.M., et al., 2012. Land use alters the resistance and resilience of soil food webs to drought. Nat. Clim. Change 2, 276–280.
Ding, G.-C., Pronk, G.J., Babin, D., Heuer, H., Heister, K., Kögel-Knabner, I., et al., 2013. Mineral composition and charcoal determine the bacterial community structure in artificial soils. FEMS Microbiol. Ecol. 86, 15–25.
Don, A., Böhme, I.H., Dohrmann, A.B., Poeplau, C., Tebbe, C.C., 2017. Microbial community composition affects soil organic carbon turnover in mineral soils. Biol. Fert. Soils 53 (4), 445–456.

Don, A., Rödenbeck, C., Gleixner, G., 2013. Unexpected control of soil carbon turnover by soil carbon concentration. Environ. Chem. Lett. 11 (4), 407–413.
Dungait, J.A.J., Hopkins, D.W., Gregory, A.S., Withmore, A.P., 2012. Soil organic matter turnover is governed by accessibility not recalcitrance. Glob. Change Biol. 18, 1781–1796.
East, R., 2013. Microbiome: soil science comes to life. Nature 501, S18–S19.
Eglinton, T.I., Aluwihare, L.I., Bauer, J.E., Druffel, E.R.M., McNichol, A.P., 1996. Gas chromatographic isolation of individual compounds from complex matrices for radiocarbon dating. Anal. Chem. 68, 904–912.
Ellerbrock, R.H., Gerke, H.H., Böhm, C., 2009. In situ DRIFT characterization of organic matter composition on soil structural surfaces. Soil Sci. Soc. Am. J. 73, 531–540.
Eusterhues, K., Rumpel, C., Kleber, M., Kögel-Knabner, I., 2003. Stabilisation of soil organic matter by interactions with minerals as revealed by mineral dissolution and oxidative degradation. Org. Geochem. 34, 1591–1600.
Fan, T.W.-M., Lane, A.N., Chekmenev, E., Wittebort, R.J., Higashi, R.M., 2004. Synthesis and physicochemical properties peptides in humic substances. J. Pept. Res. 63, 253–264.
Fan, Z., Liang, C., 2015. Significance of microbial asynchronous anabolism to soil carbon dynamics driven by litter inputs. Sci. Rep-UK 5, 9575.
Fierer, N., Bradford, M.A., Jackson, R.B., 2007. Toward an ecological classification of soil bacteria. Ecology 88, 1354–1364.
Frey, S.D., Lee, J., Melillo, J.M., Six, J., 2013. The temperature response of soil microbial efficiency and its feedback to climate. Nat. Clim. Change 3, 395–398.
Gougoulias, C., Clark, J.M., Shaw, L.J., 2014. The role of soil microbes in the global carbon cycle: tracking the below-ground microbial processing of plant-derived carbon for manipulating carbon dynamics in agricultural systems. J. Sci. Food Agric. 94, 2362–2371.
Graham, E.B., Knelman, J.E., Schindlbacher, A., Siciliano, S., Breulmann, M., Yannarell, A., et al., 2016. Microbes as engines of ecosystem function: when does community structure enhance predictions of ecosystem processes? Front. Microbiol. 7, 214.
Guggenberger, G., Frey, S.D., Six, J., Paustian, K., Elliott, E.T., 1999. Bacterial and fungal cell-wall residues in conventional and no-tillage agroecosystems. Soil Sci. Soc. Am. J. 63, 1188–1198.
Guhr, A., Borken, W., Spohn, M., Matzner, E., 2015. Redistribution of soil water by a saprotrophic fungus enhances carbon mineralization. Proc. Natl. Acad. Sci. USA 112, 14647–14651.
Jiao, N., Herndl, G.J., Hansell, D.A., Benner, R., Kattner, G., Wilhelm, S.W., et al., 2010. Microbial production of recalcitrant dissolved organic matter: long-term carbon storage in the global ocean. Nat. Rev. Microbiol. 8, 593–599.
Joergensen, R.G., Mäder, P., Fließbach, A., 2010. Long-term effects of organic farming on fungal and bacterial residues in relation to microbial energy metabolism. Biol. Fert. Soils 46, 303–307.
Kalbitz, K., Kaiser, K., Fiedler, S., Kölbl, A., Amelung, W., Bräuer, T., et al., 2013. The carbon count of 2000 years of rice cultivation. Glob. Change Biol. 19, 1107–1111.
Kallenbach, C.M., Frey, S.C., Grandy, A.S., 2016. Direct evidence for microbial-derived soil organic matter formation and its ecophysiological controls. Nat. Commun. 7, 13630.
Kästner, M., Nowak, K.M., Miltner, A., Trapp, S., Schäffer, A., 2014. Classification and modelling of non-extractable residue (NER) formation of xenobiotics in soil: a synthesis. Crit. Rev. Environ. Sci. Technol. 44, 2107–2171.
Keil, R.G., Montluçon, D.B., Prahl, F.G., Hedges, J.I., 1994. Sorptive preservation of labile organic matter in marine sediments. Nature 370, 549–552.
Kelleher, B.P., Simpson, A.J., 2006. Humic substances in soil: are they really chemically distinct? Environ. Sci. Technol. 40, 4605–4611.
Khan, K.S., Mack, R., Xiomara, C., Kaiser, M., Joergensen, R.G., 2016. Microbial biomass, fungal and bacterial residues, and their relationships to the soil organic matter C/N/P/S ratios. Geoderma 271, 115–123.

Kindler, R., Miltner, A., Richnow, H.-H., Kästner, M., 2006. Fate of gram-negative bacterial biomass in soil—mineralization and contribution to SOM. Soil Biol. Biochem. 38, 2860–2870.
Kindler, R., Miltner, A., Thullner, M., Richnow, H.-H., Kästner, M., 2009. Fate of bacterial biomass-derived fatty acids in soil and their contribution to soil organic matter. Org. Geochem. 40, 29–37.
Kleber, M., Sollins, P., Sutton, R., 2007. A conceptual model of organo-mineral interactions in soils: self-assembly of organic molecular fragments into zonal structures on mineral surfaces. Biogeochemistry 85, 9–24.
Knicker, H., 2002. The feasibility of using DCPMAS $^{15}N^{13}C$ NMR spectroscopy for a better characterization of immobilized ^{15}N during incubation of ^{13}C- and ^{15}N-enriched plant material. Org. Geochem. 33, 237–246.
Knicker, H., 2011. Soil organic N: an underrated player for C sequestration in soils? Soil Biol. Biochem. 43, 1118–1129.
Kögel-Knabner, I., 2002. The macromolecular organic composition of plant and microbial residues as inputs to soil organic matter. Soil Biol. Biochem. 34, 139–162.
Kögel-Knabner, I., 2017. The macromolecular organic composition of plant and microbial residues as inputs to soil organic matter: fourteen years on. Soil Biol. Biochem. 105, A3–A8.
Kramer, C., Gleixner, G., 2006. Variable use of plant- and soil-derived carbon by microorganisms in agricultural soils. Soil Biol. Biochem. 38, 3267–3327.
Kunhi Mouvenchery, Y., Miltner, A., Schurig, C., Kästner, M., Schaumann, G.E., 2016. Linking atomic force microscopy with nanothermal analysis to assess microspatial distribution of material characteristics in young soils. J. Plant Nutr. Soil Sci. 179, 48–59.
Kuzyakov, Y., 2010. Priming effects: interactions between living and dead organic matter. Soil Biol. Biochem. 42, 1363–1371.
Kuzyakov, Y., Blagodatskaya, E., 2015. Microbial hotspots and hot moments in soil: concept & review. Soil Biol. Biochem. 83, 184–199.
Lal, R., 2007. Soil science and the carbon civilization. Soil Sci. Soc. Am. J. 71, 1425–1437.
Lange, M., Eisenhauer, N., Sierra, C.A., Bessler, H., Engels, C., Griffiths, R.I., et al., 2015. Plant diversity increases soil microbial activity and soil carbon storage. Nat. Commun. 6, 6707.
LaRowe, D.E., van Cappellen, P., 2011. Degradation of natural organic matter: a thermodynamic analysis. Geochim. Cosmochim. Acta 75, 2030–2042.
Lederer, F., Günther, T.J., Raff, J., Flemmig, K., Pollmann, K., 2014. Eigenschaften von Bakterien aus Schwermetall-kontaminierten Halden. Biospektrum 20, 172–175.
Lehmann, J., Kleber, M., 2015. The contentious nature of soil organic matter. Nature 528, 60–68.
Lemanski, K., Scheu, S., 2014. Incorporation of ^{13}C labelled glucose into soil microorganisms of grassland: effects of fertilizer addition and plant functional group composition. Soil Biol. Biochem. 69, 38–45.
Liang, C., Balser, T.C., 2011. Microbial production of recalcitrant organic matter in global soils: implications for productivity and climate policy. Nat. Rev. Microbiol. 9, 75.
Liang, C., Cheng, G., Wixon, D., Balser, T., 2011. An Absorbing Markov Chain approach to understanding the microbial role in soil carbon stabilization. Biogeochemistry 106, 303–309.
Liang, C., David, S., Duncan, Balser, T.C., Tiedje, J.M., Jackson, R.D., 2013. Soil microbial residue storage linked to soil legacy under biofuel cropping systems in southern Wisconsin, USA. Soil Biol. Biochem. 57, 939–942.
Liang, C., Kao-Kniffin, J., Sanford, G.R., Wickings, K., Balser, T.C., Jackson, R.D., 2016. Microorganisms and their residues under restored perennial grassland communities of varying diversity. Soil Biol. Biochem. 103, 192–200.

Lobe, I., Bol, R., Ludwig, B., Du Preez, C.C., Amelung, W., 2005. Savanna-derived organic matter remaining in arable soils of the South African Highveld long-term mixed cropping: evidence from ^{13}C and ^{15}N natural abundance. Soil Biol. Biochem. 37, 1898–1909.

Ludwig, M., Achtenhagen, J., Miltner, A., Eckhardt, K.-U., Leinweber, P., Emmerling, C., et al., 2015. Microbial contribution to SOM quantity and quality in density fractions of temperate arable soils. Soil Biol. Biochem. 81, 311–322.

Lueders, T., Kindler, R., Miltner, A., Friedrich, M.W., Kaestner, M., 2006. Identification of bacterial micropredators distinctively active in a soil microbial food web. Appl. Environ. Microb. 72, 5342–5348.

Lünsdorf, H., Erb, R.W., Abraham, W.-R., Timmis, K.N., 2000. "Clay hutches": a novel interaction between bacteria and clay minerals. Environ. Microbiol. 2, 161–168.

Madigan, M.T., Bender, K.S., Buckley, D.H., Stahl, D.A., Sattley, W.M., 2017. Brock Biology of Microorganisms. Pearson Education Inc., New York.

Malik, A., Chowdhury, S., Schlager, V., Oliver, A., Puissant, J., Vazque, P.G.M., et al., 2016. Soil fungal:bacterial ratios are linked to altered carbon cycling. Front. Microbiol. 7, 1247.

Manzoni, S., Taylor, P., Richter, A., Porporato, A., Ågren, G.I., 2012. Environmental and stoichiometric controls on microbial carbon-use efficiency in soils. New Phytol. 196, 79–91.

Marschner, B., Brodowski, S., Dreves, A., Gleixner, G., Gude, A., Grootes, P.M., et al., 2008. How relevant is recalcitrance for the stabilization of organic matter in soils? J. Plant Nutr. Soil Sci. 171, 91–110.

Masoom, H., Courtier-Murias, D., Farooq, H., Soong, R., Kelleher, B.P., Zhang, C., et al., 2016. Soil organic matter in its native state: unravelling the most complex biomaterial on earth. Environ. Sci. Technol. 50, 1670–1680.

McCarty, P.L., 2007. Thermodynamic electron equivalents model for bacterial yield prediction: modifications and comparative evaluations. Biotechnol. Bioeng. 97, 377–388.

Mikutta, R., Schaumann, G.E., Gildemeister, D., Bonneville, S., Kramer, M.G., Chorover, J., et al., 2009. Biogeochemistry of mineral-organic associations across a long-term mineralogical soil gradient (0.3–4100 kyr) Hawaiian Islands. Geochim. Cosmochim. Acta 73, 2034–2060.

Miltner, A., Bombach, P., Schmidt-Brücken, B., Kästner, M., 2012. SOM genesis: microbial biomass as a significant source. Biogeochemistry 111, 41–55.

Miltner, A., Kindler, R., Knicker, H., Richnow, H.H., Kastner, M., 2009. Fate of microbial biomass-derived amino acids in soil and their contribution to soil organic matter. Org. Geochem. 40, 978–985.

Miltner, A., Richnow, H.H., Kopinke, F.D., Kastner, M., 2005. Incorporation of carbon originating from CO2 into different compounds of soil microbial biomass and soil organic matter. Isot. Environ. Health Stud. 41, 135–140.

Moinet, G.Y.K., Cieraad, E., Hunt, J.E., Fraser, A., Turnbull, M., Whitehead, D., 2016. Soil heterotrophic respiration is insensitive to changes in soil water content but related to microbial access to organic matter. Geoderma 274, 68–78.

Monod, J., 1949. The growth of bacterial cultures. Annu. Rev. Microbiol. 3, 371–394.

Mueller, C.W., Kölbl, A., Hoeschen, C., Hillion, F., Heister, K., Herrmann, A., et al., 2012. Submicron scale imaging of soil organic matter dynamics using NanoSIMS: from single particles to intact aggregates. Org. Geochem. 42, 1476–1488.

Nielsen, U.N., Ayres, E., Wall, D.H., Bardgett, R.D., 2011. Soil biodiversity and carbon cycling: a review and synthesis of studies examining diversity–function relationships. Eur. J. Soil Sci. 62, 105–116.

Or, D., Smets, B.F., Wraith, J.M., Dechesne, A., Friedman, S.P., 2007. Physical constraints affecting bacterial habitats and activity in unsaturated porous media: a review. Adv. Water Resour. 30, 1505–1527.

Otto, B., Schlosser, D., Reisser, W., 2010. First description of a laccase-like enzyme in soil algae. Arch. Microbiol. 192, 759–768.

Plaza, C., Courtier-Murias, D., Fernández, J.M., Polo, A., Simpson, A.J., 2013. Physical, chemical, and biochemical mechanisms of soil organic matter stabilization under conservation tillage systems: a central role for microbes and microbial by-products in C sequestration. Soil Biol. Biochem. 57, 124–134.

Poeplau, C., Don, A., 2013. Sensitivity of soil organic carbon stocks and fractions to different land-use changes across Europe. Geoderma 192, 189–201.

Potts, M., 1994. Desiccation tolerance of prokaryotes. Microbiol. Rev. 58, 755–805.

Pronk, G.J., Heister, K., Ding, G.-C., Smalla, K., Kögel-Knabner, I., 2012. Development of biogeochemical interfaces in an artificial soil incubation experiment: aggregation and formation of organo-mineral associations. Geoderma 189-190, 585–594.

Pronk, G.J., Heister, K., Kögel-Knabner, I., 2013. Is turnover and development of organic matter controlled by mineral composition? Soil Biol. Biochem. 67, 235–244.

Pronk, G.J., Heister, K., Vogel, C., Babin, D., Bachmann, J., Ding, G.-C., et al., 2017. Interaction of minerals, organic matter, and microorganisms during biogeochemical interface formation as shown by a series of artificial soil experiments. Biol. Fert. Soils 53 (1), 9–22.

Rein, A., Adam, I.K.U., Miltner, A., Brumme, K., Kästner, M., Trapp, S., 2016. Impact of bacterial activity on turnover of insoluble hydrophobic substrates (phenanthrene and pyrene): model simulations for prediction of bioremediation success. J. Hazard. Mater. 305, 105–114.

Rethemeyer, J., Kramer, C., Gleixner, G., John, B., Yamashita, T., Flessa, H., et al., 2005. Transformation of organic matter in agricultural soils: radiocarbon concentration versus soil depth. Geoderma 128, 94–105.

Rossi, F., De Philippis, R., 2016. Exocellular polysaccharides in microalgae and cyanobacteria: chemical features, role and enzymes and genes involved in their biosynthesis. In: Borowitzka, M.A. (Ed.), The Physiology of Microalgae, Developments and Applied Phycology.

Rossi, F., Potrafka, R.M., Pichel, F.G., De Philippis, R., 2012. The role of the exopolysaccharides in enhancing hydraulic conductivity of biological soil crusts. Soil Biol. Biochem. 46, 33–40.

Rousk, J., Bååth, E., 2011. Growth of saprotrophic fungi and bacteria in soil. FEMS Microb. Ecol. 78, 17–30.

Rousk, J., Frey, S.D., Bååth, E., 2012. Temperature adaptation of bacterial communities in experimentally warmed forest soils. Glob. Change Biol. 18, 3252–3258.

Santruckova, H., Bird, M.I., Elhottova, D., Novak, J., Picek, T., Simek, M., et al., 2005. Heterotrophic fixation of CO_2 in soil. Microb. Ecol. 49, 218–225.

Sara, M., Sleytr, U.B., 2000. S-layer proteins. J. Bacteriol. 182, 859–868.

Schaeffer, A., Nannipieri, P., Kästner, M., Schmidt, B., Botterweck, J., 2015. From humic substances to soil organic matter-microbial contributions: in honour of Konrad Haider and James P. Martin for their outstanding research contribution to soil science. J. Soils Sedim. 15, 1865–1881.

Schaumann, G.E., Kunhi Mouvenchery, Y., 2012. Potential of AFM-nanothermal analysis to study the microscale thermal characteristics in soils and natural organic matter (NOM). J. Soils Sedim. 12, 48–62.

Schimel, J., Balser, T.C., Wallenstein, M., 2007. Microbial stress-response physiology and its implications for ecosystem function. Ecology 88, 1386–1394.

Schimel, J., Schaeffer, S.M., 2012. Microbial control over carbon cycling in soil. Front. Microbiol. 3, 1–11.

Schimel, J.P., Weintraub, M.N., 2003. The implications of exoenzyme activity on microbial carbon and nitrogen limitation in soil: a theoretical model. Soil Biol. Biochem. 35, 549–563.

Schmidt, M.W.I., Skjemstad, J.O., Jäger, C., 2002. Carbon isotope geochemistry and nanomorphology of soil black carbon: black chernozemic soils in central Europe originate from ancient biomass burning. Global Biogeochem. Cycles 16, 1123.

Schmidt, M.W.I., Torn, M.S., Abiven, S., Dittmar, T., Guggenberger, G., Janssens, I.A., et al., 2011. Persistence of soil organic matter as an ecosystem property. Nature 478, 49–56.
Schmidt, O., 2007. Indoor wood-decay basidiomycetes: damage, causal fungi, physiology, identification and characterization, prevention and control. Mycol. Prog. 6, 261–279.
Schulze, E.D., Freibauer, A., 2005. Carbon unlocked from soils. Nature 437, 205–206.
Schurig, C., Smittenberg, R.H., Berger, J., Kraft, F., Woche, S.K., Göbel, M.O., et al., 2013. Microbial cell-envelope fragments and the formation of soil organic matter: a case study from a glacier forefield. Biogeochemistry 113, 595–612.
Schweigert, M., Herrmann, S., Miltner, A., Fester, T., Kästner, M., 2015. Fate of ectomycorrhizal fungal biomass in a soil bioreactor system and its contribution to soil organic matter formation. Soil Biol. Biochem. 88, 120–127.
Schwertmann, U., 1988. Occurrence and formation of iron oxides in various pedoenvironments. In: Stucki, J.W., Goodman, B.A., Schwertmann, U. (Eds.), Iron in Soils and Clay Minerals. Springer, Dordrecht, Netherlands.
Seifert, A.-G., Roth, V.-N., Dittmar, T., Gleixner, G., Breuer, L., Houska, T., et al., 2016. Comparing molecular composition of dissolved organic matter in soils and stream water: influence of land use and chemical characteristics. Sci. Total Environ. 571, 142–152.
Simkins, S., Alexander, M., 1984. Models for mineralization kinetics with the variables of substrate concentration and population density. Appl. Environ. Microbiol. 47, 1299–1306.
Simpson, A.J., Simpson, M.J., Smith, E., Kelleher, B.P., 2007a. Microbially derived inputs to soil organic matter: are current estimates too low? Environ. Sci. Technol. 41, 8070–8076.
Simpson, A.J., Song, G., Smith, E., Lam, B., Novotny, E.H., Hayes, M.H.B., 2007b. Unraveling the structural components of soil humin by use of solution-state nuclear magnetic resonance spectroscopy. Environ. Sci. Technol. 41, 876–883.
Sollins, P., Homann, P., Caldwell, B.A., 1996. Stabilization and destabilization of soil organic matter: mechanisms and controls. Geoderma 74, 65–105.
Solomon, D., Lehmann, J., Harden, J., Wang, J., Kinyangi, J., Heymann, K., et al., 2012. Micro- and nano-environments of carbon sequestration: multi-element STXM-NEXAFS spectromicroscopy assessment of microbial carbon and mineral associations. Chem. Geol. 329, 53–73.
Stevenson, F.J., 1994. Humus Chemistry, Genesis, Composition, Reactions. John Wiley & Sons, New York, NY.
Strickland, M.S., Lauber, C., Fierer, N., Bradford, M.A., 2009. Testing the functional significance of microbial community composition. Ecol. Lett. 90, 441–451.
Sutton, R., Sposito, G., 2005. Molecular structure in soil humic substances: the new view. Environ. Sci. Technol. 39, 9009–9015.
Tecon, R., Or, D., 2016. Bacterial flagellar motility on hydrated rough surfaces controlled by aqueous film thickness and connectedness. Sci. Rep. 6, 19409.
Trapp, S., Libonati Brock, A., Nowak, K.M., Kästner, M., 2017. Prediction of the formation of biogenic non-extractable residues during degradation of environmental chemicals from biomass yields. Environ. Sci. Technol., in revision.
Trivedi, P., Anderson, I.C., Singh, B.K., 2013. Microbial modulators of soil carbon storage: integrating genomic and metabolic knowledge for global prediction. Trends Microbiol. 21, 641–651.
Tscherko, D., Rustemeier, J., Richter, A., Wanek, W., Kandeler, E., 2003. Functional diversity of the soil microflora in primary succession across two glacier forelands in the Central Alps. Eur. J. Soil Sci. 54, 685–696.
van der Heijden, M.G.A., Bardgett, R.D., van Straalen, N.M., 2008. The unseen majority: soil microbes as drivers of plant diversity and productivity in terrestrial ecosystems. Ecol. Lett. 11, 296–310.

von Lützow, M., Kögel-Knabner, I., Ekschmitt, K., Flessa, H., Guggenberger, G., Matzner, E., et al., 2007. SOM fractionation methods: relevance to functional pools and to stabilization mechanisms. Soil Biol. Biochem. 39, 2183–2207.

von Lützow, M., Kögel-Knabner, I., Ekschmitt, K., Matzner, E., Guggenberger, G., Marschner, B., Flessa, H., 2006. Stabilization of organic matter in temperate soils: mechanisms and their relevance under different soil conditions: a review. Eur. J. Soil Sci. 57, 426–445.

von Lützow, M., Kögel-Knabner, I., Ludwig, B., Matzner, E., Flessa, H., Ekschmitt, K., et al., 2008. Stabilization mechanisms of organic matter in four temperate soils: development and application of a conceptual model. J. Plant Nutr. Soil Sci. 171, 111–124.

Vrede, K., Heldal, M., Norland, S., Bratbak, G., 2002. Elemental composition (C, N, P) and cell volume of exponentially growing and nutrient-limited bacterioplankton. Appl. Environ. Microb. 68, 2965–2971.

Waksman, S.A., 1925. What is humus? P. Natl. Acad. Sci. USA 11, 463–468.

Wieder, W.R., Allison, S.D., Davidson, E.A., Georgiou, K., Hararuk, O., He, Y., et al., 2015. Explicitly representing soil microbial processes in earth system models. Glob. Biogeochem. Cyc. 29, 1782–1800.

Wieder, W.R., Bonan, G.B., Allison, S.D., 2013. Global soil carbon projections are improved by modelling microbial processes. Nat. Clim. Change 3, 909–912, Reid.

Wieder, W.R., Grandy, A.S., Kallenbach, C.M., Bonan, G.B., 2014. Integrating microbial physiology and physio-chemical principles in soils with the Microbial-Mineral Carbon Stabilization (MIMICS) model. Biogeosciences 11, 3899–3917.

Woche, S.K., Goebel, M.-O., Mikutta, R., Schurig, C., Kaestner, M., Guggenberger, G., et al., 2017. Decreasing soil wettability can be explained by surface elemental composition: an XPS study. Sci. Rep. 7, 42877.

Worrich, A., Stryhanyuk, H., Musat, N., König, S., Banitz, T., Centler, F., et al., 2017. Mycelium-mediated transfer of water and nutrients stimulates bacteria activity in dry and oligotrophic environments. Nat. Commun. 8, 15472.

Young, I.M., Crawford, J.W., 2004. Interactions and self-organization in the soil-microbe complex. Science 304, 1634–1637.

Zang, H., Wang, J., Kuzyakov, Y., 2016. N fertilization decreases soil organic matter decomposition in the rhizosphere. Appl. Soil Ecol. 108, 47–53.

CHAPTER 6

Microbial Control of Soil Carbon Turnover

Yilu Xu*, Balaji Seshadri*, Binoy Sarkar, Cornelia Rumpel†, Donald Sparks‡, Nanthi S. Bolan***
*University of Newcastle, Callaghan, NSW, Australia
**The University of Sheffield, Sheffield, United Kingdom
†Institute of Ecology and Environment Paris, Thiverval-Grignon, France
‡University of Delaware, Newark, DE, United States

Chapter Outline

INTRODUCTION

The terrestrial biotic C accounts for c. 37.33% of total soil organic C content (up to 1 m belowground) (Guo and Gifford, 2002; Jobbágy and Jackson, 2000; Stockmann et al., 2013). As such, the mineralization of terrestrial C accelerates greenhouse gas (CO_2) accumulation in the atmosphere (Houghton, 2007; Lal, 2013). Terrestrial C sequestration can partially compensate CO_2 accumulation caused by anthropogenic activities, such as deforestation and extensive land management practices (1.6 Pg C year^{-1}, Lal, 2010), land-use changes (1.29 Pg C year^{-1}, Nakicenovic et al., 2000), fossil fuel combustion and industrial activities in general (8.0 Pg C year^{-1}, Lal, 2010). Photosynthesis and heterotrophic microbial respiration are the two major processes regulating terrestrial C balance. Therefore, soil C sink can be increased either by increasing organic C inputs or reducing litter and soil organic matter breakdown (Luo et al., 2003). Jastrow et al. (2007)

The Future of Soil Carbon
http://dx.doi.org/10.1016/B978-0-12-811687-6.00006-7

proposed the "residence time" (RT) as a decisive measurement of soil C storage capacity. Soil microorganisms are the driving factors behind soil organic C dynamics and decomposition, and soil microbial properties (such as microbial activity, microbial biomass, and microbial diversity) are recognized as sensitive indicators of soil health and quality (Nannipieri et al., 2003) (Fig. 6.1). There are considerable gaps in understanding soil microbiota and its impact on the biological mechanisms regulating C exchange between atmospheric, terrestrial, and aquatic reservoirs, and the way microbial communities interact with climate perturbation and environmental factors to affect soil C dynamics (Bardgett et al., 2008; Ganie et al., 2016; Gong et al., 2009). As such, ecosystem research that couples approaches ranging from macroscale to molecular scale interactions is necessary to understand the role of specific soil microorganisms and mechanisms contributing to C storage (King, 2011). In this chapter, we highlight the recent research and findings about microbial mediating soil C dynamic, especially under environmental parameter variation. We also address certain microbial related C degradation features, such as priming effect (PE), and modern techniques to detecting microbiota and modeling strategy. We then evaluate research focusing on how soil microbiota regulates C loading and organic matter decomposition,

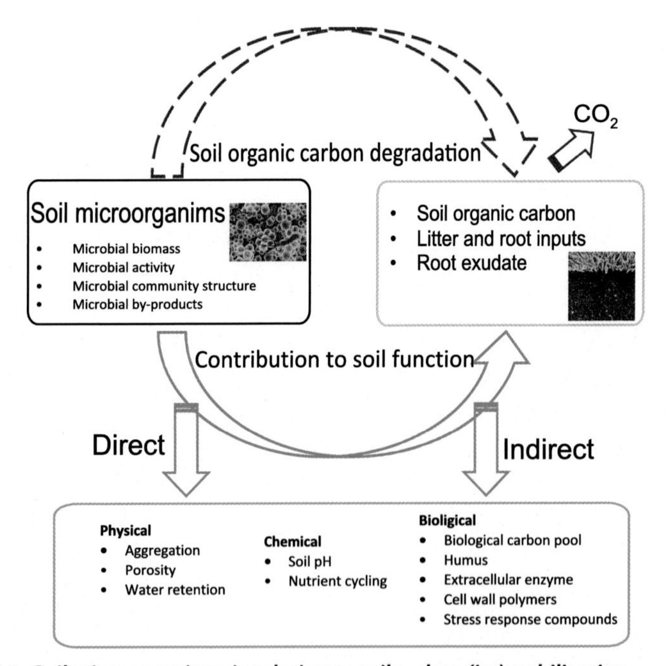

Figure 6.1 ***Soil microorganisms in relation to soil carbon (im)mobilization.***

studying their implications for soil C sequestration and long-term management. We also suggest some future directions in these research areas.

SOIL MICROORGANISMS IN RELATION TO C CYCLING PROCESSES

Soil C Pools and Microbial Accessibility

C turnover rate after addition of litter or exogenous organic matter depends on several factors, including the chemical and physical properties of the materials, their capacity for adsorption on soil minerals, and their suitability for microbial degradation (Kuzyakov et al., 2003). There are three major routes for plants to transfer organic C into soil systems: (1) aboveground plant litter and leachates (i.e., dissolved organic C washed by infiltrating rainfall), (2) below ground root exudates, and (3) dead roots (Brimecombe et al., 2007; Six et al., 2006). Such materials originating from root cells and tissues, mucilages, volatiles, and soluble lysates and exudates (releasing from live and/or damaged cells) collectively form the rhizodeposits in due course of time (Dennis et al., 2010; Gougoulias et al., 2014). Although the relative magnitude of organic input is time- and location-specific, the C-containing compounds can be separated into two categories depending on their molecular structures. In the first category, there would be low molecular weight organics, such as sugar, amino acids, alcohols, organic acids, and oligosaccharides, while the second category includes polymers like starch, lignin, cellulose, and hemicellulose (Gougoulias et al., 2014). Low molecular weight organics can be directly used by microbes within seconds to hours, while high molecular weight compounds need to be hydrolyzed before microbial take up (Jones and Darrah, 1994). As such, soil organic C consists of a continuum of organic matter compounds ranging from labile to recalcitrant C pools. The various labile C fractions are key indicators of soil productivity and health. Generally, the labile C pool decreases right after entering into the soil due to diffusion and convection processes (Raynaud, 2010), and is more easily degraded by microbes, leading to short mean residence time (MRT) in the range of days/weeks/months (Cheng et al., 2007; Ogle et al., 2012). Labile C compounds provide energy to soil biota, and is thus rapidly mineralized, if not stabilized through interactions with clay mineral surfaces, or physically protected within soil aggregates (Krull et al., 2003; McLauchlan and Hobbie, 2004). The chemically recalcitrant C pool consists of less decomposable organic matter consisting of aromatic or aliphatic compounds. Although chemically recalcitrant, this pool is also degraded by the soil microbial biomass, its degradation occurs comparatively slowly, and a part of this recalcitrant pool may survive in soils for decades contributing to soil properties, such as cation exchange capacity (CEC) (Bot and Benites, 2005).

Microbial activity is highly related to the availability of fresh organic C inputs in the soil surface (Fang and Moncrieff, 2005; Ghiorse and Wilson, 1988) and in deeper soil horizons (Sanaullah et al., 2016). The microbial biomass C pool includes the living microbial biomass C and the dead microbial biomass. Notably, the dead microorganisms become a readily available C source to the living microorganisms, and the necromasses can represent a huge amount of C in soil. As such, microbes and microbial products contribute in a large proportion to the stabilized organic matter pool (Miltner et al., 2012) with MRTs exceeding decades. It is well established that soil environment (e.g., soil texture, pH, moisture, and temperature) can influence microbial turnover rates of soil organic C (Davidson et al., 1998; Fierer and Jackson, 2006) (Table 6.1). Moreover, both soil physical properties (e.g., temperature and pore aeration) and organic C accessibility lead to different microbial C utilization patterns (Cotrufo et al., 2013). Ekschmitt et al. (2008) indicated that turnover rates of soil organic C depend not only on organic C concentration, but also on factors, such as substrate properties (e.g., substrate quality, composition, and initial C:N:P ratio), organic C distribution in the soil matrix and activity, biomass and composition of soil biota.

Soil Properties in Relation to Microorganisms

Among many other physical (e.g., texture, structure, moisture content, aeration level) and chemical (e.g., pH, electrical conductivity [EC], clay contents and types) soil properties, the CEC can reflect the overall soil fertility. It can interact with polycations, which can act as a bridge between soil aggregates and soil organic matter (Tisdall, 1996). Polyvalent cations, such as Ca^{2+}, Al^{3+}, and Fe^{3+}, can also lead to the formation of aggregates with organic C occluded in them. This interaction reduces the microbial degradation of soil organic C (Fig. 6.1). It is well established that the negative surface charge on clay particles increases by increasing soil pH (Bronick and Lal, 2005; Singh et al., 2017b). In variable charged soils, soil pH can significantly affect the adsorption of dissolved organic C as well as metals (Appel and Ma, 2002; Bolan et al., 1999; Singh et al., 2016, 2017a). Since pH affects activity and composition of microbial communities and also the charges of surface-reactive soil particles, changing soil pH may have a complex effect on microbial reactions and organic C turnover.

Table 6.1 Selected references on environmental factors and the influences on soil microorganisms

Factor of influence	Laboratory/field conditions	Treatments	Microbial activity and composition response	References
Soil temperature	Laboratory incubation	Temperature control	Microbial respiration increased with temperature	Wei et al. (2014)
	Laboratory column incubation	Temperature control	Emissions of CO_2 and CH_4 at 23°C were an average of 2.4 and 6.6 times larger, respectively, than those at 10°C	Moore and Dalva (1993)
	Laboratory chamber incubation	Temperature control	Microbial respiration increased with temperature up to 32°C	Fang and Moncrieff (2001)
Water content	United states		Microbial communities consistently, exhibited the uniform distribution pattern regardless of soil water content	Zhou et al. (2002)
	Laboratory column incubation	Water table position control	Static water table depths of 0, 10, 20, 40, and 60 cm, CO_2 emissions showed a positive, linear relation with depth, whereas CH_4 emissions revealed a negative, logarithmic relation with depth	Moore and Dalva (1993)
	Laboratory incubation	Drying- rewetting frequency	Microbial respiration decreased with drying-rewetting frequency	Fierer and Schimel (2002)
	Laboratory chamber incubation	Soil moisture control	Microbial respiration response to soil moisture content between 20 and 50 vol.% is not so obvious	Fang and Moncrieff (2001)

(Continued)

Table 6.1 Selected references on environmental factors and the influences on soil microorganisms (*cont.*)

Factor of influence	Laboratory/field conditions	Treatments	Microbial activity and composition response	References
Soil pH	United Kingdom	Amendments with lime for pH gradient	The composition of the bacterial communities was closely defined by soil pH, Fungal community composition was less strongly affected by pH than bacteria	Rousk et al. (2010)
	United States		Soil pH as predictor of soil bacteria community, the overall phylogenetic diversity of the bacterial communities was at peak diversity in soils with near-neutral pH	Lauber et al. (2009)
	United Kingdom	pH control	Archaeal amoA gene and transcript abundance decreased with increasing soil pH, while bacterial amoA gene abundance was generally lower and transcripts increased with increasing pH	Nicol et al. (2008)
	Germany		Fungal/bacterial biomass index increased slightly with increasing pH, and microbial community was to a large extent determined on pH	Bååth and Anderson (2003)
Organic carbon content	United States		Microbial communities in low-carbon surface soils showed remarkably uniform distributions, and all species were equally abundant	Zhou et al. (2002)
Soil heavy metal content	Laboratory incubation	Amendments with heavy metals	Heavy metal pollution severely decreases the functional diversity of the soil microbial community and impairs specific pathways of nutrient cycling	Kandeler et al. (1996)

Microbial Adaptation to Elevated Temperature

Elevated temperature can stimulate microbial activity, and it is thus expected that soil organic C decreases upon global warming (Kirschbaum, 2004). However, this hypothesis was derived from laboratory incubation experiments with disturbed soil samples. Meanwhile, the higher CO_2 concentration may improve plant productivity and alter nutrient and organic matter inputs into the soil. Incorporation of microbial properties (composition, activity, and biomass) into C dynamics models is a prerequisite to build up a theoretical framework to understand microbial mechanisms of mediating global C cycle. However, because of the variations of the ecosystems and models, as well as curve fitting techniques, feedback of soil microbiota to temperature is a major uncertainty (Davidson and Janssens, 2006; Knorr et al., 2005). The other problem is that the experiment length may blur the results and eliminate the microbial resilience in response to environmental stress. A hypothesis on microbial acclimation to global warming suggests that soil respiration (mainly microbial respiration) keeps dynamically changing to the seasonal change in temperature (Fenner et al., 2005). The initial stimulation of soil CO_2 efflux in the early stage could be only transitory due to the depletion of labile C pool (Eliasson et al., 2005; Kirschbaum, 2004). However, some responses of stable C pools could also be observed at this stage (Lefèvre et al., 2014; Leifeld and Fuhrer, 2005). Hartley et al. (2007) proposed that the substrate availability affects the response of microbial heterotrophic respiration to warming. Therefore, the heterogeneity of substrate C also needs to be considered when investigating microbial regulating soil C and simulating models. Fierer et al. (2005) reported that when the organic C content in litter declined, decomposition was more sensitive to temperature change. In another study, Knorr et al. (2005) demonstrated that nonlabile organic C was more sensitive to temperature change than labile organic C. Interestingly, Pries et al. (2017) argued that there was no significant differentiation in microorganisms in terms of accessing different soil organic matter after 2½ years of warming. This indicated the potential C release due to subsoil depletion in the background of increasing temperature. The composition of soil microbial communities may also change with temperature variation (Dinh et al., 2014; Zogg et al., 1997). Bardgett et al. (1999) found a fast growth of Gram-positive bacteria in the first succession stages under an elevated atmospheric temperature, but no direct effect on slow growing microorganisms, such as fungi and actinomycetes. Although the bibliography on soil heterotrophic respiration is extensive, only a few models include the overall response of soil microbial activity to climate change. In this context, Allison et al. (2010) calculated the C use efficacy (CUE), defined as the organic C incorporated into microbial biomass, to better simulate the process of soil-organic C response to global warming. Attempts to quantify the effects of natural climate gradients on soil C and microbial activity based on laboratory and field experiments can often be vulnerable to uncertainty. This is because soil organic C pool consists of a variety of C compounds rather than having a uniform composition (Qafoku, 2015). Future modeling research should therefore address microbial processes more explicitly, for example, by including PE, enzymatic activities, and C use efficiency under the changing climate, and incorporate them into the C budget calculations.

Soil Microbial Activity in Response to Drought and Rewetting Cycle

Soil moisture deficit or surplus can cause soil physical and chemical changes, such as gas and solution diffusion, water retention, and pore space alteration, and these changes can affect soil microbial properties. Microbial respiration may reach a plateau at an optimum moisture level because of the diffusion limitation of substrates with the consequently reduced substrate availability to microbes (Moyano et al., 2013). In addition, there may be a "Birch Effect" occurring, which refers to the flush of soil CO_2 production because of a sudden rewetting after a long period of drought (Xiang et al., 2008). A long period of drought leads to inhibitory affect in soil microorganisms, while the rewetting increase microbial activity with sufficient organic accumulation during the whole drought period. Thus, there can be a close connection between increased water movement and simulated microbial activity/organic C depletion. Especially in the arid and semi-arid ecosystems, because of microbial sensitivity to water impulse, the stimulation of soil organic C decomposition by a sudden precipitation can lead to a cascade of responses (Carbone et al., 2011). Miller et al. (2005) reported that there was a 60% increase in CO_2 efflux due to a drought-rewetting event compared to soils at constant and optimal water content. In the models reported by Yuste et al. (2005) and Li et al. (2006), the rewetting pulse alone accounted for 10%−14% of the annual CO_2 fluxes.

In order to understand how soil microorganisms respond to drought stress and rewetting, it is essential to study how the increased intracellular concentration of solutes may prevent microbial dehydration. Due to the existence of different C allocation strategies among microbial species (i.e., different CUE), bacteria tend to produce amino compounds, such as proline, glutamine, and glycine betaine (Csonka, 1989), while fungi generate more polyols, such as glycerol, erythritol, and mannitol (Witteveen and Visser, 1995). Fierer and Schimel (2003) proposed that the initial production of CO_2 after rewetting may be a result of the mineralization of cytoplasmic solutes, which maintains the equilibrium between intracellular and soil water potentials, following the depletion of easily available organic C. However, the physiological processes required for this phenomenon in both bacteria and fungi can be energy intensive. Killham and Firestone (1984) estimated that osmotic stress can reduce more than a half microbial CUE compared to unstressed conditions. Multiple drought and rewetting cycles thus may strongly influence soil C flows. The magnitude of CO_2 release on rewetting varies between 7% and 20% of the size of microbial biomasses (Killham and Firestone, 1984). Major concerns about the consequences of repeating drought followed by rewetting are: (1) sudden big flux of CO_2 emitting from soil, and (2) uncertainty about the recovery of microbial biomasses following the rewetting (Fierer and Schimel, 2003). Schimel et al. (2007) found it difficult to conclude whether soil drought and rewetting slowed down or accelerated microbial decomposition of organic C, but suggested that this stress could shift the microbial allocation of degradable C substrates. This hypothesis was supported by the work of Xiang et al. (2008), who showed an increasing microbial use of substrates upon drying–wetting cycles related to physical release of protected C. In conclusion, there is a need of integrating soil drought-rewetting processes to both microorganism physiology and community composition as kinetic components regarding soil C dynamics and model development.

MICROBIAL COMMUNITY COMPOSITION AND SOIL C DYNAMICS

Because of the complex interactions among soil microorganisms and environmental factors, identification of microbial parameters that are primarily responsible for soil C dynamics is challenging. However, the succession of microbial communities and their structural change may be significantly related to the changes occurring during soil organic matter decomposition (Bai et al., 2016; Sanaullah et al., 2016). This may explain even more than 2/3 of soil C reactions (Louis et al., 2016). Fungi and bacteria are two major groups of microorganisms responsible for soil C decomposition. These groups thus significantly relate with most of the soil quality parameters, and are responsible for more than 90% of the soil functions (McGuire and Treseder, 2010). However, there are significant differences between fungi and bacteria in terms of C use preferences. The general understanding is that, microbial C use is related to the ratio of soil bacterial to fungal populations due to the difference in their C utilization abilities. Fungi favor acidic pH environment, while bacterial cells prefer neutral or slightly alkaline conditions (Ganie et al., 2016; Rousk et al., 2009). Compared to bacteria, fungi promote condensation type biochemical reactions, and can more easily grow on complex substrate sources, such as cellulose, tannin-protein, and lignitic materials (Hanson et al., 2008). This leads to the trait differences of extracellular enzymes they produce in order to decompose target C substrates. For example, fungi produce more phenoloxidases, laccases, and peroxidases than bacteria. On the other hand, bacteria produce more lipases and cellulases than fungi (Jastrow et al., 2007). Although both fungi and bacteria can degrade saccharides, it has been proved that when the colonization of fresh litter happens, development of fungal population is more rapid than bacteria (Gessner and Chauvet, 1994; Hieber and Gessner, 2002). In addition, fungi may slightly modify the soil structure through the production of glomalin or glomalin-related soil proteins that act as a "glue" for soil aggregate formation (Rillig and Mummey, 2006). In addition to the habitat differences between fungal and bacterial species, Fierer et al. (2003) studied the relationship between C sources and microorganisms, and suggested that the dominant microbial species might change along the soil profil due to variation in organic C availability. Ekschmitt et al. (2008) reported that fungi were located mostly in the upper 20 cm of soil profile, while bacterial habitats could be extended to more than 1.4 m soil depth. In addition, Gram-positive bacteria and actinomycetes tend to inhabit deeper soil layers with limited C availability. Conversely, Gram-negative bacteria, fungi, and protozoa have higher abundance in the surface soil. These differences consequently affect C allocation pathway and soil C dynamics. A model of Fontaine and Barot (2005) showed that the soil C pool is highly related to microbial community size and composition, and would continuously increase if nutrients were sufficient before reaching a threshold.

When soil microbiota faces environmental disadvantage, drought, for example, physiological modification is their primary short-term acclimation mechanism, while the shift of microbial community structure is generally the response to a long-term stress (Schimel et al., 2007). A significant research effort has been directed to understand microbial community composition shifts in relation to environmental stresses. There have been numerical results in terms of microbial species response to drought-rewetting cycles (Göransson et al., 2013). Fungi and bacteria generally have different preferences in decomposing various organic matter compounds, and play specific functional roles in the ecosystem. Soil moisture stress could result in different substrate depletion and nutrient utilization patterns by microorganisms. Schimel et al. (2007) found Gram-positive bacteria and fungi more drought-tolerant than Gram-negative bacterial because of their enhanced biochemical ability of decompose soil organic C. Williams (2007) and Gordon et al. (2008) reported bacteria were more dominant in the rewetting cycle than fungi. However, there were reports indicating almost no difference among microbial groups due to rewetting (Butterly et al., 2009; Hamer et al., 2007). As such, drought stress might not always be related to the equilibrium of microbial community shift (Balser and Firestone, 2005). The magnitude of fungal activity is influenced by environmental parameters, such as temperature and root development patterns of plants (Bell et al., 2009; Vandenkoornhuyse et al., 2002). Moreover, fungal function and community structure are both closely dependent on soil chemical parameters. The variation of fungi population due to living condition changes will affect soil C degradation in return.

In addition, microorganisms find a natural balance between ecological selection and evolutionary adaptation in order to survive and bounce back from environmental stresses (Schimel et al., 2007). Categorized by ecological strategy, soil microbial community can be generalized into aggregated groups that share similar ecological function traits (Blagodatskaya et al., 2004; Chen et al., 2016). K-strategies (oligotrops) have higher C use efficiency yet slow growth rate in contrast to r-strategist that grow fast in rich substrate conditions with lower C use efficiency. Blagodatskaya et al. (2004) suggested that microbial community shift to species with better stress tolerance, known as K-strategies, rather than to r-strategy species. However, there is limited research trying to explain drought-rewetting scenario with microbial r-K growth strategy. Moreover, the relative influence of drought-rewetting on microbial CUE are highly site-dependent, especially in the context of changes in long-term soil C pools (Balogh et al., 2011).

The investigation of soil microfunctions in relation to microbial community composition is more challenging under field conditions than in laboratory-scale microcosm experiments because the former involve highly complex and unpredictable environmental factors. Modern molecular biology–based techniques are powerful in investigating the complex microbial community compositions under such situations (Table 6.2). Techniques, such as cultivation (plating) and phospholipid fatty acid (PLFA) analysis, can be used to study the abundance of broad/large microbial

Table 6.2 Techniques to identify microbial community structure

Analyzing communities approaches	Advantages	Current disadvantages	References
Plate counts	Fast, inexpensive	Unculturable microorganisms not detected; bias towards fast growing individuals and fungal species with large quantities of spores	Kirk et al. (2004)
Community level physiological profiling (CLPP)/ Sole-carbon-source utilization (SCSU) pattern	Fast, highly reproducible, relatively inexpensive; option of using bacterial, fungal plates, or site specific carbon sources	Only represents culturable fraction of community; favors fast growing organisms; only represents those organisms capable of utilizing available carbon sources and metabolic diversity instead of in situ diversity; sensitive to inoculum density	Kirk et al. (2004)
Cell membrane lipid fatty acids (Phospholipid fatty acids-PLFA)	Detect changes in microbial composition; direct extraction from soil; culturing of microorganisms is not required; distinguish broad taxonomic groups (such as Gram-negative and positive bacteria, actinomycetes, fungi, etc., and some narrow taxa, such as methanotrophic bacteria)	Difficult in connecting specific microbial populations with ecosystem dynamics; more material is required when fungal spores are used; can be influenced and confused by external factors or other microorganisms	Frostegård et al. (1993); Schimel and Gulledge (1998); Kirk et al. (2004)
Substrate-use patterns and kinetics	Detect changes in bacteria composition in response to environment changes	Difficult in relating to in situ bacterial community function	Lidstrom (1996); Schimel and Gulledge (1998); Kirk et al. (2004)

Mol.% guanine plus cytosine (GC content)	Quantitative techniques; not influenced by Polymerase chain reaction (PCR) biases; includes all DNA extracted and rare microbial member	Requires large quantities of DNA, coarse level of resolution; dependent on lysing and extraction efficiency	Kirk et al. (2004)
Nucleic acid re-association and hybridization technique	In situ specific study with total DNA extracted or RNA; not influenced by PCR biases	Lack of sensitivity; sequences need to be in high copy number for detection, dependent on lysing and extraction efficiency	Amann et al. (1995); Holmes et al. (1995); Teske et al. (1996); Schimel and Gulledge (1998); Kirkby et al. (2013)
DNA microarrays and DNA hybridization	Same but more specified than nucleic acid hybridization; reproducible; thousands of genes can be analyzed	Only detect the most abundant species; need to culture organisms; only accurate in low diversity systems	Kirk et al. (2004)
Denaturing and temperature gradient gel electrophoresis (DGGE and TGGE)	Reliable, reproducible, and repaid to specific genes; large number of samples can be analyzed simultaneously	Medium resolution; does not provide present species identification; PCR biases; influenced by lying and extraction efficiency, and sampling; one band can represent more than one species (co-migration); only detects dominate species	Kirk et al. (2004)
Single strand conformation polymorphism (SSCP)	Same as DGGE/TGGE; no GC clamp or gradient	PCR biases; some ssDNA can form more than one stable conformation	Kirk et al. (2004)

(Continued)

Table 6.2 Techniques to identify microbial community structure (*cont.*)

Analyzing communities approaches	Advantages	Current disadvantages	References
qPCR	Quantitative rRNA/rDNA ratios; detect active members of microbial community; can be used in a specific gene as well as a broad scale	Influenced by extraction, lysing efficiency, and choice of restriction; type of *Taq* can increase variability	Teske et al. (1996); Schimel and Gulledge (1998)
Restriction fragment length polymorphism (RFLP)	Detect structural changes in microbial community	PCR biases; banding patterns often too complex	Kirk et al. (2004)
Terminal restrict fragment length polymorphism (T-RFLP)	Simpler banding patterns than RFLP; reproducible, automated; large number of samples can be analyzed; can compare microbial communities differences	PCR biases; influenced by extraction; lysing efficiency and choice of restriction; type of *Taq* can increase variability	Kirk et al. (2004)
Ribosomal intergenic spacer analysis (RISA)/automated ribosomal intergenic spacer analysis (ARISA)/amplified ribosomal DNA restriction analysis (ARDRA)	Highly reproducible community profiles	PCR biases; required large sample of DNA (for RISA)	Kirk et al. (2004)
Pyrosequencing/ metagenomics/	Reproducible, reliable, and rapid; large number of samples can be analyzed to detect microbial species in situ	Very expensive; datasets can be very big	Lane et al. (1985); Hugenholtz et al. (1998)

groups (e.g., bacteria and fungi) (Torsvik and Øvreås, 2002). However, polymerase chain reaction (PCR)/quantitative polymerase chain reaction (qPCR) and genetic fingerprinting techniques, such as denaturant gradient gel electrophoresis (DGGE), amplified rDNA restriction analysis (ARDRA), terminal restriction fragment length polymorphism (T-RFLP), and ribosomal intergenic spacer analysis (RISA), should preferably be used to obtain precise and in-depth information of microbial community structures under complex soil environmental conditions. Moreover, a combination of two or more of these techniques provides an overview of the dominant soil microbiota. However, due to the limitation of specifically targeted microbial communities (<1000), some microbial species may not be able to be accessed or detected (Van Elsas and Boersma, 2011). Other techniques, such as stable isotopic probing and brdU methods are suitable for detecting active communities in situ, and microarrays can provide information microorganism diversity at phylogenetic or functional levels (Van Elsas and Boersma, 2011). Molecular technique supports the study of microbial C use beyond the concept as organic decomposer. Since the particular functional genes may tend to cluster in the genome, and the microbial dispersal limitation may constrain microbial recolonization, bioengineering approaches by using modified bacteria have also been attempted to enhance CO_2–C sequestration in soils. Some promising bioengineering approaches include the formation of dolomite [$CaMg(CO_3)_2$] by genetically modified bacteria in certain hypersaline ecosystems (King, 2011), and incorporation of C into membranes of modified living microorganisms through small macrocyclic Cate receptors (Brooks et al., 2006; Tossell, 2009). Whitman et al. (1998) predicted that 1.59×10^{31} bioengineered bacterial could potentially sequester around 1 Gt of C, which is about 12% of global annual anthropogenic CO_2 emission.

The earlier discussions indicate that soil abiotic conditions, fresh organic matter additions, and other anthropogenic interferences may directly or indirectly alter the microbial community composition with possible effects on C dynamics in soils. Anthropogenic disturbances, such as tillage and grazing may trigger higher C loss from soils due to their influence on physical soil parameters and also the soil microbial community structure (Ingram et al., 2008; Jackson et al., 2003). Other factors, like fresh organic matter inputs, may also alter the microbial community structure by influencing the C utilization patterns (either indigenous or added C) of microorganisms. For instance, >3.6 mg C g^{-1} as root exudates resulted in a negative PE in soils, and significantly shifted the fungal:bacterial population ratio (De Graaff et al., 2010). Schutter and Dick (2001) also reported that microbial richness was positively related to the substrate utilization potentials of the soil microorganisms. Anderson et al. (2011), by using the TRFLP fingerprint techniques, found that certain microbial species that were related to recalcitrant C decomposition were promoted in the presence of biochar, which was followed by an indigenous organic C depletion. Yu et al. (2005) in a study on the effects of invading plant species, suggested that exotic weeds could replace the native plants by changing the native soil microbial community and their interactions with the plants. This ultimately would alter the C dynamics in soils, too. In addition, environmental stresses caused by heavy metal(loid)s may lead to reduction in the microbial community diversity and activity in soils (Wang et al., 2007). However, the attempt of using soil microflora and microbial activity as indicators of soil heavy metal(loid)s bioavailability should underscore the sensitivity of microorganisms to the concerned toxicity (Bolan et al., 2010). Despite the uncertainties in parameter calibrations and elicit assumptions, an increased number of researches have recently begun to incorporate microbial community shifts in modeling and simulation studies (Sinsabaugh et al., 2013). For example, in light of the concept of "multiple C pools" models in Wu et al. (2013), they divided terrestrial C pools into eight parts, including 2 biotic C pools as surface microbial and soil microbial pools. This was in light of the fact that microbial population developed different features under different substrate (e.g., easily decomposable or recalcitrant C) availabilities and their decomposition stages (e.g., residence age) (Berg and McClaugherty, 2008). However, Louis et al. (2016) suggested that a universal applicability and predication through these models may be challenging, mainly because: (1) fixed parameters in these models are suitable in certain time and space, and (2) there are many discrepancy for species and functional diversity of microorganisms. McGuire and Treseder (2010) proposed several mechanisms that could influence the incorporation of microbial community diversity into model settings.

Although there has been a debate around the old microbiological citation "Everything is everywhere, but, the environment selects" from Baas Becking (1934) and De Wit and Bouvier (2006), microbial diversity is still critical in terms of supporting soil stability and resilience ability (Griffiths and Philippot, 2013), and it is highly related to soil physiochemical properties and function, and consequently influences organic C decomposition (Baumann et al., 2013).

PRIMING EFFECT

PE of organic matter mineralization refers to the enhancement or retardation of organic matter decomposition as a result of the addition of fresh C substrates into the soil. Most of the previous studies reported on PE were laboratory incubation–based experiments with significant manipulation of the real environmental conditions. Nevertheless, there is a slowly growing consensus in considering PE as a natural process, which is induced by a pulse or continuous input of fresh organic matter (Kuzyakov, 2010). Rhizosphere and detritusphere are considered as the hot spots for PE because these regions are more concentrated in labile C than bulk soils due to the occurrence of root exudates and leaf litters (Kuzyakov, 2010). Continuous organic matter inputs in these areas enable a higher microbial biomass and C turnover rate than in the bulk soil. Hütsch et al. (2002) reported that 64%–86% of the "root-borne C" was rapidly used by soil microbes, while 2%–5% were incorporated to the soil matrix and protected from future degradation. It is commonly believed that fresh organic matter addition leads to two major phenomena: (1) enhancement of microbial activity, and (2) uneven growth of microbial populations (Fontaine et al., 2003). Fresh organic matter inputs induce microbial community succession (Nottingham et al., 2009). Certain microbial species can be triggered by a dramatic increase in populations based on the characteristics of the fresh C supply, which indicate a strong C substrate preference for soil microbes (Cardon et al., 2001; Hagedorn et al., 2003; Nottingham et al., 2009). If the previously starving microbial species are able to use the new C supply, their population composition is supposed to experience significant growth. The r- and K-strategy theory provides information to understand the microbial succession with organic substrate alteration. Generally, the r-strategy microbes are more adapted to fresh and easily available organic compounds under nutrient-favorable conditions, while the K-strategy species are expected to decompose polymerized compounds at a late succession period (Blagodatskaya et al., 2009). The growth pattern difference is due to the different energy allocation strategy, leading to the C use discrepancy (Fig. 6.2). The r-strategists prefer a population expansion, while K-strategists focus on extracellular enzyme production and defense from predation (Fontaine et al., 2003; Tate, 1995). Kuzyakov (2010) termed the CO_2 released during the first 3 days after fresh organic amendment input as apparent PEs, and suggested that this CO_2 flux indicated the accelerated microbial turnover rate or C pool alteration. The value of real PE was relatively lower than the apparent PE, which increased with the incubation period and possessed a

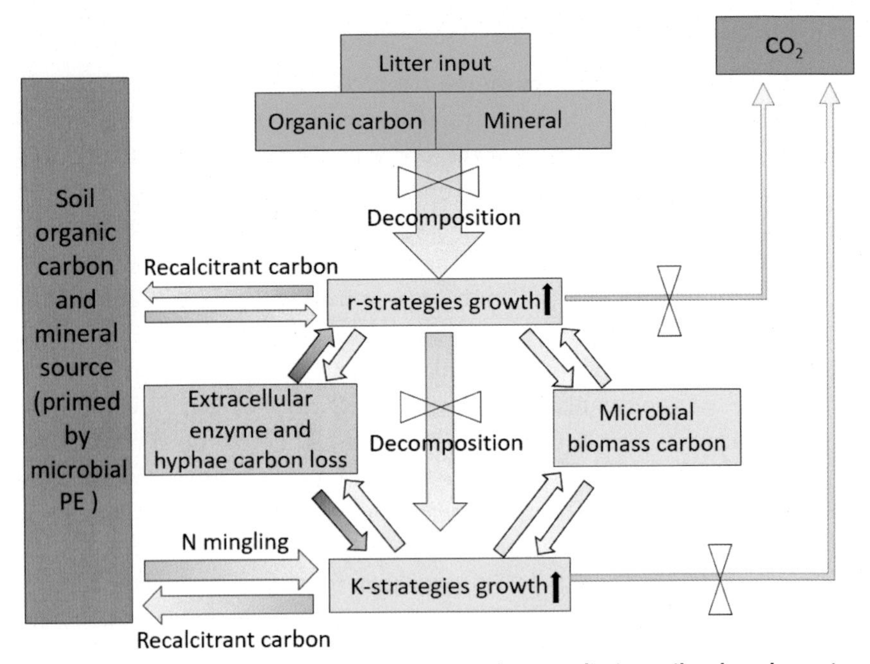

Figure 6.2 ***Concept r- and K-strategy microorganisms mediating soil carbon dynamics.*** The *arrows* indicate the carbon flow direction and volume (by *arrow bar* thickness). The sum of extracellular enzyme and hyphae carbon loss and microbial biomass carbon is the carbon sequestrated by microorganisms.

large portion of the total CO_2 flux in the later incubation stage (Blagodatsky et al., 2010; Kuzyakov, 2010). As such, soil microorganisms with fast growth rate, that is, r-strategists, are the major response for the apparent PE, while the later stage microbes, that is, K-strategists, produce real PE. Fontaine et al. (2003) suggested that the K-strategists play a more important role in maintaining nutrient availability and fertility of soils because their extensive growth of mycelia enables quick release of N from soil organic matter. Microorganisms regulate the direction and magnitude of PE, and the microbial community composition along with microbial abundance have essential influences on PE (Blagodatskaya and Kuzyakov, 2008). The biochemical properties of freshly applied organic matter alter the microbial population and biomass, consequently controlling the variation of PE and CO_2 production. Use of isotopically labeled C compounds to separate CO_2 released from added C source and native organic C is an essential technique to distinguish between real PE and apparent PE.

Most of the studies on PE showed a positive PE after fresh organic input, meaning an enhancement of soil organic C degradation (Cheng, 2009; Zhu et al., 2014). Fontaine et al. (2004) reported that soil humus was depleted by 174 mg C kg^{-1} because of cellulose application. As a result of positive PE, a decrease in soil mineral nutrient contents can be expected due to the stimulated microbial activity, which in turn may affect soil nutrient availability (Fontaine and Barot, 2005). However, negative PE and no obvious C mineralization were also observed in biochar amended soils (Liang et al., 2010; Zimmerman et al., 2011). The quantity of organic inputs may also mediate the PE direction. De Graaff et al. (2010) found that negative PE occurred at a higher liable C input rate (>3.6 mg C g^{-1}), while positive PE took place at a lower C input rate (0.7 mg C g^{-1} soil). The characteristics of fresh organic materials, such as polymerization and C/N ratio (Brant et al., 2006; Khalil et al., 2005), and the inherent soil factors, such as pH and soil type (Khalil et al., 2005) may have significant impacts on microbial population size and composition, and in turn on the magnitude of the PE. In addition, because specific enzymes are involved in the degradation of certain organic substrates, the enzyme type, quantity, and activity in soils might significantly change with the congruent substrate depletion processes (Fontaine et al., 2003). Fontaine et al. (2004) demonstrated that specific enzymes were produced with cellulose induced C decomposition, and this was the primary driver of positive PE in the system.

The plant species affect PE by changing the litter input quality and quantity, as well as root exudates and phenology (Pascault et al., 2013). An intensified PE was observed in an increased forest litter fall accompanied with rising temperature (Sayer et al., 2011), leading to significant soil organic C decomposition (Zhu and Cheng, 2011). Chen et al. (2014) suggested that the fresh organic C inputs served as a "primer" for native C decomposition, while the PE was intensely regulated by mineral nitrogen supply. Crow et al. (2009) found a low PE and slow C turnover rates in soils with high nutrient availability. Similarly, Fontaine et al. (2011) suggested that microbes tend to decompose more indigenous soil organic matter and release more nutrients with a high PE performance in soils with limited nutrient availability. With the application of pyrolyzed C, Lu et al. (2014) noticed a negative PE in a sandy loam soil. Fang et al. (2015) suggested that the pyrolysis temperature may have a high impact on the microbial degradation ability, for instance, the biochar produced at more than 550°C led to a greater local C stabilization than that produced at lower temperature. However, no universal consensus on the magnitude of PE nor a fully satisfied mechanism to explain the process yet exist in the literature (Fontaine et al., 2011; Kuzyakov et al., 2000).

In conclusion, fresh organic matter application may alter the degradation of native organic matter and microbiota, causing soil PEs. However, the variety of litter quality and soil physicochemical properties could lead to different PE. Soil microorganisms are the regulators of PE. Microbial community composition and characteristics, including metabolic activity and enzyme production, shift, depending on the fresh amendments. Although the "hot spot" area, such as rhizosphere is the most active place for PE, the input affects soil organic matter in deeper layer along soil profile. Fontaine et al. (2007) observed an increase in deeper organic C decomposition after fresh organic C application to soils. Other environmental factors, such as elevated soil temperature and frequently disturbed fields (i.e., tillage, drought) influence soil PE by controlling microbial functions. Future research attention should therefore be aimed at the following aspects in order to understand PE: (1) evaluating long-term PE effect with pulse or continuous supply of fresh organic matter to encompass the timing and changes of C supply, (2) incorporating PE with conditions in the real ecosystems, especially in a scenario of temperature, and (3) upscale to across ecosystem models; integrated models in the purpose of estimating global C turnover to evaluate directions and magnitude of C flow.

INCORPORATION OF MICROBIAL PARAMETERS INTO SOIL C MODELING

C was constantly transformed between organic and inorganic forms among biotic and abiotic C reservoirs, keeping a continuous flow. Soil microorganisms are the driving force for soil organic C decomposition as well as for biological C pool buildup in soils. These "double roles" of microorganisms makes the interpretation of soil microbial functions and mechanisms influencing C dynamics really challenging. It is noteworthy that most of the current established models are based on the concept of understanding soil as a potential sink or source of CO_2 for atmosphere (Manzoni and Porporato, 2009; Wieder et al., 2013). Because soil microbes are the main driving force for organic matter decomposition, including microbial communities into the models is necessary to improve their predictive capacity. However, incorporation of microorganisms and microbial features in the models also need to consider the feedback from environmental condition changes. Modeling the response of microorganisms to environmental changes is complicated due to the ecosystem diversity. This is one of the reasons why most of the models concerning C turnover and microorganisms are restricted to certain soil conditions or specific period (Lützow et al., 2006). To address that, long-term and large-scale field-based studies are required in the future. However, the challenge may still exist even if there is a "one fit all" model because the model can smudge results from environmental factors as they also function as indirect influence. In addition, scaling up (both in time and space) is also important in terms of ecological modeling, uncertainties, such as land histories and spatial variations, must be considered (Luo et al., 2010). Singh et al. (2010) suggested that both reductionist approach, for individual taxa or communities in response of each environmental variable and multifactorial approach, for trophic interactions are crucial in models interpreting ecological processes.

There are a variety of models regarding soil C dynamics including microorganisms as kinetic parameters. Miltner et al. (2012) developed a model based on the concept that the microbial residues are the source for the microbial biomass. This model explained and addressed soil microorganism not only as C decomposer, but also as a significant contributor to soil organic C formation. In addition, microbial species, especially fungi and bacteria represent most of the microbial functions in soil C degradation, and have been considered as important parameters in current models. Clemmensen et al. (2013) suggested the important role of fungi in the rhizosphere, and their model revealed a great contribution to a boreal forest from roots and roots-associated microorganisms. Resat et al. (2012) demonstrated that the bacterial community members are highly connected to microbial biochemical processes. Moreover, microbial function groups are also another important microbial parameter to be include while modeling. Waring et al. (2013) suggested that the functional groups can replace the broad bacterial and fungal diversity in terms of their different influences on soil C degradation. For instance, microbial enzymes and extracellular polysaccharides are the initial approaches that link microbial C use and soil functions (Schimel and Schaeffer, 2012).

In conclusion, the incorporation of soil microbial parameters into global C dynamic modeling can be powerful because of the information they can provide. However, the uncertainty about the accuracy of the models is challenging (Schimel, 2013).

CONCLUSIONS AND FUTURE RESEARCH DIRECTIONS

The earlier discussions suggest that while environmental parameters (e.g., temperature, moisture, pH, CEC, soil nutrient contents) significantly affect soil microbial activity and community composition, interactions between fresh and stabilized organic matter also influence the soil microbial parameters.

Soil organic C stability is largely influenced by the accessibility of the substrates to soil microbiota in different fractions. While modern molecular techniques (e.g., TGGE and 16sRNA) can be used to harness information on soil microbial taxa, the next step could be to answer how to link their functional and physiological capabilities with across the ecosystems and along environmental gradients. Manipulation of microorganisms using molecular and gene technologies have potentials to reveal the role of soil microorganism in terms of soil C sequestration, but significant future research is needed to understand the concerned mechanisms and also the risk of adding genetically modified microorganisms to the natural environment. Clearly, our fundamental knowledge of soil microorganisms is not qualified enough to give a mechanical explanation to present hitherto unknown microbial mediating soil C dynamics. Some progress in this regard can be made by focusing future research on the following aspects:

1. Introduction of soil measurement methods for a better understanding of soil functions in the perspective of microbiota, such as the function of soil structure as microbial habitat.
2. Modeling soil microsystem for integrating ecosystem and environmental gradients in a global level.
3. Understanding the functional implication of high diversity of microbial community and microbial succession stages in relation to C turnover.
4. Use of molecular techniques and identification on C-cycle related microbial functional genes and taxa.
5. Understanding enzyme types and activities in relation to soil organic matter decomposition in various microbe-plant systems.

Soil microorganisms are essential for the rejuvenation of C depleted soils. Microbiological models and research have already proved the ecological importance to protect microbial diversity and functions in terms of soil resilience and stability. However, there are still many questions that remain to be answered. For instance,

- Would global climate change affect microorganisms due to their spatial distribution?
- What will be the long-term effect of environmental stress on microorganisms?
- How far does the acclimation ability support microorganisms in enduring those stresses?
- Could soil microbiota be a reliable indicator in modeling or environmental assessment approaches?

Addressing the aforementioned research aspects will need interdisciplinary cooperation as well as modified measuring techniques. However, we hold high confidence for a better understanding of soil microbial features and their important functional roles in regulating the C dynamics, especially with the help of the emerging molecular analytical tools.

ACKNOWLEDGMENTS

Yilu Xu is thankful to the University of Newcastle and Department of Education and Training, Government of Australia, for awarding her PhD Scholarship. This research was partly supported by an Australian Research Council Discovery-Project (DP140100323).

REFERENCES

Allison, S.D., Wallenstei, M.D., Bradford, M.A., 2010. Soil-carbon response to warming dependent on microbial physiology. Nature Geosci. 3, 336–340.

Amann, R.I., Ludwig, W., Schleifer, K.-H., 1995. Phylogenetic identification and in situ detection of individual microbial cells without cultivation. Microbiol. Rev. 59, 143–169.

Anderson, C.R., Condron, L.M., Clough, T.J., Fiers, M., Stewart, A., Hill, R.A., Sherlock, R.R., 2011. Biochar induced soil microbial community change: Implications for biogeochemical cycling of carbon, nitrogen and phosphorus. Pedobiologia 54, 309–320.

Appel, C., Ma, L., 2002. Concentration, pH, and surface charge effects on cadmium and lead sorption in three tropical soils. J. Environ. Qual. 31, 581–589.

Baas Becking, L.G.M., 1934. Geobiologie of inleiding tot de milieukunde. W.P. Van Stockum & Zoon (Dutch), The Hague, the Netherlands.

Bååth, E., Anderson, T.-H., 2003. Comparison of soil fungal/bacterial ratios in a pH gradient using physiological and PLFA-based techniques. Soil Biol. Biochem. 35, 955–963.

Bai, Z., Liang, C., Bodé, S., Huygens, D., Boeckx, P., 2016. Phospholipid ^{13}C stable isotopic probing during decomposition of wheat residues. Appl. Soil Ecol. 98, 65–74.

Balogh, J., Pintér, K., Fóti, S., Cserhalmi, D., Papp, M., Nagy, Z., 2011. Dependence of soil respiration on soil moisture, clay content, soil organic matter, and CO_2 uptake in dry grasslands. Soil Biol. Biochem. 43, 1006–1013.

Balser, T.C., Firestone, M.K., 2005. Linking microbial community composition and soil processes in a California annual grassland and mixed-conifer forest. Biogeochemistry 73, 395–415.

Bardgett, R.D., Kandeler, E., Tscherko, D., Hobbs, P.J., Bezemer, T.M., Jones, T.H., Thompson, L.J., 1999. Below-ground microbial community development in a high temperature world. Oikos 85, 193–203.

Bardgett, R.D., Freeman, C., Ostle, N.J., 2008. Microbial contributions to climate change through carbon cycle feedbacks. ISME J. 2, 805–814.

Baumann, K., Dignac, M.-F., Rumpel, C., Bardoux, G., Sarr, A., Steffens, M., Maron, P.-A., 2013. Soil microbial diversity affects soil organic matter decomposition in a silty grassland soil. Biogeochemistry 114, 201–212.

Bell, C.W., Acosta, -Martinez, V., McIntyre, N.E., Cox, S., Tissue, D.T., Zak, J.C., 2009. Linking microbial community structure and function to seasonal differences in soil moisture and temperature in a Chihuahuan desert grassland. Microbial Ecol. 58, 827–842.

Berg, B., McClaugherty, C., 2008. Plant Litter: Decomposition, Humus Formation, Carbon Sequestration, second ed. Springer.

Blagodatskaya, E., Kuzyakov, Y., 2008. Mechanisms of real and apparent priming effects and their dependence on soil microbial biomass and community structure: critical review. Biol. Fert. Soils 45, 115–131.

Blagodatskaya, E., Ermolaev, A., Myakshina, T., 2004. Ecological strategies of soil microbial communities under plants of meadow ecosystems. Biol. Bull. Russ. Acad. Sci. 31, 620–627.

Blagodatskaya, E., Blagodatsky, S., Anderson, T.H., Kuzyakov, Y., 2009. Contrasting effects of glucose, living roots and maize straw on microbial growth kinetics and substrate availability in soil. Eur. J. Soil Sci. 60, 186–197.

Blagodatsky, S., Blagodatskaya, E., Yuyukina, T., Kuzyakov, Y., 2010. Model of apparent and real priming effects: linking microbial activity with soil organic matter decomposition. Soil Biol. Biochem. 42, 1275–1283.

Bolan, N.S., Naidu, R., Syers, J., Tillman, R., 1999. Surface charge and solute interactions in soils. Adv. Agron. 67, 87–140.

Bolan, N.S, Naidu, R., Choppala, G., Park, J., Mora, M.L., Budianta, D., Panneerselvam, P., 2010. Solute interactions in soils in relation to the bioavailability and environmental remediation of heavy metals and metalloids. International Symposium. Challenges to Soil Degradation Towards Sustaining Life and Environment (special issue, Tokyo Metropolitan University Symposium Series No. 2) 53, 1–18.

Bot, A., Benites, J., 2005. The importance of soil organic matter: key to drought-resistant soil and sustained food production. No. 80. Food and Agriculture Org.

Brant, J.B., Sulzman, E.W., Myrold, D.D., 2006. Microbial community utilization of added carbon substrates in response to long-term carbon input manipulation. Soil Biol. Biochem. 38, 2219–2232.

Brimecombe, M.J., De Leij, F., Lynch, J.M., 2007. Rhizodeposition and microbial populations. The Rhizosphere Biochemistry and Organic Susbstances at the Soil-Plant Interface. CRC Press, Boca Raton, Florida, 73-109.

Bronick, C.J., Lal, R., 2005. Soil structure and management: a review. Geoderma 124, 3–22.

Brooks, S.J., Gale, P.A., Light, M.E., 2006. Anion-binding modes in a macrocyclic amidourea. Chem. Commun. 41, 4344–4346.

Butterly, C., Bünemann, E., McNeill, A., Baldock, J., Marschner, P., 2009. Carbon pulses but not phosphorus pulses are related to decreases in microbial biomass during repeated drying and rewetting of soils. Soil Biol. Biochem. 41, 1406–1416.

Carbone, M.S., Still, C.J., Ambrose, A.R., Dawson, T.E., Williams, A.P., Boot, C.M., et al., 2011. Seasonal and episodic moisture controls on plant and microbial contributions to soil respiration. Oecologia 167, 265–278.

Cardon, Z., Hungate, B.A., Cambardella, C., Chapin, F., Field, C., Holland, E., Mooney, H., 2001. Contrasting effects of elevated CO_2 on old and new soil carbon pools. Soil Biol. Biochem. 33, 365–373.

Chen, R., Senbayram, M., Blagodatsky, S., Myachina, O., Dittert, K., Lin, X., et al., 2014. Soil C and N availability determine the priming effect: microbial N mining and stoichiometric decomposition theories. Global Change Biol. 20, 2356–2367.

Chen, Y., Chen, G., Robinson, D., Yang, Z., Guo, J., Xie, J., et al., 2016. Large amounts of easily decompostion carbon stored in subtropical forest subsoil are associated with r-strategies-dominate soil microbes. Soil Biol. Biochem. 95, 233–242.

Cheng, W., 2009. Rhizosphere priming effect: its functional relationships with microbial turnover, evapotranspiration, and C-N budgets. Soil Biol. Biochem. 41, 1795–1801.

Cheng, L., Leavitt, S., Kimball, B., Pinter, P., Ottman, M.J., Matthias, A., et al., 2007. Dynamics of labile and recalcitrant soil carbon pools in a sorghum free-air CO_2 enrichment (FACE) agroecosystem. Soil Biol. Biochem. 39, 2250–2263.

Clemmensen, K., Bahr, A., Ovaskainen, O., Dahlberg, A., Ekblad, A., Wallander, H., et al., 2013. Roots and associated fungi drive long-term carbon sequestration in boreal forest. Science 339, 1615–1618.

Cotrufo, M.F., Wallenstein, M.D., Boot, C.M., Denef, K., Paul, E., 2013. The Microbial Efficiency Matrix Stabilization (MEMS) framework integrates plant litter decomposition with soil organic matter stabilization: do labile plant inputs form stable soil organic matter? Global Change Biol. 19, 988–995.

Crow, S.E., Lajtha, K., Bowden, R.D., Yano, Y., Brant, J.B., Caldwell, B.A., Sulzman, E.W., 2009. Increased coniferous needle inputs accelerate decomposition of soil carbon in an old-growth forest. Forest Ecol. Manage. 258, 2224–2232.

Csonka, L.N., 1989. Physiological and genetic responses of bacteria to osmotic stress. Microbiol. Rev. 53, 121–147.
Davidson, E.A., Janssens, I.A., 2006. Temperature sensitivity of soil carbon decomposition and feedbacks to climate change. Nature 440, 165–173.
Davidson, E., Belk, E., Boone, R.D., 1998. Soil water content and temperature as independent or confounded factors controlling soil respiration in a temperate mixed hardwood forest. Global Change Biol. 4, 217–227.
De Graaff, M.A., Classen, A.T., Castro, H.F., Schadt, C.W., 2010. Labile soil carbon inputs mediate the soil microbial community composition and plant residue decomposition rates. New Phytol. 188, 1055–1064.
De Wit, R., Bouvier, T., 2006. "Everything is everywhere, but, the environment selects"; what did Baas Becking and Beijerinck really say? Environ. Microbiol. 8, 755–758.
Dennis, P.G., Miller, A.J., Hirsch, P.R., 2010. Are root exudates more important than other sources of rhizodeposits in structuring rhizosphere bacterial communities? FEMS Microbiol. Ecol. 72, 313–327.
Dinh, N.T., Hatta, K., Kwon, S.H., Rollon, A.P., Nakasaki, K., 2014. Changes in the microbial community during the acclimation stages of the methane fermentation for the treatment of glycerol. Biomass Bioenergy 68, 240–249.
Ekschmitt, K., Kandeler, E., Poll, C., Brune, A., Buscot, F., Friedrich, M., et al., 2008. Soil-carbon preservation through habitat constraints and biological limitations on decomposer activity. J. Plant Nutr. Soil Sci. 171, 27–35.
Eliasson, P.E., McMurtrie, R.E., Pepper, D.A., Strömgren, M., Linder, S., Ågren, G.I., 2005. The response of heterotrophic CO_2 flux to soil warming. Global Change Biol. 11, 167–181.
Fang, C., Moncrieff, J.B., 2001. The dependence of soil CO_2 efflux on temperature. Soil Biol. Biochem. 33, 155–165.
Fang, C., Moncrieff, J.B., 2005. The variation of soil microbial respiration with depth in relation to soil carbon composition. Plant Soil 268, 243–253.
Fang, Y., Singh, B., Singh, B.P., 2015. Effect of temperature on biochar priming effects and its stability in soils. Soil Biol. Biochem. 80, 136–145.
Fenner, N., Freeman, C., Reynolds, B., 2005. Observations of a seasonally shifting thermal optimum in peatland carbon-cycling processes: implications for the global carbon cycle and soil enzyme methodologies. Soil Biol. Biochem. 37, 1814–1821.
Fierer, N., Jackson, R.B., 2006. The diversity and biogeography of soil bacterial communities. Proc. Natl. Acad. Sci. U.S.A. 103, 626–631.
Fierer, N., Schimel, J.P., 2002. Effects of drying-rewetting frequency on soil carbon and nitrogen transformations. Soil Biol. Biochem. 34, 777–787.
Fierer, N., Schimel, J.P., 2003. A proposed mechanism for the pulse in carbon dioxide production commonly observed following the rapid rewetting of a dry soil. Soil Sci. Soc. Am. J. 67, 798–805.
Fierer, N., Schimel, J.P., Holden, P.A., 2003. Variations in microbial community composition through two soil depth profiles. Soil Biol. Biochem. 35, 167–176.
Fierer, N., Craine, J.M., McLauchlan, K., Schimel, J.P., 2005. Litter quality and the temperature sensitivity of decomposition. Ecology 86, 320–326.
Fontaine, S., Mariotti, A., Abbadie, L., 2003. The priming effect of organic matter: a question of microbial competition? Soil Biol. Biochem. 35, 837–843.
Fontaine, S., Barot, S., 2005. Size and functional diversity of microbe populations control plant persistence and long-term soil carbon accumulation. Ecol. Lett. 8, 1075–1087.
Fontaine, S., Bardoux, G., Benest, D., Verdier, B., Mariotti, A., Abbadie, L., 2004. Mechanisms of the priming effect in a savannah soil amended with cellulose. Soil Sci. Soc. Am. J. 68, 125–131.
Fontaine, S., Barot, S., Barré, P., Bdioui, N., Mary, B., Rumpel, C., 2007. Stability of organic carbon in deep soil layers controlled by fresh carbon supply. Nature 450, 277–280.
Fontaine, S., Henault, C., Aamor, A., Bdioui, N., Bloor, J.M.G., Maire, V., et al., 2011. Fungi mediate long term sequestration of carbon and nitrogen in soil through their priming effect. Soil Biol. Biochem. 43, 86–96.

Frostegård, Å., Bååth, E., Tunlio, A., 1993. Shifts in the structure of soil microbial communities in limed forests as revealed by phospholipid fatty acid analysis. Soil Biol. Biochem. 25, 723–730.
Ganie, M.A., Mukhtar, M., Dar, M.A., Ramzan, S., 2016. Soil microbiological activity and carbon dynamics in the current climate change scenarios: a review. Pedosphere 26, 577–591.
Gessner, M.O., Chauvet, E., 1994. Importance of stream microfungi in controlling breakdown rates of leaf litter. Ecology 75, 1807–1817.
Ghiorse, W.C., Wilson, J.T., 1988. Microbial ecology of the terrestrial subsurface. Adv. Appl. Microbiol. 33, 107–172.
Gong, W., Yan, X., Wang, J., Hu, T., Gong, Y., 2009. Long-term manure and fertilizer effects on soil organic matter fractions and microbes under a wheat-maize cropping system in northern China. Geoderma 149, 318–324.
Göransson, H., Godbold, D.L., Jones, D.L., Rousk, J., 2013. Bacterial growth and respiration responses upon rewetting dry forest soils: impact of drought-legacy. Soil Biol. Biochem. 57, 477–486.
Gordon, H., Haygarth, P.M., Bardgett, R.D., 2008. Drying and rewetting effects on soil microbial community composition and nutrient leaching. Soil Biol. Biochem. 40, 302–311.
Gougoulias, C., Clark, J.M., Shaw, L.J., 2014. The role of soil microbes in the global carbon cycle: tracking the below-ground microbial processing of plant-derived carbon for manipulating carbon dynamics in agricultural systems. J. Sci. Food Agric. 94, 2362–2371.
Griffiths, B.S., Philippot, L., 2013. Insights into the resistance and resilience of the soil microbial community. FEMS Microbiol. Rev. 37, 112–129.
Guo, L.B., Gifford, R., 2002. Soil carbon stocks and land use change: a meta analysis. Global Change Biol. 8, 345–360.
Hagedorn, F., Spinnler, D., Siegwolf, R., 2003. Increased N deposition retards mineralization of old soil organic matter. Soil Biol. Biochem. 35, 1683–1692.
Hamer, U., Unger, M., Makeschin, F., 2007. Short Communication Impact of air-drying and rewetting on PLFA profiles of soil microbial communities. J. Plant Nutr. Soil Sci. 170, 259–264.
Hanson, C.A., Allison, S.D., Bradford, M.A., Wallenstein, M.D., Treseder, K.K., 2008. Fungal taxa target different carbon sources in forest soil. Ecosystems 11, 1157–1167.
Hartley, I.P., Heinemeyer, A., Ineson, P., 2007. Effects of three years of soil warming and shading on the rate of soil respiration: substrate availability and not thermal acclimation mediates observed response. Global Change Biol. 13, 1761–1770.
Hieber, M., Gessner, M.O., 2002. Contribution of stream detrivores, fungi, and bacteria to leaf breakdown based on biomass estimates. Ecology 83, 1026–1038.
Holmes, A.J., Owens, N.J., Murrell, J.C., 1995. Detection of novel marine methanotrophs using phylogenetic and functional gene probes after methane enrichment. Microbiology 141, 1947–1955.
Houghton, R., 2007. Balancing the global carbon budget. Annual Rev. Earth Planet. Sci. 35, 313–347.
Hugenholtz, P., Goebel, B.M., Pace, N.R., 1998. Impact of culture-independent studies on the emerging phylogenetic view of bacterial diversity. J. Bacteriol. 180, 4765–4774.
Hütsch, B.W., Augustin, J., Merbach, W., 2002. Plant rhizodeposition-an important source for carbon turnover in soils. J. Plant Nutr. Soil Sci. 165, 397–407.
Ingram, L., Stahl, P., Schuman, G., Buyer, J., Vance, G., Ganjegunte, G., Welker, J., Derner, J., 2008. Grazing impacts on soil carbon and microbial communities in a mixed-grass ecosystem. Soil Sci. Soc. Am. J. 72, 939–948.
Jackson, L., Calderon, F., Steenwerth, K., Scow, K., Rolston, D., 2003. Responses of soil microbial processes and community structure to tillage events and implications for soil quality. Geoderma 114, 305–317.

Jastrow, J.D., Amonette, J.E., Bailey, V.L., 2007. Mechanisms controlling soil carbon turnover and their potential application for enhancing carbon sequestration. Clim. Change 80, 5–23.

Jobbágy, E.G., Jackson, R.B., 2000. The vertical distribution of soil organic carbon and its relation to climate and vegetation. Ecol. Appl. 10, 423–436.

Jones, D., Darrah, P., 1994. Amino-acid influx at the soil-root interface of *Zea mays* L. and its implications in the rhizosphere. Plant Soil 163, 1–12.

Kandeler, F., Kampichler, C., Horak, O., 1996. Influence of heavy metals on the functional diversity of soil microbial communities. Biol. Fertil. Soils 23, 299–306.

Khalil, M., Hossain, M., Schmidhalter, U., 2005. Carbon and nitrogen mineralization in different upland soils of the subtropics treated with organic materials. Soil Biol. Biochem. 37, 1507–1518.

Killham, K., Firestone, M., 1984. Proline transport increases growth efficiency in salt-stressed *Streptomyces griseus*. Appl. Environ. Microbiol. 48, 239–241.

King, G.M., 2011. Enhancing soil carbon storage for carbon remediation: potential contributions and constraints by microbes. Trends Microbiol. 19, 75–84.

Kirk, J.L., Beaudette, L.A., Hart, M., Moutoglis, P., Klironomos, J.N., Lee, H., Trevors, J.T., 2004. Methods of studying soil microbial diversity. J. Microbiol. Methods 58, 169–188.

Kirkby, C.A., Richardson, A.E., Wade, L.J., Batten, G.D., Blanchard, C., Kirkegaard, J.A., 2013. Carbon-nutrient stoichiometry to increase soil carbon sequestration. Soil Biol. Biochem. 60, 77–86.

Kirschbaum, M.U., 2004. Soil respiration under prolonged soil warming: are rate reductions caused by acclimation or substrate loss? Global Change Biol. 10, 1870–1877.

Knorr, W., Prentice, I., House, J., Holland, E., 2005. Long-term sensitivity of soil carbon turnover to warming. Nature 433, 298–301.

Krull, E.S., Baldock, J.A., Skjemstad, J.O., 2003. Importance of mechanisms and processes of the stabilisation of soil organic matter for modelling carbon turnover. Funct. Plant Biol. 30, 207–222.

Kuzyakov, Y., 2010. Priming effects: interactions between living and dead organic matter. Soil Biol. Biochem. 42, 1363–1371.

Kuzyakov, Y., Friedel, J., Stahr, K., 2000. Review of mechanisms and quantification of priming effects. Soil Biol. Biochem. 32, 1485–1498.

Kuzyakov, Y., Raskatov, A., Kaupenjohann, M., 2003. Turnover and distribution of root exudates of *Zea mays*. Plant Soil 254, 317–327.

Lal, R., 2010. Soils as source and sink of environmental carbon dioxide. Molecular Environmental Soil Science at the Interfaces in the Earth's Critical Zone. Springer, Berlin, Heidelberg, pp. 11-12.

Lal, R., 2013. Soil carbon management and climate change. Carbon Manage. 4, 439–462.

Lane, D.J., Pace, B., Olsen, G.J., Stahl, D.A., Sogin, M.L., Pace, N.R., 1985. Rapid determination of 16S ribosomal RNA sequences for phylogenetic analyses. Proc. Natl. Acad. Sci. 82, 6955–6959.

Lauber, C.L., Hamady, M., Knight, R., Fierer, N., 2009. Pyrosequencing-based assessment of soil pH as a predictor of soil bacterial community structure at the continental scale. Appl. Environ. Microbiol. 75, 5111–5120.

Lefèvre, R., Barré, P., Moyano, F.E., Christensen, B.T., Bardoux, G., Eglin, T., et al., 2014. Higher temperature sensitivity for stable than for labile soil organic carbon: evidence from incubations of long-term bare fallow soils. Global Change Biol. 20, 633–640.

Leifeld, J., Fuhrer, J., 2005. The temperature response of CO_2 production from bulk soils and soil fractions is related to soil organic matter quality. Biogeochemistry 75, 433–453.

Li, X., Meixner, T., Sickman, J.O., Miller, A.E., Schimel, J.P., Melack, J.M., 2006. Decadal-scale dynamics of water, carbon and nitrogen in a California chaparral ecosystem: DAYCENT modeling results. Biogeochemistry 77, 217–245.

Liang, B., Lehmann, J., Sohi, S.P., Thies, J.E., O'Neill, B., Trujillo, L., et al., 2010. Black carbon affects the cycling of non-black carbon in soil. Organ. Geochem. 41, 206–213.
Lidstrom, M.E., 1996. Environmental molecular biology approaches: promises and pitfalls. Microbiology of Atmospheric Trace Gases. Springer, Berlin, Heidelberg, pp. 121-134.
Louis, B.P., Maron, P.-A., Viaud, V., Leterme, P., Menasseri-Aubry, S., 2016. Soil C and N models that integrate microbial diversity. Environ. Chem. Lett. 14, 331–344.
Lu, W., Ding, W., Zhang, J., Li, Y., Luo, J., Bolan, N., Xie, Z., 2014. Biochar suppressed the decomposition of organic carbon in a cultivated sandy loam soil: a negative priming effect. Soil Biol. Biochem. 76, 12–21.
Luo, Y., White, L.W., Canadell, J.G., DeLucia, E.H., Ellsworth, D.S., Finzi, A., et al., 2003. Sustainability of terrestrial carbon sequestration: a case study in Duke Forest with inversion approach. Global Biogeochem. Cycles 17 (1).
Luo, Z., Wang, E., Sun, O.J., 2010. Soil carbon change and its responses to agricultural practices in Australian agro-ecosystems: a review and synthesis. Geoderma 155, 211–223.
Lützow, M.V., Kögel-Knabner, I., Ekschmitt, K., Matzner, E., Guggenberger, G., Marschner, B., Flessa, H., 2006. Stabilization of organic matter in temperate soils: mechanisms and their relevance under different soil conditions: a review. Eur. J. Soil Sci. 57, 426–445.
Manzoni, S., Porporato, A., 2009. Soil carbon and nitrogen mineralization: theory and models across scales. Soil Biol. Biochem. 41, 1355–1379.
McGuire, K.L., Treseder, K.K., 2010. Microbial communities and their relevance for ecosystem models: decomposition as a case study. Soil Biol. Biochem. 42, 529–535.
McLauchlan, K.K., Hobbie, S.E., 2004. Comparison of labile soil organic matter fractionation techniques. Soil Sci. Soc. Am. J. 68, 1616–1625.
Miller, A.E., Schimel, J.P., Meixner, T., Sickman, J.O., Melack, J.M., 2005. Episodic rewetting enhances carbon and nitrogen release from chaparral soils. Soil Biol. Biochem. 37, 2195–2204.
Miltner, A., Bombach, P., Schmidt-Brücken, B., Kästner, M., 2012. SOM genesis: microbial biomass as a significant source. Biogeochemistry 111, 41–55.
Moore, T., Dalva, M., 1993. The influence of temperature and water table position on carbon dioxide and methane emissions from laboratory columns of peatland soils. Eur. J. Soil Sci. 44, 651–664.
Moyano, F.E., Manzoni, S., Chenu, C., 2013. Responses of soil heterotrophic respiration to moisture availability: an exploration of processes and models. Soil Biol. Biochem. 59, 72–85.
Nakicenovic, N., Alcamo, J., Grubler, A., Riahi, K., Roehrl, R., Rogner, H.H., Victor, N., 2000. Special Report on Emissions Scenarios (SRES), A Special Report of Working Group III of the Intergovernmental Panel on Climate Change. Cambridge University Press.
Nannipieri, P., Ascher, J., Ceccherini, M.T., Landi, L., Pietramellara, G., Renella, G., 2003. Microbial diversity and soil functions. Eur. J. Soil Sci. 54, 655–670.
Nicol, G.W., Leininger, S., Schleper, C., Prosser, J.I., 2008. The influence of soil pH on the diversity, abundance and transcriptional activity of ammonia oxidizing archaea and bacteria. Environ. Microbiol. 10, 2966–2978.
Nottingham, A.T., Griffiths, H., Chamberlain, P.M., Stott, A.W., Tanner, E.V., 2009. Soil priming by sugar and leaf-litter substrates: a link to microbial groups. Appl. Soil Ecol. 42, 183–190.
Ogle, S.M., Swan, A., Paustian, K., 2012. No-till management impacts on crop productivity, carbon input and soil carbon sequestration. Agric. Ecosyst. Environ. 149, 37–49.
Pascault, N., Ranjard, L., Kaisermann, A., Bachar, D., Christen, R., Terrat, S., et al., 2013. Stimulation of different functional groups of bacteria by various plant residues as a driver of soil priming effect. Ecosystems 16, 810–822.
Pries, C.E.H., Castanha, C., Porras, R., Torn, M., 2017. The whole-soil carbon flux in response to warming. Science 355, 1420–1423.
Qafoku, N.P., 2015. Chapter two-climate-change effects on soils: accelerated weathering, soil carbon, and elemental cycling. Adv. Agron. 131, 111–172.

Raynaud, X., 2010. Soil properties are key determinants for the development of exudate gradients in a rhizosphere simulation model. Soil Biol. Biochem. 42, 210–219.

Resat, H., Bailey, V., McCue, L.A., Konopka, A., 2012. Modeling microbial dynamics in heterogeneous environments: growth on soil carbon sources. Microbial Ecol. 63, 883–897.

Rillig, M.C., Mummey, D.L., 2006. Mycorrhizas and soil structure. New Phytol. 171, 41–53.

Rousk, J., Brookes, P.C., Bååth, E., 2009. Contrasting soil pH effects on fungal and bacterial growth suggest functional redundancy in carbon mineralization. Appl. Environ. Microbiol. 75, 1589–1596.

Rousk, J., Bååth, E., Brookes, P.C., Lauber, C.L., Lozupone, C., Caporaso, J.G., Knight, R., Fierer, N., 2010. Soil bacterial and fungal communities across a pH gradient in an arable soil. ISME J. 4, 1340–1351.

Sanaullah, M., Chabbi, A., Maron, P.-A., Baumann, K., Tardy, V., Blagodatskaya, E., et al., 2016. How do microbial communities in top-and subsoil respond to root litter addition under field conditions? Soil Biol. Biochem. 103, 28–38.

Sayer, E.J., Heard, M.S., Grant, H.K., Marthews, T.R., Tanner, E.V., 2011. Soil carbon release enhanced by increased tropical forest litterfall. Nat. Clim. Change 1, 304–307.

Schimel, J.P., 2013. Soil carbon: Microbes and global carbon. Nat. Clim. Change 3, 867.

Schimel, J.P., Gulledge, J.M., 1998. Microbial community structure and global trace gases. Global Change Biol. 4, 745–758.

Schimel, J.P., Schaeffer, S.M., 2012. Microbial control over carbon cycling in soil. Front. Microbiol. 3, 1–11.

Schimel, J.P., Balser, T.C., Wallenstein, M., 2007. Microbial stress-response physiology and its implications for ecosystem function. Ecology 88, 1386–1394.

Schutter, M., Dick, R., 2001. Shifts in substrate utilization potential and structure of soil microbial communities in response to carbon substrates. Soil Biol. Biochem. 33, 1481–1491.

Singh, B.K., Bardgett, R.D., Smith, P., Reay, D.S., 2010. Microorganisms and climate change: terrestrial feedbacks and mitigation options. Nat. Rev. Microbiol. 8, 779–790.

Singh, M., Sarkar, B., Biswas, B., Churchman, J., Bolan, N.S., 2016. Adsorption-desorption behavior of dissolved organic carbon by soil clay fractions of varying mineralogy. Geoderma 280, 47–56.

Singh, M., Sarkar, B., Biswas, B., Bolan, N.S., Churchman, G.J., 2017a. Relationship between soil clay mineralogy and carbon protection capacity as influenced by temperature and moisture. Soil Biol. Biochem. 109, 95–106.

Singh, M., Sarkar, B., Hussain, S., Ok, Y.S., Bolan, N.S., Churchman, G.J., 2017b. Influence of physico-chemical properties of soil clay fractions on the retention of dissolved organic carbon. Environ. Geochem. Health 28, 1–16.

Sinsabaugh, R.L., Manzoni, S., Moorhead, D.L., Richter, A., 2013. Carbon use efficiency of microbial communities: stoichiometry, methodology and modelling. Ecol. Lett. 16, 930–939.

Six, J., Frey, S., Thiet, R., Batten, K., 2006. Bacterial and fungal contributions to carbon sequestration in agroecosystems. Soil Sci. Soc. Am. J. 70, 555–569.

Stockmann, U., Adams, M.A., Crawford, J.W., Field, D.J., Henakaarchchi, N., Jenkins, M., et al., 2013. The knowns, known unknowns and unknowns of sequestration of soil organic carbon. Agric. Ecosyst. Environ. 164, 80–99.

Tate, R.L., 1995. Soil Microbiology. Wiley and Sons, New York.

Teske, A., Wawer, C., Muyzer, G., Ramsing, N.B., 1996. Distribution of sulfate-reducing bacteria in a stratified fjord (Mariager Fjord, Denmark) as evaluated by most-probable-number counts and denaturing gradient gel electrophoresis of PCR-amplified ribosomal DNA fragments. Appl. Enrivon. Microbiol. 62, 1405–1415.

Tisdall, J., 1996. Formation of soil aggregates and accumulation of soil organic matter. Struct. Org. Matter Storage Agric. Soils, 57–96.

Torsvik, V., Øvreås, L., 2002. Microbial diversity and function in soil: from genes to ecosystems. Curr. Opin. Microbiol. 5, 240–245.

Tossell, J.A., 2009. Catching CO_2 in a bowl. Inorgan. Chem. 48, 7105–7110.

Van Elsas, J., Boersma, F., 2011. A review of molecular methods to study the microbiota of soil and the mycosphere. Eur. J. Soil Biol. 47, 77–87.

Vandenkoornhuyse, P., Husband, R., Daniell, T., Watson, I., Duck, J., Fitter, A., Young, J., 2002. Arbuscular mycorrhizal community composition associated with two plant species in a grassland ecosystem. Mol. Ecol. 11, 1555–1564.

Wang, Y., Shi, J., Wang, H., Lin, Q., Chen, X., Chen, Y., 2007. The influence of soil heavy metals pollution on soil microbial biomass, enzyme activity, and community composition near a copper smelter. Ecotoxicol. Environ. Saf. 67, 75–81.

Waring, B.G., Averill, C., Hawkes, C.V., 2013. Differences in fungal and bacterial physiology alter soil carbon and nitrogen cycling: insights from meta-analysis and theoretical models. Ecol. Lett. 16, 887–894.

Wei, G., Zhou, Z., Guo, Y., Dong, Y., Dang, H., Wang, Y., Ma, J., 2014. Long-term effects of tillage on soil aggregates and the distribution of soil organic carbon, total nitrogen, and other nutrients in aggregates on the semi-arid loess plateau, China. Arid Land Res. Manage. 28, 291–310.

Whitman, W.B., Coleman, D.C., Wiebe, W.J., 1998. Prokaryotes: the unseen majority. Proc. Natl. Acad. Sci. 95, 6578–6583.

Wieder, W.R., Bonan, G.B., Allison, S.D., 2013. Global soil carbon projections are improved by modelling microbial processes. Nat. Clim. Change 3, 909–912.

Williams, M.A., 2007. Response of microbial communities to water stress in irrigated and drought-prone tallgrass prairie soils. Soil Biol. Biochem. 39, 2750–2757.

Witteveen, C.F., Visser, J., 1995. Polyol pools in *Aspergillus niger*. FEMS Microbiol. Lett. 134, 57–62.

Wu, T., Li, W., Ji, J., Xin, X., Li, L., Wang, Z., et al., 2013. Global carbon budgets simulated by the Beijing climate center climate system model for the last century. J. Geophys. Res. 118, 4326–4347.

Xiang, S.-R., Doyle, A., Holden, P.A., Schimel, J.P., 2008. Drying and rewetting effects on C and N mineralization and microbial activity in surface and subsurface California grassland soils. Soil Biol. Biochem. 40, 2281–2289.

Yu, X., Yu, D., Lu, Z., Ma, K., 2005. A new mechanism of invader success: exotic plant inhibits natural vegetation restoration by changing soil microbe community. Chin. Sci. Bull. 50, 1105–1112.

Yuste, J.C., Janssens, I., Ceulemans, R., 2005. Calibration and validation of an empirical approach to model soil CO_2 efflux in a deciduous forest. Biogeochemistry 73, 209–230.

Zhou, J., Xia, B., Treves, D.S., Wu, L.-Y., Marsh, T.L., O'Neill, R.V., et al., 2002. Spatial and resource factors influencing high microbial diversity in soil. Appl. Environ. Microbiol. 68, 326–334.

Zhu, B., Cheng, W., 2011. Rhizosphere priming effect increases the temperature sensitivity of soil organic matter decomposition. Global Change Biol. 17, 2172–2183.

Zhu, B., Gutknecht, J.L., Herman, D.J., Keck, D.C., Firestone, M.K., Cheng, W., 2014. Rhizosphere priming effects on soil carbon and nitrogen mineralization. Soil Biol. Biochem. 76, 183–192.

Zimmerman, A.R., Gao, B., Ahn, M.-Y., 2011. Positive and negative carbon mineralization priming effects among a variety of biochar-amended soils. Soil Biol. Biochem. 43, 1169–1179.

Zogg, G.P., Zak, D.R., Ringelberg, D.B., White, D.C., MacDonald, N.W., Pregitzer, K.S., 1997. Compositional and functional shifts in microbial communities due to soil warming. Soil Sci. Soc. Am. J. 61, 475–481.

CHAPTER 7

Recycling of Organic Wastes to Soil and Its Effect on Soil Organic Carbon Status

Heribert Insam, María Gómez-Brandón, Judith Ascher-Jenull
University of Innsbruck, Innsbruck, Austria

Chapter Outline

INTRODUCTION

We bet that maintenance of soil organic C (SOC) has a lot to do with environmental politics, even big global politics. Let us start with this example: the Russian–Chinese friendship in the 1950s resulted in a forestry cooperation. The original climax tropical monsoon forest in the Guangdong province had been destroyed due to human overuse, resulting in severe soil erosion. The suggestion by the Russian advisors was to reclaim the land in southern China by planting highly productive *Pinus massoniana* and *Eucalyptus exserta* trees, a good idea, indeed. In a second stage, these forests were replaced by another artificially mixed forest. However, after a few years the trees ceased to grow. Initially, to supply the forest with nutrients, circular trenches were dug, and in each one domestic waste material was placed. Presumably, this waste material had been predominantly organic, and it delivered nutrients to the trees for years. In 1989 as a result of a joint European–Chinese project the immediate reason for the cessation of growth was discovered: local

The Future of Soil Carbon
http://dx.doi.org/10.1016/B978-0-12-811687-6.00007-9

people had, for years, meticulously removed all the litter, and along with it, the carbon and nutrients (Insam, 1990). Compared to a nearby natural forest, the SOC content had dropped from 3.10% to 0.77%. Within 6 years of protection from litter removal, the organic C content recovered to 1.37% (Ding et al., 1992). This story is an example of how organic matter can be replenished by appropriate management measures, and how SOC reminds us about politics.

Concern About Soil Organic C Loss

How far the concern about SOC loss dates back is not easy to resolve. We know, however, it is not only the nutrients that are responsible for soil fertility and plant health; let us remember the dust bowl of the 1930s that for a decennium threw farmers of the Midwest in poverty. John Steinbeck deplored in his novel *The Grapes of Wrath*: "*And then the dispossessed were drawn west- from Kansas, Oklahoma, Texas, New Mexico; from Nevada and Arkansas, families, tribes, dusted out, tractored out. Car-loads, caravans, homeless and hungry; twenty thousand and fifty thousand and a hundred thousand and two hundred thousand... Like ants scurrying for work, for food, and most of all for land.*" Then, in the early decades of excessive mineral fertilization it was realized that organic matter might have some irreplaceable value for maintaining soil fertility. That time, agricultural policy seems to have failed and we should have learned from it. However, having a look at a recent book published by the International Fertilizer Association (Reetz, 2016), which devotes a meager 3 of more than 100 pages to organic fertilizers, shows a certain bias by the industry toward mineral fertilizers. Legal difficulties for marketing novel fertilizers based on organic sources proliferate, and these difficulties may be due, in fact, to extensive lobbying by those incentivized to sell mineral fertilizers. For this reason, the authors attempt some lobbying themselves for organic fertilizers and soil conditioners.

Carbon Sequestration

SOC can reach soil through two main pathways, directly through plant input (primary production) and through external inputs like products from organic waste (secondary production, if the wastes are not agricultural ones). In addition, some input may be expected from atmospheric deposition (Fig. 7.1). SOC models like the CENTURY (Parton et al., 1987) and the Rothamsted Carbon Model (RothC) (Coleman and Jenkinson, 1996) allow for estimating C balances in soils. The RothC is a model for the

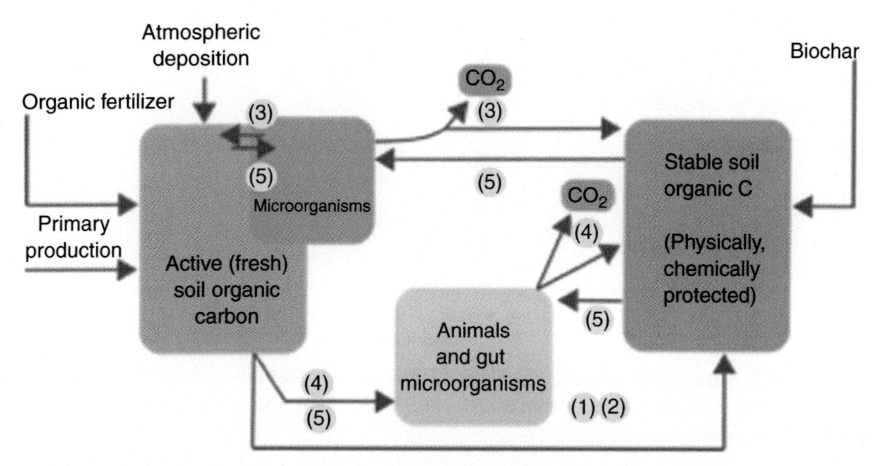

Figure 7.1 ***How carbon is sequestered into soil.*** (1, 2) Root and shoot debris are formed into soil organic matter (SOM); (3) microorganisms respire C compounds and their biomass eventually becomes part of the SOM pool; (4) plant matter is ingested by animals and residues are transferred to the SOM pool; (5) microorganisms (and soil animals) thrive on SOM and form new biomass.

turnover of organic C in nonwaterlogged topsoil that allows for the effects of soil type, temperature, soil moisture, and plant cover on the turnover process. RothC was originally developed and parameterized to model the turnover of organic C in arable topsoil from the Rothamsted long-term field experiments and was later extended to model turnover in grassland and in woodland and to operate in different soils and under different climates. RothC is designed to run in two modes: "forward" in which known inputs are used to calculate changes in soil organic matter (SOM) and "inverse," when inputs are calculated from known changes in SOM. External input may be fed into these models, which, in general, are very suitable predictors of SOC dynamics.

In a metastudy comprising 21 long-term experimental sites with various management regimes, distributed all over the North American subcontinent, Insam et al. (1989) and Insam (1990) found a distinct relationship of climate and the organic C balance. As the best climatic predictor, they found the precipitation/evaporation ratio (P/E-ratio) (Fig. 7.2). Their C equilibrium model described the C_{mic}-to-C_{org}-ratio as approaching the minimum when the PE-ratio was around 1 (mean annual precipitation equaling pan evaporation). Any measured C_{mic}-to-C_{org}-ratio deviating from the model equilibrium line would indicate an imbalance in the SOC status. In particular, C_{mic}-to-C_{org}-ratio higher than the prediction for a certain climate would

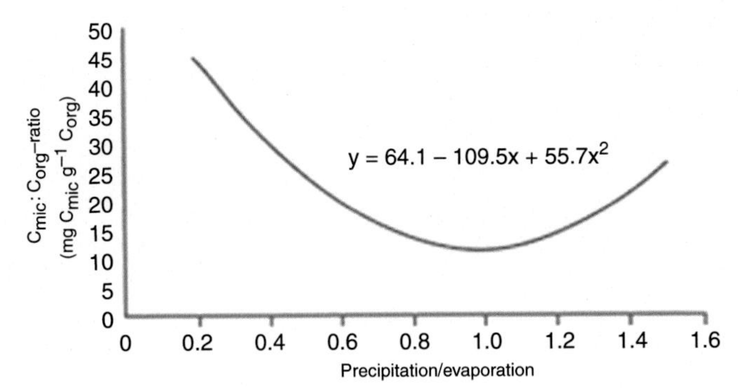

Figure 7.2 ***Baseline for organic C balance.*** If the C_{mic}-C_{org}-ratio is above the equilibrium line, a soil is gaining carbon, otherwise it is losing carbon under the prevailing climatic condition *(after Insam, 1990)*.

Table 7.1 mg C_{mic} g^{-1} C_{org} for the two main groups of fertilization and cropping practice

Mineral fertilizer		Manure	
Monoculture	**Crop rotation**	**Monoculture**	**Crop rotation**
16.6 ± 0.4 (66)	19.9 ± 0.3 (102)	18.0 ± 0.5 (48)	22.5 ± 0.7 (54)
18.3		20.3	

Mean ± standard deviation (in parentheses, number of observations, *n*).
Source: From Insam, H., Parkinson, D., Domsch, K.H., 1989. Influence of macroclimate on soil microbial biomass. Soil Biol. Biochem. 21, 211–221.

indicate that the soil would be in the status of C gain, or below the equilibrium line a C loss would be indicated. Comparing differently managed soils, it was shown that both fertilization practice (mineral fertilizer versus manure) and cropping practice (monoculture versus crop rotation) had a significant impact on SOC status (Table 7.1). Extra organic input (manure) increased C_{mic}-to-C_{org}-ratio from 18.3 to 20.3, an indication of increasing C accumulation.

ORGANIC WASTES AND PRETREATMENT OPTIONS

Domestic waste composition, collection, and treatment have considerably changed all over the world since the onset of the aforementioned afforestation experiment. Nowadays, in many countries organic waste is source-separately collected, or there is at least a mechanical sorting of the original waste to obtain an organic fraction. Wastewater treatment with resulting aerobic or anaerobic sludges is implemented in many places, and the residual slurries are often in a quality that may allow for utilization for agricultural, horticultural, or forestry purposes. The impact of manure-based biogas fermentation residues (digestates) on soil fertility has been recently reviewed by Insam et al. (2015). Agricultural wastes have long been used for fertilizing soils, or for maintaining SOM. However, due to an industrialization of agriculture, wastes like manure, liquid manure, or straw are often found in centralized operations where an ultimate use on soils is impossible.

The use of appropriate management technologies, involving the stabilization of the waste prior to use or disposal, could mitigate the health and environmental risks associated with the local overproduction and the application of excessive amounts of organic waste (Gómez-Brandón and Podmirseg, 2013). Composting and vermicomposting have been used, either separately or in combination with each other, for processing wastes from different origins under aerobic conditions (Gómez-Brandón and Domínguez, 2014). Unlike composting, vermicomposting depends on the joint action of earthworms and microorganisms and does not involve a thermophilic phase as classical composting does. During these biological processes organic wastes are transformed into a safer and more stabilized product, with benefits for both agriculture and the environment, thereby resulting in a more balanced nutrient mix and increased nutrient bioavailability for plants in comparison with the untreated waste (Gómez-Brandón and Domínguez, 2014). Indeed, composted materials have been shown to provide manifold benefits when used as soil amendments, as they increase SOM levels, soil porosity, and aggregate stability. They may also lead to an increase in soil microbial biomass and activity (Ros et al., 2006a,b), which could be attributed to the activation of the indigenous soil microbiota by the supply of C-rich organic compounds contained in the composts. Furthermore, C addition to soil seems to select for specific microbial groups that feed primarily on organic compounds therefore also leading to changes in microbial community composition (Carrera et al., 2007; Ros et al., 2006b).

Anaerobic digestion (AD) has also become an important technology for recycling wastes from different origins due to the fact that the available fossil fuel reserves are decreasing, and the biogas produced by AD can be utilized as an ecofriendly energy source (Insam et al., 2015). However, the sustainability of the production of biogas depends on the proper use of the digested material, which must be treated, disposed of, or reused properly, to avoid any negative environmental impacts (Insam and Wett, 2008; Insam et al., 2015). The use of the digested material as an organic fertilizer in agriculture seems to be an optimal option for its use, because it contains significant amounts of residual organic C and nutrients for plants (Alburquerque et al., 2012a,b). According to Odlare et al. (2011) the C in biogas residues is more easily degradable than that of composts because the mineralization is less efficient under anaerobic conditions. Consequently, when biogas residues are applied to soil, their C is expected to be more rapidly metabolized, leading to an increase in soil microbial biomass in the short term (Insam et al., 2015). However, Gómez-Brandón et al. (2016) found that the addition of digestates was not accompanied by a significant long-term increase in soil microbial biomass and activity assessed as substrate-induced and basal respiration, compared to the unamended soil and raw manure, after 15 and 60 days of incubation.

Overall, AD appears not to negatively affect the SOC status in the long term (Insam et al., 2015) compared to composting. Nonetheless, as pointed out by de la Fuente et al. (2013) a posttreatment of digestates via liquid–solid separation or composting, might further increase their C sequestration potential. All in all, this will help to maintain the organic matter in the soil after the use of biogas residues as organic amendments and ultimately, it will lead to positive effects on crop yields and soil microbial functions (Abubaker et al., 2012; Insam et al., 2015). Future research dealing with longer term studies and varying the source and application rate of digestates will contribute to gain knowledge into the agronomic effects of biogas residues into soil and their impact in SOC.

Agricultural, Horticultural, and Silvicultural Sources

An overview of the various sources of organic wastes and their potential use as biofertilizers after appropriate treatment and quality-check/risk assessment is given in Fig. 7.3. The constraints and the potential for the use of these organic wastes in agriculture will be discussed, bearing in mind that organic farming—based on recycling of organic wastes/by-products—is considered as the backbone of sustainable agriculture with the chief objectives: (1) to adopt ecofriendly and modern techniques to recycle rather than discard precious organic material; (2) to reduce the environmental impact, health issues, and high cost of landfilling (on average 80 € T^{-1} landfilling tax in EU) or diverse storage (e.g., random piles); (3) reduce the environmental impact of chemical fertilizers; and (4) produce high amounts of high-quality products. All in all, there is still the urgent need of finding and/or applying sustainable solutions for modern agriculture and economies (Hofer, 2009).

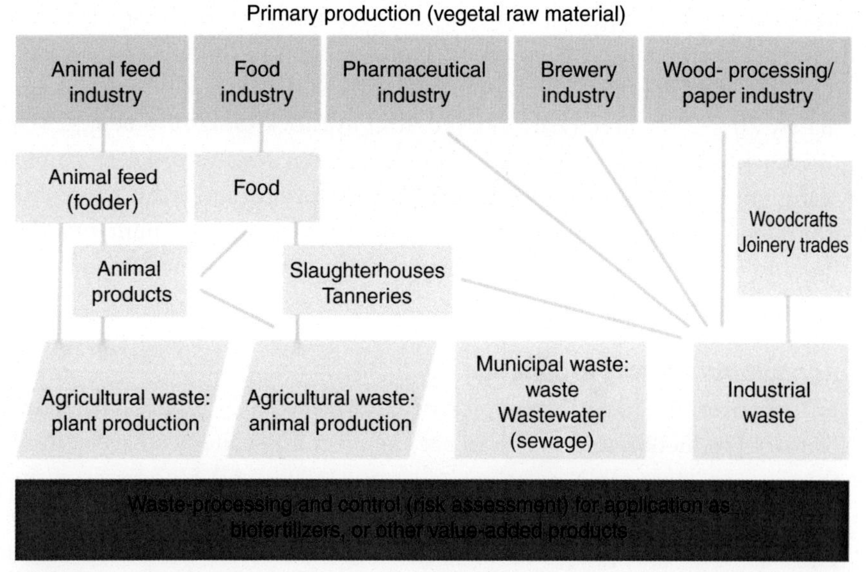

Figure 7.3 ***Different sources of wastes as potential biofertilizers or other value-added products.***

Animal Production

Large amounts of livestock products, for example, manure, litter, compost, and wastewater, are readily available due to an increased growth of confinement livestock and poultry operations. Their use in agriculture for crop and forage production is convenient for multiple aspects: cost-effective strategy to beneficially manage large amounts of by-products from animal production; supply of plant nutrients (especially N and P), ideal for contrasting the N and P limiting crop production; increase of SOM with benefits for plant growth and yield. Anyway, in order to guarantee the environmentally friendly use, adequate storage, use and land application recommendations have to be strictly followed, so as to prevent/avoid any potential public health and environmental impact (Sorathiya et al., 2014), for example, contamination of (1) surface and groundwater by excessive N (mobile nitrate); excessive P (stimulation of aquatic plant growth with risk of eutrophication); organic matter associated with manure (reduction of dissolved oxygen in water bodies); heavy metals (e.g., Cu, Zn, As); harmful pathogens (e.g., bacteria and viruses); antibiotics; and (2) air quality due to odors associated with animal production, as a result of decomposition of organic material, such as feed, manure, and mortalities; among the most frequent noxious gases resulting from anaerobic decomposition of complex liquid biological wastes are CH_4, H_2S, CO_2, and NH_4 (Gerber et al., 2005). In fact, livestock waste is a major source of pollution, pathogens, odor, and greenhouse gases (GHG). Of the global GHG production, 16% is methane, of which approximately 20%–25% directly are derived from livestock (Sorathiya et al., 2014). Livestock waste can be recycled in many modern ways in order to combat rising energy prices, promote sustainable agriculture, and reduce the environmental threats from traditional livestock waste management practices (Sorathiya et al., 2014), and most importantly used for sequestering C into the organic soil pool.

Agricultural Primary Production

Johansson et al. (2010) have constructed a database of global agricultural primary production to estimate its net energy content and postharvest losses, focusing on sustainable recycling of crop residues and bioorganic wastes for efficient biofuel production. This database enables calculating the energy content of crops and residues globally, and for the member states of the European Union. Within the context of global change, the mitigation of GHG emission has been placed at the top of the global issues to be addressed. Considering agricultural soils as potential sinks for atmospheric CO_2 due to their C sequestration potential, sustainable cropping management has become crucial for the global C budget (Liu et al., 2014; Wang et al., 2016).

The annual SOC sequestration rate of cropping systems with recommended management was estimated to be 0.4–0.8 Pg C globally, accounting for 33.3%–100% of the total potential of soil C sequestration worldwide (Liu et al., 2014). Straw return is among the most convenient recommended management strategies, capable of increasing SOC sequestration in croplands. Liu et al. (2014) performed a meta-analysis on the net global warming potential balance of straw C input (straw-return practice) to manage C sequestration in agricultural systems, based on the dataset derived from >170 field studies, by calculating the response ratios of SOC concentrations, GHGs emission, nutrient contents, and other important soil properties to straw addition. These authors concluded that straw return significantly increased (1) both the total and active SOC concentrations; (2) CO_2 emissions in both upland and paddy systems, (3) CH_4 emissions only in rice paddy soils; and (4) N_2O emission in paddy soils, while a decline was recorded in upland soils. The aforementioned information about C budget effects (i.e., straw return increased C sink in upland soils but increased C source in paddy soils due to enhanced CH_4 emission), points out the importance to differentiate between upland and paddy soils for future agroecosystem models and cropland management. Overall, the potential of straw-return practice to improve SOC accumulation, soil quality, including soil nutrient status, and crop yield has led to (1) positive correlations between macroaggregates and crop yield with increasing SOC concentration; (2) significant effects on SOC dynamics of the straw-C input rate and clay content; and (3) a significant positive relationship between annual SOC sequestered and duration, suggesting a soil C saturation after 12 years under straw return.

Recently, Wang et al. (2016) used the soil C model RothC to assess the critical C input for maintaining SOC stocks in wheat systems at global scale as a baseline for assessing soil C dynamics in croplands under changing scenarios (management practices vs. climate). The critical C input was estimated to be 2.0 Mg C ha^{-1} yr^{-1}, with a large spatial variability depending on local soil and climatic conditions. Due to the higher soil C stocks present in wheat systems of central United States and Western Europe, higher C inputs are required. The authors suggested a metamodel driven by current SOC level, mean annual temperature, precipitation, and soil clay content to be capable to effectively estimate the critical C input, so as to reduce or reverse C loss in agricultural soils.

Wastes From Forestry/Wood Industries

The forest-products industry based on the use of wood as raw materials, for example, lumber, furniture, paper products, and pulp, can be divided into (1) primary wood-products industry (from lumber production to the manufacture of finished products (Burton et al., 2003; Pentti et al., 2002); and (2) the value-added or secondary forest-products industry processing raw or semiprocessed materials (e.g., production of pallets and light furniture), both generating large amounts of wood wastes, that is, unsuitable material for the production of wood products, including also low-quality raw material (e.g., bark, small chips, saw dust, wood edges) (Murphy et al., 2007; Top, 2015). Wood wastes, classified into bark, coarse and fine waste, are potential biomass resources (Skog et al., 2011) for either (1) *energy applications* (combustion in wood burners or larger biomass boilers to cover energy demands of homes or industrial enterprises; at an industrial scale, forest residues and waste wood can be converted into advanced biofuels or intermediates through various thermochemical pathways (http://biofuelstp.eu/forest.html); or (2) *nonenergy applications* (production of composite boards and wood pulp; land reclamation; animal bedding material; landscaping; agricultural mulch; and landfilling (Murphy et al., 2007; Top, 2015).

The technological platform "European Biofuels" (https://ec.europa.eu/energy/en/topics/renewable-energy/biofuels) offers insights into the state of the art of advanced technologies of bioenergy and biofuels production from different types of forestry sources: (1) residues from harvest operations that are left in the forest after stem wood removal, such as branches, foliage, roots; (2) complementary fellings, which describe the difference between the maximum sustainable harvest level and the actual harvest needed to satisfy round wood demand; (3) wood wastes from, for example, construction or demolition wastes; and (4) waste from manufacturing of wood-based products. Furthermore, this web-platform also highlights the environmental and commercial benefits of harvesting biomass to both maintain and optimize forest health (e.g., removing diseased and hazardous trees) and to provide valuable feedstock for bioenergy production.

Importantly, some forest residues must be left in situ to provide ecological benefits, for example, to provide various habitats and as such preserve biodiversity, guarantee soil nutrient status, and improve overall soil quality. In fact, deadwood plays an important role in the functioning of forest ecosystems and their structure, contributing to maintaining the biodiversity and natural regeneration in forests, as well as influencing nutrient cycling and overall C storage (Gómez-Brandón et al., 2017; Petrillo et al., 2016).

Domestic Wastes in Their Broader Sense

Separately Collected Organic Part of the Solid Waste

Source-separate collection (Fig. 7.4) has been a major push for proper recycling programs for organic matter in many countries. As an example of a straightforward strategy, the Austrian Compost Ordinance (2001) offers an unprecedented end-of waste-option for organic waste that underwent precisely defined treatment, that is, composting. Beyond conventional agriculture, the end-product compost may even be used in organic agriculture as long it meets certain additional specifications, for example, related to heavy-metal contents. Since 2001, nearly 100% of Austrian organic waste is composted, or treated in analogy by AD. No data are available about how this has changed overall organic matter content of Austrian soils but based on the data we know from various models, the impact must have been considerable. Per capita, the annual production is 111 kg organic waste, or more than 4 mio T per year for the country. At an estimated water content of 80%, this is 800.000 T per year organic dry matter of which around 400.000 T remain after the composting process. With this amount, it may

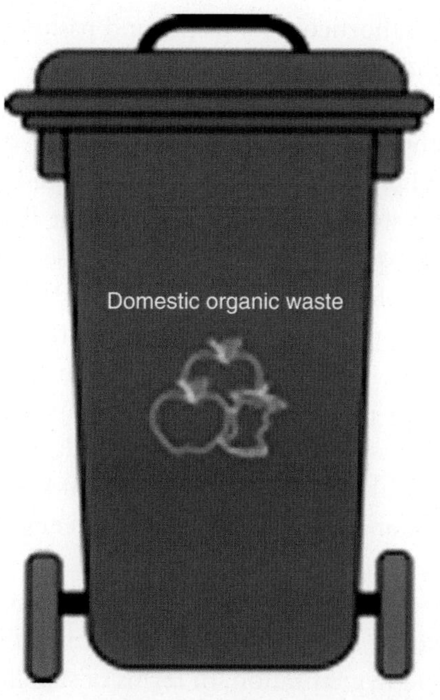

Figure 7.4 ***Source-separate collection is a prerequisite for good quality recycling.***

be estimated that each of the 1.3 Mio ha of agricultural land in Austria has, per year, received more than 3 T of additional high-quality organic matter, which is, in the long term, a considerable amount! Nonsource-separate collection of organic waste would not provide a recycling product of a quality good enough to allow for long-term agricultural use.

Waste Water and Sewage Sludge

As a result of rapidly increasing population, urbanization, and industrialization, wastewater production and sewage sludge generation have increased manifold (Usman et al., 2012). While sewage sludge was regarded in the past as a waste product due to its expected high level of contaminants (e.g., harmful pollutants, pathogens, heavy metals, and synthetic materials) and therefore mainly incinerated, dumped on occasion or landfill, nowadays, due to escalating high cost of mineral fertilizers, there is an increasing trend of processing and using sewage sludge in sustainable agriculture. In fact, sewage sludge/biosolids, the by-products of municipal and industrial wastewater treatment, are rich in organic matter, macro- and micronutrients, and as such potential and low-cost sources as fertilizer and soil conditioner for food, vegetable crops, horticultural plants, and pasture, which in most cases can be beneficially recycled (Usman et al., 2012).

Among the most effective treatment processes of sewage sludges are pasteurization; mesophilic AD; thermophilic aerobic digestion; composting (windows or aerated piles); lime stabilization of liquid sludge; liquid storage; and dewatering and storage. Among the different types of sludge used in Europe there are (1) liquid sludge (resulting from sedimentation of screened sewage; 2%–7% dry solid containing up to 75% OM); (2) untreated sewage cake (formed by dewatering of liquid sludge with soil-similar consistency and farm yard manure-similar microbiological characteristics); (3) conventionally treated sludge (digestion); (4) enhanced treated sludge (heat treatment of dried, lime pasteurized, or digested sludge); (5) composted sludge (odorless friable soil-like material rich in P); and (6) lime-treated sludge (lime amendment of undigested sludge cake generates friable products with high pH) (Usman et al., 2012).

Compost as a product from both source-separate collected organic waste as well as sewage sludge needs to make sure to meet the basic requirements for both stability and hygienization. In terms of stability it is recommended to use mature compost, if possible of Rottegrad V (Table 7.2). The more mature a compost is, the more humified is the organic matter and the longer will the half-life of the carbon be.

Danon et al. (2008) showed that after maturation, the microbial community composition of composts is still changing while no changes in chemical properties can be found. An additional *curing phase* that improves the benefit of composts is still disputed. Plant-disease suppressiveness, one of the main properties of good composts, has not been shown to improve with prolonged curing, but it may be possible that the mobilization of soil indigenous organic matter through addition of fresh composts (also known as priming effect) is reduced. This could mean an additional benefit for the long-term SOM status.

Table 7.2 Rottegrad of composts is measured by a self-heating test in Dewar-flasks

Rottegrad	T_{max} in °C	Product
I	>60	Fresh compost
II	50–60	Fresh compost
III	40–50	Mature compost
IV	30–40	Mature compost
V	<30	Mature compost

Organic Fraction of the Waste (Collected in Bulk, and Then Separated)

Municipal or household wastes, generated from several sources with variable human activities, and from different socioeconomic areas all over the world, are highly heterogeneous in nature (Miezah et al., 2015; Valkenburg et al., 2008) having variable, source-dependent physical characteristics (e.g., food waste, yard waste, wood, plastics, papers, metals, leather, rubbers, inert materials, batteries, paint containers, textiles, construction, and demolishing materials), which makes their utilization as raw materials challenging. Therefore, prior to figuring out any appropriate treatment process, there is the need of source sorting/waste fractionation, so as to obtain qualiquantitative data on the various waste fractions (organic versus nonorganic wastes). In fact, effective recycling relies on effective sorting. Among the most common methods for waste sorting adopted by waste disposal companies in Europe there are (1) trommel separators/drum screens; (2) Eddy current separators; (3) induction sorting; (4) near infrared sensors (NIR); (5) X-ray technologies; and (6) manual sorting (https://waste-management-world.com/a/waste-sorting-a-look-at-the-separation-and-sorting-techniques-in-todayrsquos-european-market).

Industrial Wastes

Wastes From the Food Industry

Among food-processing industry-derived wastes the main waste streams are fruit-and-vegetable wastes, wastes from dairy, olive oil, fermentation industries, meat, poultry, and seafood by-products (Kosseva, 2009, 2013) along with vast volumes of aqueous wastes (Grismer et al., 2002). The problem with the utilization of such wastes is often their centralized availability, incurring high transportation costs. Often, the production of biomethane, and a solid–liquid separation of the digestate make sense for its further utilization, for example, after a stabilization process by composting.

One example from the food industry is olive oil production. Due to different ways of extraction of oil from olives, different waste streams are known. Recently, the trend points toward the production of *alperujo*, wet solid lignocellulosic material also containing mashed olive stones. Alperujo may be composted and yields good quality in terms of nutrient content, stabilized and nonphytotoxic organic matter, and low heavy-metal contents best suited as a soil amendment (Alburquerque et al., 2011). While AD of olive mill effluents has proven feasible in combination with cattle manure, household waste, sewage sludge, poultry manure, wine-grape and slaughterhouse wastewater, and laying hen litter, there are only few studies that have shown the codigestability of two-phase olive mill wastes (TPOMW, alperujo); for example, Goberna et al. (2010) who found cofermentation of cattle manure with alperujo a suitable treatment option yielding biogas and fertilizer.

Wastes From the Pharmaceutical Industry

Herbal pharmaceutical industries generate large volumes of wastewater containing alkaloids, plant extracts, heavy-metal ions, and toxic solutes (Vanerkar et al., 2015). This wastewater must not be discharged, but treatment efforts are necessary. A feasible option is the treatment with microalgae that may then be used for biomethanization, and composting as a posttreatment.

Organic wastes from the pharmaceutical industry often are based on the biomass of fungi or bacteria that are used for metabolite production. The study by Ceccherini et al. (2007) on how to "dis-activate" fungal biomass, that is, to degrade DNA (antibiotics encoding genes)—the by-product of antibiotics production (Cephalosporine from *Cephalosporium* sp.)—provided the basis for obtaining a European patent (EP 1 529 766 B1). Overall, it can be seen as a practical example of successful biowaste processing for environmentally safe agricultural application as biofertilizer, known to have beneficial effects on the biological, chemical, and physical properties of the soil (Aescht and Foissner, 1992; Haselwandter et al., 1988; Nannipieri et al., 2017a,b), and to avoid the risk of potential genetic exchange (horizontal gene transfer) in soil via natural transformation (Pietramellara et al., 2006, 2009).

It is beyond the scope of this review to address numerous further specific organic industrial wastes from tanneries, slaughterhouses, bakeries, fruit processors, breweries, and so on. Fact is, if there is no other added value that can be generated from such wastes, like bulk chemicals, or feed protein, stabilized organic matter would always be a beneficial product aiding soil functional properties.

WHICH OPTIONS FOR WASTE UTILIZATION DO WE HAVE?

The utilization of by-products and waste materials from animal production (meat, poultry, and fish processing industries) has been recently reviewed by Jayathilakan et al. (2012). Furthermore, ecofriendly and modern methods of livestock waste recycling for enhancing farm profitability and reducing/avoiding public health and environmental concerns has been critically reviewed by Sorathiya et al. (2014). As mentioned earlier, there is a broad spectrum of available biomass resources for energetic use deriving from different industries or household (Fig. 7.3). Organic residues, by-products, and waste can be classified in herbaceous biomass (e.g., straw, landscape conservation material), wood (e.g., forest residual wood, industrial residual wood) and other biomass (e.g., excrements, organic industrial waste) (Hofer, 2009). In order to estimate their energetic biomass potential, not only the "technical potential" (the total amount of available biomass evaluated by considering technical restrictions), but also structural, environmental, and legal restrictions have to be considered (Hofer, 2009). The forest biomass potential includes the wood not used as a raw material, that is, firewood and forest residual wood (Hofer, 2009).

Next to recycling organic wastes for energetic purposes, there is the well-targeted production of plant biomass for green energy, estimating also the required amount of biomass to replace fossil oil (Henry, 2010) and the bioenergy production potential of global biomass plantations (Beringer et al., 2011). The example of targeted plant cultivation in Central Western Europe (e.g., Federal Republic of Germany) shows the cultivation of different types of plants for different purposes: (1) thermochemical conversion (mixed cultivation of lignocellulosic plants for producing solid biofuels); (2) physical–chemical conversion (e.g., rape-seed cultivation); and (3) biochemical conversion (two-culture system for producing substrates for biogas- and ethanol production) (Hofer, 2009).

A team of Swedish researchers (Röös et al., 2016) recently developed a promising model for designing and assessing the sustainability of "fair" diets, based on the globally available arable land per capita. The authors concluded that the proposed concept of using "ecological leftovers for livestock production" is capable to design diets using food produced on (Swedish) agricultural land that satisfy both nutritional and environmental recommendations. Anyway, although their model is advantageous with respect to current diets, the production of these "fair diets," including a drastically reduced consumption of meat, still results in environmental impacts that cause several planetary boundaries to be transgressed, but is suggested to provide a promising basis for sustainable livestock consumption.

Grape marc compost, as another example, is an excellent fertilizer and due to its high tannin contents, an excellent plant growth promoting and disease suppressive agent that, at the same time, is best suited to build up SOM (Carmona et al., 2012). Various other industrial wastes like those from the pulp and paper industry, tanneries, breweries, and other operations are available, however, their reutilization are not covered in this chapter.

QUALITY CRITERIA FOR THE USE OF WASTES

If waste materials are eventually converted into products for use in agriculture, horticulture, or forestry it has carefully to be evaluated which valuables (amount and binding forms of micro-, macronutrients, carbon compounds) and pollutants they contain. The pollutants of concern are heavy metals, toxins, antibiotics, and xenobiotics. In particular, because of heavy-metal loads, sewage sludges have been poorly reputed for years, their qualities in terms of lowered heavy-metal loads, however, are increasing.

GLOBAL CHANGE ASPECTS

The Marrakech Accords allow biospheric, including soil, C sinks and sources in forests, croplands and pastures to be included to meet emission reduction targets for the first commitment period of the Kyoto Protocol. It is estimated that European Union croplands lose 78 Tg (C) per year, thus there is significant potential to decrease the flux of C to the atmosphere from cropland, and for cropland management to sequester soil C, relative to the amount of C stored in cropland soils at present. The biological potential for C storage in European (EU 15) cropland is of the order of 90–120 Tg (C) per year, with a range of options available that include reduced and zero tillage, set-aside, perennial crops, deep rooting crops, and, importantly, efficient use of organic amendments (animal manure, sewage sludge, cereal straw, compost), improved rotations, irrigation, bioenergy crops, extensification, organic farming, and conversion of arable land to grassland or woodland. For socioeconomic and other constraints, a realistically achievable potential is estimated to be about 20% of the biological potential. If C sequestration in croplands is to be used in helping to meet emission reduction targets for the first commitment period of the Kyoto Protocol, the changes in soil C content must be measurable and verifiable, which is considered difficult within a 5-year commitment period. Soil C sequestration is a riskier long-term strategy for climate mitigation than direct reduction of C emissions. However, improved agricultural management often has a range of other environmental and economic benefits in addition to climate mitigation potential, and this may make attempts to improve soil C storage attractive as part of integrated sustainability policies (Marmo, 2008; Smith and Falloon, 2005; Wiesmeier et al., 2014).

SOIL SUSTAINABILITY

Apart from the important effects on mitigation of climate change, we should still keep in mind the beneficial effects of high organic matter pools in soils that aid the maintenance of a high microbiological diversity, high water and nutrient holding capacity, better textural properties of the soils that altogether improve the resistance and resilience toward disturbances. Increasing SOM may always be seen in the context of making agricultural, horticultural, and forest operations more sustainable. SOM models should help in quantifying the effect of organic matter management by returning C from wastes to fields.

REFERENCES

Abubaker, J., Risberg, K., Pell, M., 2012. Biogas residues as fertilizers e effects on wheat growth and soil microbial activities. Appl. Energy 99, 126–134.

Aescht, E., Foissner, W., 1992. Effects of mineral and organic fertilizers on the microfauna in a high-altitude reafforestation trial. Biol. Fertil. Soils 13, 17–24.

Alburquerque, J.A., de la Fuente, C., Bernal, M.P., 2011. Improvement of soil quality after "alperujo" compost application to two contaminated soils characterised by differing heavy metal solubility. J. Environ. Manage. 92 (3), 733–741.

Alburquerque, J.A., de la Fuente, C., Bernal, M.P., 2012a. Chemical properties of anaerobic digestates affecting C and N dynamics in amended soils. Agric. Ecosyst. Environ. 160, 15–22.

Alburquerque, J.A., de la Fuente, C., Ferrer-Costa, A., Carrasco, L., Cegarra, J., Abad, M., Bernal, M.P., 2012b. Assessment of the fertiliser potential of digestates from farm and agroindustrial residues. Biomass Bioenergy 40, 181–189.

Austrian Compost Ordinance, 2001. Verordnung des Bundesministers für Land- und Forstwirtschaft, Umwelt und Wasserwirtschaft über Qualitätsanforderungen an Komposte aus Abfällen (Kompostverordnung; BGBl. II Nr. 292/2001). https://www.ris.bka.gv.at/GeltendeFassung.wxe?Abfrage=Bundesnormen&Gesetzesnummer=20001486.

Beringer, T., Lucht, W., Schaphoff, S., 2011. Bioenergy production potential of global biomass plantations under environmental and agricultural constraints. GCB Bioenergy 3, 299–312.

Burton, J.P., Messier, C., Smith, D.W., Wiktor, L.A., 2003. Towards Sustainable Management of the Boreal Forest. NRC Research Press, Ottawa, Ontario, Canada.

Carmona, E., Moreno, M.T., Avilés, M., Ordovás, J., 2012. Use of grape marc compost as substrate for vegetable seedlings. Sci. Hortic. Amsterdam 137, 69–74.

Carrera, L.M., Buyer, J.S., Vinyard, B., Abdul-Baki, A.A., Sikora, L.J., Teasdale, J.R., 2007. Effects of cover crops, compost, and manure amendments on soil microbial community structure in tomato production systems. Appl. Soil Ecol. 37, 247–255.

Ceccherini, M.T., Ascher, J., Pietramellara, G., Mocali, S., Nannipieri, P., 2007. The effect of pharmaceutical waste-fungal biomass, treated to degrade DNA, on the composition of eubacterial and ammonia oxidizing populations of soil. Biol. Fertil. Soils 44, 299–306.

Coleman, K.W., Jenkinson, D.S., 1996. RothC-26. 3 – A model for the turnover of carbon in soil. Powlson, D.S., Smith, P., Smith, J. (Eds.), Evaluation of Soil Organic Matter Models

Using Existing Long-term Datasets, NATO ASI Series, 1 (38), Springer-Verlag, Heidelberg, pp. 237–246.
Danon, M., Franke-Whittle, I.H., Insam, H., Chen, Y., Hadar, Y., 2008. Molecular analysis of bacterial community succession during prolonged compost curing. FEMS Microbiol. Ecol. 65, 133–144.
de la Fuente, C., Alburquerque, J.A., Clemente, R., Bernal, M.P., 2013. Soil C and N mineralisation and agricultural value of the products of an anaerobic digestion system. Biol. Fertil. Soils 49, 313–322.
Ding, M.M., Yi, W.M., Liao, L.Y., Martens, R., Insam, H., 1992. Effect of afforestation on microbial biomass and activity in soils of tropical China. Soil Biol. Biochem. 24, 865–872.
Gerber, P., Chilonda, P., Franceschini, G., Menzi, H., 2005. Geographical determinants and environmental implications of livestock production intensification in Asia. Bioresour. Technol. 96, 263–276.
Goberna, M., Schoen, M.A., Sperl, D., Wett, B., Insam, H., 2010. Mesophilic and thermophilic co-fermentation of cattle excreta and olive mill wastes in pilot anaerobic digesters. Biomass Bioenergy 34, 340–346.
Gómez-Brandón, M., Podmirseg, S.M., 2013. Biological waste treatment. Waste Manage. Res. 31, 773–774.
Gómez-Brandón, M., Domínguez, J., 2014. Recycling of solid organic wastes through vermicomposting: microbial community changes throughout the process and use of vermicompost as a soil amendment. Crit. Rev. Environ. Sci. Technol. 44, 1289–1312.
Gómez-Brandón, M., Fernández-Delgado Juárez, M., Zangerle, M., Insam, H., 2016. Effects of digestate on soil chemical and microbiological properties: a comparative study with compost and vermicompost. J. Hazard. Mater. 302, 267–274.
Gómez-Brandón, M., Ascher-Jenull, J., Bardelli, T., Fornasier, F., Fravolini, G., Arfaioli, P., et al., 2017. Physico-chemical and microbiological evidence of exposure effects on *Picea abies*–coarse woody debris at different stages of decay. Forest Ecol. Manage. 391, 376–389.
Grismer, M.E., Ross, C.C., Valentine, Jr., G.E., Smith, B.M., Walsh, Jr., J.L., 2002. Food-processing wastes. Water Environ. Res. 74 (4), 377–384.
Haselwandter, K., Krismer, R., Holzmann, H., Reid, C.P.P., 1988. Hydroxamate siderophore content of organic fertilizers. J. Plant Nutr. 11, 959–967.
Henry, R.J., 2010. Evaluation of plant biomass resources available for replacement of fossil oil. Plant Biotechnol. J. 8, 288–293.
Hofer, R., 2009. History of the sustainability concept renaissance of renewable resources. Sustainable Solutions for Modern Economies. RSC Publishing, RSC Green Chemistry Seriesdoi: 10.1039/9781847552686-00001, 1-11.
Insam, H., Parkinson, D., Domsch, K.H., 1989. Influence of macroclimate on soil microbial biomass. Soil Biol. Biochem. 21, 211–221.
Insam, H., 1990. Are the soil microbial biomass and basal respiration governed by the climatic regime? Soil Biol. Biochem. 22, 525–533.
Insam, H., Wett, B., 2008. Control of GHG emission at the microbial community level. Waste Manage. 28, 699–706.
Insam, H., Gómez-Brandón, M., Ascher, J., 2015. Manure-based biogas fermentation residues-friend or foe of soil fertility. Soil Biol. Biochem. 84, 1–14.
Jayathilakan, K., Sultana, K., Radhakrishna, K., Bawa, A.S., 2012. Utilization of byproducts and waste materials from meat, poultry and fish processing industries: a review. J. Food Sci. Technol. 49, 278–293.
Johansson, K., Liljequist, K., Ohlander, L., Aleklett, K., 2010. Agriculture as provider of both food and fuel. AMBIO 39, 91–99.
Kosseva, M.R., 2009. Processing of food wastes. Adv. Food Nutr. Res. 58, 57–136.

Kosseva, M.R., 2013. Food industry wastes: assessment and recuperation of commodities. In: Taylor, S.L. (Ed.), Chapter 3 – Sources, Characterization and Composition of Food Industry Wastes. Academic Press, pp. 37–60.

Liu, C., Lu, M., Cui, J., Li, B., Fang, C., 2014. Effects of straw carbon input on carbon dynamics in agricultural soils: a meta-analysis. Global Change Biol. 20, 1366–1381.

Marmo, L., 2008. EU strategies and policies on soil and waste management to offset greenhouse gas emissions. Waste Manage. 28, 685–689.

Miezah, K., Obiri-Danso, K., Kádár, Z., Fei-Baffoe, B., Mensah, M.Y., 2015. Municipal solid waste characterization and quantification as a measure towards effective waste management in Ghana. Waste Manage. 46, 15–27.

Murphy, J.A., Smith, P.M., Wiedenbeck, J., 2007. Wood residue utilization in Pennsylvania: 1988 vs 2003. Forest Prod. J. 57, 101–106.

Nannipieri, P., Ascher, J., Ceccherini, M.T., Landi, L., Pietramellara, G., Renella, G., 2017a. Landmark paper: microbial diversity and soil functions. Eur. J. Soil Sci. 68, 1–26.

Nannipieri, P., Ascher-Jenull, J., Ceccherini, M.T., Giagnoni, L., Pietramellara, G., Renella, G., 2017b. Landmark papers: No. 6. Eur. J. Soil Sci. 68, 2–5, Reflections by Nannipieri, P., Ascher, J., Ceccherini, M.T., Landi, L., Pietramellara, G., Renella, G., 2003. Microbial diversity and soil functions: reflections. Eur. J. Soil Sci. 54, 655–670.

Odlare, M., Arthurson, V., Pell, M., Svensson, K., Nehrenheim, E., Abubaker, J., 2011. Land application of organic waste-effects on the soil ecosystem. Appl. Energy 88, 2210–2218.

Parton, W.J., Schimel, D.S., Cole, C.V., Ojima, D.S., 1987. Analysis of factors controlling soil organic matter levels in Great Plains grasslands. Soil Sci. Soc. Am. J. 51, 1173–1179.

Pentti, H., Anssi, N., Andreas, O., Markku, T., Johanna, V., 2002. Forest Related Perspectives for Regional Development in Europa. Research report 13. Brill: Academic Publishers, Leiden, The Netherlands.

Petrillo, M., Cherubini, P., Fravolini, G., Ascher, J., Schärer, M., Synal, H.-A., et al., 2016. Time since death and decay rate constants of Norway spruce and European larch deadwood in subalpine forests determined using dendrochronology and radiocarbon dating. Biogeosciences 13, 1537–1552.

Pietramellara, G., Ceccherini, M.T., Ascher, J., Nannipieri, P., 2006. Persistence of transgenic and not transgenic extracellular DNA in soil and bacterial transformation. Biol. Forum 99, 37–68.

Pietramellara, G., Ascher, J., Borgogni, F., Ceccherini, M.T., Guerri, G., Nannipieri, P., 2009. Extracellular DNA in soil and sediment: fate and ecological relevance. Biol. Fertil. Soils 45, 219–235.

Reetz, Jr., H.F., 2016. Fertilizers and Their Efficient Use, first ed. International Fertilizer Industry Association (IFA), France.

Röös, E., Patel, M., Spångberg, J., Carlsson, G., Rydhmer, L., 2016. Limiting live-stock production to pasture and by-products in a search for sustainable diets. Food Policy 58, 1–13.

Ros, M., Pascual, J.A., Garcia, C., Hernandez, M.T., Insam, H., 2006a. Hydrolase activities, microbial biomass and bacterial community in a soil after long-term amendment with different composts. Soil Biol. Biochem. 38, 3443–3452.

Ros, M., Klammer, S., Knapp, B.A., Aichberger, K., Insam, H., 2006b. Long term effects of soil compost amendment on functional and structural diversity and microbial activity. Soil Use Manage. 22, 209–218.

Skog, K., Lebow, P., Dykstra, D., Miles, P., Stokes, B.J., Perlack, R.D., et al., 2011. Forest biomass and wood waste resources. U.S. Billion-Ton Update: Biomass Supply for a Bioenergy and Bioproducts Industry: Chapter 32011Oak Ridge National Laboratory, Oak Ridge, Tenn, 16-51.

Smith, P., Falloon, P., 2005. Carbon sequestration in European croplands. SEB Experimental Biology Series, 47-55.

Sorathiya, L.M., Fulsounder, A.B., Tyagi, K.K., Patel, M.D., Singh, R.R., 2014. Eco-friendly and modern methods of livestock waste recycling for enhancing farm profitability. Int. J. Recycl. Org. Waste Agric. 3, 50.

Top, Y., 2015. Waste generation and utilisation in micro-sized furniture-manufacturing enterprises in Turkey. Waste Manage. 35, 3–11.

Usman, K., Khan, S., Ghulam, S., Khan, M.U., Khan, N., Khan, M.A., et al., 2012. Sewage sludge: an important biological resource for sustainable agriculture and its environmental implications. Am. J. Plant Sci. 3, 1708–1721.

Valkenburg, C., Walton, C.W., Thompson, B.L., Gerber, M.A., Jones, S., Stevens, D.J., 2008. Municipal solid waste (MSW) to liquid fuels synthesis. Volume 1: Availability of Feedstock and Technol. PNNL 18144. Pacific Northwest National Laboratory, Richland, WA.

Vanerkar, A.P., Fulke, A.B., Lokhande, S.K., Giripunje, M.D., Satyanarayan, S., 2015. Recycling and treatment of herbal pharmaceutical wastewater using *Scenedesmus quadricuada*. Curr. Sci. India 108, 979–983.

Wang, G., Luo, Z., Han, P., Chen, H., Xu, J., 2016. Critical carbon input to maintain current soil organic carbon stocks in global wheat systems. Sci. Rep.-UK 6, 19327.

Wiesmeier, M., Hübner, R., Spörlein, P., Geuß, U., Hangen, E., Reischl, A., et al., 2014. Carbon sequestration potential of soils in southeast Germany derived from stable soil organic carbon saturation. Glob. Change Biol. 20, 653–665.

CHAPTER 8

Soil Erosion and C Losses: Strategies for Building Soil Carbon

Felipe Bastida, Teresa Hernández, Carlos García
CEBAS-CSIC, Campus Universitario de Espinardo, Murcia, Spain

Chapter Outline

SOIL EROSION: RATES AND C DYNAMICS

Soil erosion is one of the biggest threats to European soils. Erosion degrades soil ecosystems and can lead to reduced crop yields, threatening food security and farmers' incomes. Water eroding soil can transport agricultural pollutants into waterways, and also, wind-eroding soil may transport soil particles and pollution to other sites. With the erosion, soil (and C) can be lost (Berhe et al., 2007; Lal, 2003).

Protecting European soils from erosion is, therefore, a priority under the European Commission's soil protection thematic strategy. Under the Good Agricultural and Environmental Condition (GAEC) requirements, member states are obliged to prevent soil erosion and maintain soil organic matter through minimal soil cover maintenance, minimum land management reflecting site-specific conditions to soil loss, and maintenance of soil organic matter level (Borrelli et al., 2016).

The susceptibility of a soil to water erosion is primarily determined by the erosive potential of the rainfall, the slope of the land surface and position

The Future of Soil Carbon
http://dx.doi.org/10.1016/B978-0-12-811687-6.00008-0

of the soil in the catchment, and the vegetative cover on the soil surface (FAO, 2015). Typical soil erosion rates by water have been calculated for representative agroecological conditions. Hilly croplands under conventional agriculture and orchards without additional soil cover in temperate climate zones are subject to erosion rates up to 10–20 tonnes ha^{-1} $year^{-1}$, while average rates are often < 10 tonnes ha^{-1} $year^{-1}$. Erosion rates on hilly croplands in tropical and subtropical areas may reach values up to 50–100 tonnes ha^{-1} $year^{-1}$ whereas average rates are around 10–20 tonnes ha^{-1} $year^{-1}$. These high rates are due to the combination of an erosive climate (high-intensity rainfall) and slope gradients, which are generally steeper than those on cultivated land in the temperate zones. In addition, in these areas the incidence of erosion is due to the combination of a high population pressure with low-intensity agriculture, leading to the cultivation of marginal steeplands. Rangelands and pasturelands in hilly tropical and subtropical areas may suffer erosion rates similar to those of tropical croplands. Due to the lack of field boundaries, which often act as barriers for sediment and runoff and promote infiltration, these rangelands may also be particularly vulnerable to gully formation. This may not affect topsoil so much but may make land inaccessible and hence unusable. Rangelands and pasturelands in temperate areas are characterized by erosion rates, which are generally much lower being usually below 1 tonnes ha^{-1} $year^{-1}$. These rangelands are less intensively used and better managed than (sub-) tropical rangelands (FAO, 2015).

Soil erosion enhances soil degradation with a long-term decline in soil productivity and in the capacity of soil to carry out ecosystem functions (Lal, 2001). Half of the topsoil on the planet has been lost in the last 150 years. The effects of soil erosion go beyond the loss of fertile land because it has led to increased pollution and sedimentation in streams and rivers, clogging these waterways and causing declines in the abundance of fish and other species. The sustainable land use can help to reduce the impacts of agriculture and livestock, preventing soil degradation and erosion and the loss of valuable land to desertification.

Data from the European Environment Agency shows that 105 million ha, or 16% of Europe's total land area (excluding Russia), were affected by water erosion in 1990s (Fig. 8.1). About 42 million ha of land were estimated to be affected by wind erosion, of which around 1 million ha were categorized as being severely affected. The last models indicate that almost 20% of the EU-27 land area is subjected to soil loss in excess of 10 tonnes ha^{-1} yr^{-1}. The increased aridity that will be enhanced by climate change in many regions of the planet, will make finer-textured soils more

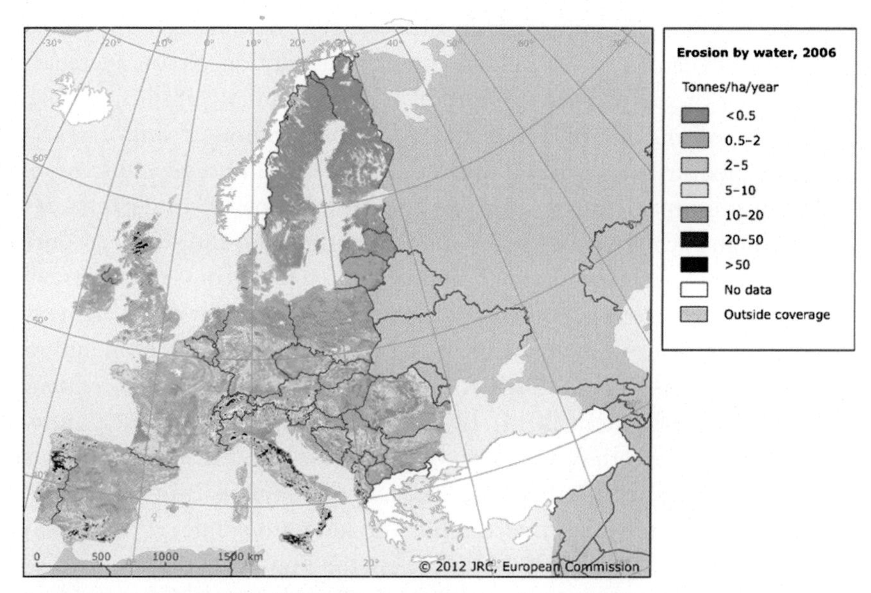

Figure 8.1 ***Estimated soil erosion by water in Europe.*** *(http://www.eea.europa.eu/data-and-maps/indicators/soil-erosion-by-water-1/assessment)*

vulnerable to erosion, especially if accompained by a decrease in soil organic matter levels.

The total land area affected by severe erosion (including water plus wind erosion) is 132 Mha in Europe, 20 Mha in Oceania, 267 Mha in Africa, 407 Mha in Asia, 93 Mha in South America, 50 Mha in Central America, and 78 Mha in North America (Lal, 2003). There are also several regions with high rates of erosion such Himalayan–Tibetean ecosystem in South Asia, the Loess Plateau in China, the subhumid and semiarid regions of sub-Saharan Africa, highlands of Central America, the Andean region, Haiti and the Caribbean area (Lal, 2003; Scherr and Yadav, 1996).

Other chapters have discussed the importance of the organic C in the ecosystem services provided by soil. As mentioned earlier, one of the main consequences of soil erosion is the loss of soil organic matter in upper horizons. Because it is well established that the soil organic matter is related to soil quality (Bastida et al., 2006), the loss of soil C is often recognized as a cause of soil degradation (Lal, 2003). The impacts of erosion in soil degradation, productivity, and food security are widely known, however, its effects on the C dynamics and on the emission of greenhouse gases are poorly known (Lal, 2003). Soil erosion is a four-stage process involving detachment, breakdown, transport and redistribution, and deposition of sediments.

The soil organic C is influenced in all these steps. Soil erosion impacts soil organic C dynamics by its influence on: (1) disruption of soil aggregates; (2) removal of soil organic C by ruoff water or dust storms; (3) mineralization of soil organic matter in situ; (4) mineralization of soil organic C displaced, redistribution of soil organic C over the landscape and transportation of soil organic matter in rivers and dust storms; (5) reaggregation of soil through formation of organi–mineral complexes; and (6) deep burial of C-enriched sediments in depositional sites, flood plains, reservoirs, and ocean floor (Lal, 2003). Consequently, the cycle of C (and other elements) is strongly influenced by soil erosion, which affects their fluxes in and out of the soil system, storage, distribution within the soil matrix, and residence time in soil (Berhe et al., 2014). Every year, soil erosion distributes around 75 Gt of topsoil and the amount of total C removed by erosion on the earth is around 4–6 Pg yr^{-1} (Berhe et al., 2007). Interestingly, part of the removed C can be easily mineralized, accounting for an erosion-induced emission of 0.8–1.2 Pg C yr^{-1} (Lal, 2003). Nevertheless, the impacts of erosion in C dynamics are highly dependent on the type of landform position considered. For instance, Berhe et al. (2008) observed that the redistribution of soil C with erosion and deposition resulted in a considerable increase in C storage downslope. Moreover, these authors found that the eroded C was replaced by new photosynthate C and that the rate of organic C decomposition in the depositional setting was slower to that in the eroding position, suggesting that erosion can be a C sink process (Berhe et al., 2008).

CAUSES AND IMPACT OF SOIL EROSION AND C LOSS

As the human population has expanded, more and more land has been cleared for agriculture and other pursuits that degrade the soil and make erosion more likely to occur. When agriculture fields replace natural vegetation, topsoil is exposed and can dry out. The diversity, activity, and biomass of microbial communities, all relates to soil fertility, can decrease, and nutrients may wash out. As already mentioned soil can be blown away by the winds or washed away by rains, if it is not covered by plants (Albaladejo et al., 2000, 2008).

The conversion of natural ecosystems to pasture can lead to high rates of erosion and loss of topsoil, nutrients, and organic C. Overgrazing can reduce groundcover, enabling erosion and compaction of soil, with the consequent reduction in plant growth and water penetration. Under certain conditions the loss of fertile soil can lead to desertification, with huge economic costs for nations where deserts are growing.

In addition to its on-site effects, the soil that is detached by accelerated water or wind erosion may be transported to considerable distances, rising "off-site problems." For example, the movement of sediment and agricultural pollutants into watercourses can lead to the silting-up of dams, disruption of lake ecosystems, and contamination of drinking water (Lal, 2001). In some cases, increased downstream flooding may also occur due to the reduced capacity of eroded soil to absorb water.

Strategies of soil restoration that can reduce soil erosion and C losses include the promotion of sustainable agriculture practices. Farming is the world's largest industry, employing a billion people who produce more than $1 trillion of food annually. We should promote the use and development of sustainable agriculture that preserves and restores critical habitats, helps to protect watersheds, and improves the health of soil and water. In addition, we should reduce deforestation, which is a factor causing soil erosion (García-Franco et al., 2015). Given the extent of deforestation around the world, zero net deforestation is impossible to be obtained. However, actions should be taken in this context as done by some nations like Paraguay, which reduced the rate of deforestation by 85% after the 2004 Zero Deforestation Law.

CLIMATE CHANGE AND C DYNAMICS

One the most serious environmental problems is how climate change will impact the overall C dynamics in the planet. Soil, as one of the major soil organic C compartment in the planet, may be severely impacted by climate change but how and how much are not known. For instance, soil respiration due to microbial and root-respired CO_2 is the second-largest terrestrial carbon flux (Schimel, 1995) and it is affected by warming (Bond-Lamberty and Thomson, 2010).

Northern peatlands are rich in soil organic matter and contain one third of total soil organic C in the planet. If the decomposition of these soil organic C is accelerated by climate change-derived factors, a marked impact on the global climate will occur. For example, methane emissions from permafrost in high-latitude regions can increase due to warming, permafrost thaw, and CO_2 fertilization (Koven et al., 2011). Dorrepaal et al. (2007) observed that 1°C warming increased total ecosystem CO_2 released rates from peat of about 60% and such effect was sustained for 8 years. About 69% of the respired CO_2 came from C in permafrost peat. Therefore, climate warming may accelerate the decline of soil organic C in permafrost with long-term positive feedbacks to the global climate. Nevertheless, the impacts of warming on C cycling depended on the temperature range on which warming occurs in the Swiss Central Alps (Ferrari et al., 2016). In Illinois, the combined elevated CO_2 and warming increased respiration from soil microbial community by 20%, while warming reduced respiration from roots and rhizosphere by 25% (Black et al., 2017).

Variations in temperature and precipitacion may alter both biotic and abiotic factors that control C immobilization in arid and semiarid soils (Evans and Wallenstein, 2012). The positive microbial community feedbacks in response to elevated CO_2 concentration and warming can accelerate microbial decomposition and lead to soil C losses (Nie et al., 2013). At the global level, the effects of temperature in the decomposition of SOM is controversial (Giardina and Ryan, 2000), because other studies have pointed that global emissions of CO_2 will increase as a consequence of OM decomposition by raising temperatures (Jones et al., 2005), whereas, in contrast, Shen et al. (2009) suggested that dryland soils will likely sequestrate CO_2 with an increase in future precipitation but release C with the decrease in future precipitation, as also discussed by Albaladejo et al. (2013).

Episodic water availability clearly affects erosion and element cycling in arid and semiarid ecosystems (Austin et al., 2004; Gebauer and Ehleringer, 2000). High temperatures and erratic moisture inputs generate a pulsed pattern on biological activities (Collins et al., 2008), thus affecting the C and N turnover; so organic matter tends to accumulate during dry periods when plant and microbial growth are restricted. Moreover, drought affects the quality and composition of humic acids in semiarid soils but these effects can be ameliorated by organic amendments (Hueso et al., 2012).

The low rates of organic matter accumulation in arid and semiarid soils are due to several factors, including microclimate, low net primary productivity, microbial and abiotic degradation, and the presence of invertebrates. Accumulated microbial and plant necromass, lysis of live microbial cells, release of solutes and exposure of previously protected organic matter may explain the additional mineralization during wetting of dry soils (Borken and Matzner, 2009). Nevertheless, soil processes in arid lands are principally controlled by water availability; but the photodegradation of aboveground litter and the overriding importance of spatial heterogeneity can modulate biotic responses to water availability (Austin, 2011). Liu et al. (2009) suggested that soil water availability is more important than temperature in regulating soil microbial respiration and microbial biomass in a semiarid temperate steppe. In this sense, it has been observed that organic matter

(OM) stocks can be preserved by the increased duration and intensity of drought periods (Conant et al., 2000). Conversely, the OM content decreased with aridity under Mediterranean conditions (Lavee et al., 1998). Recently, Zhang et al. (2016) revealed that semiarid regions of the Eurasian Steppe could act as a net carbon source under increased temperatures, unless preciptiation increases concurrently.

A major concern is also the impact of agriculture in Asia. Paddy rice is the main cropping system in Southeast Asia. The reduced water availability in some regions has forced to swiths into a less irrigation intensive upland cropping system. Simulation results suggest that approximately 2.8–3.4 t C ha^{-1} $year^{-1}$ residue incorporation after harvest is needed to achieve stable SOC stocks after switching to a lower irrigation manegement in a padding rice agroecosystem (Kraus et al., 2016).

LAND USE IN DEPLETED ORGANIC C SOILS

Arid and semiarid soils are located in extensive agricultural regions such California, Israel, southeast Spain, Greece, and southern Italy, and hence may suffer the consequences of a human overexplotation. These ecosystems have less vegetation and hence lower organic C accumulation than boreal or tropical areas, but they are estimated to contain 20% of the global soil C pool (organic plus inorganic). Lal (2004) suggested that the predicted amounts of C in drylands are 159–191 billion of tons ranging from 35 to 42 (tons C ha^{-1}). If we compare these ranges with those estimated for boreal (247–344 tons C ha^{-1}), tropical (121–113 t tons C ha^{-1}) and tundra (121–127 tons C ha^{-1}) ecosystems, dry soils are C depleted on carbon, both for "natural" or "anthropogenic reasons." The increase of the relative SOC contents can increase crop yield and soil quality to ensure food security by 2025 (Lal, 2006).

The organic matter content of soil depends on physical, chemical, biochemical, and biological properties; for example, SOM can be physically protected into aggregates against microbial mineralization (Six et al., 2002; von Lutzow et al., 2006). Changes on both land uses and climate may alter C stocks in soil, causing C losses and soil degradation.

Human impact on soil erosion occurs through agriculture and this impact is critical in areas with low soil organic C content. For instance, tillage can increase soil erosion (Lal, 2001; Quine et al., 1999). However, the direct influences of agriculture on soil erosion and on global C cycle are controversal at a planetary scale. van Oost et al. (2007) suggested an erosion-sink of atmospheric C equivalent to approximately 26% of the C transported by erosion. These authors estimated a global sink of 0.12 pentagrams of C year^{-1} resulting from erosion in agroecosystems.

Adequate land-use management helps to control global stocks of organic C in drylands and prevent soil desertification (Albaladejo et al., 2013; de Baets et al., 2012). However, despite the extensive bibliography on the effects of land use on organic C stocks, there are still some discrepancies. For instance, the agriculture use of native arid and semiarid soils can decrease stocks of C in these soils (Abule et al., 2005; Girmay et al., 2008; Pérez-Quezada et al., 2011; Steffens et al., 2008). Disturbance by shrub removal and/or livestock grazing significantly reduced the amount of SOC in an Australian semiarid woodland (Daryanto et al., 2013). However, other studies did not find a significant effect of land management on SOC contents (Seddaiu et al., 2013; Steffens et al., 2009). As stated by Booker et al. (2013), the C inputs to arid and semiarid soils are most often controlled by abiotic factors not easily changed by management or vegetation. In this sense, photodegradation, which is highly intense in arid ecosystems, exerts a dominant control on aboveground litter decomposition (Austin and Vivanco, 2006). Losses through photochemical reactions may represent a short circuit in the C cycle, with a substantial fraction of carbon fixed in plant biomass being lost directly to the atmosphere without cycling through soil organic matter pools (Austin and Vivanco, 2006).

Refforestation may impact C balances, increase soil C stocks, and prevent erosion and desertification in many arid and semiarid regions (Hu et al., 2013; Li et al., 2012; Ringrose et al., 1998; Su et al., 2010). In general, arid and semiarid soils show a positive relationship between SOC content and plant cover (García et al., 2002, 2005). Nevertheless, the spatial heterogeneity of plant cover in semiarid shrub lands is the principal cause of the spatial heterogeneity of SOM content, which is associated with the development of island of fertility under shrubs (Schlesinger et al., 1996).

THE IMPORTANCE OF MICROBIAL COMMUNITIES IN SOILS WITH LOW ORGANIC C CONTENT

Drylands occupy 6.31×10^9 ha or 47% of the earth's land area (UNCED, 1992) and are distributed among four climates: hyperarid (1.0×10^9 ha), arid (1.62×10^9 ha), semiarid (2.37×10^9 ha), and dry subhumid (1.32×10^9 ha). Arid and semiarid or subhumid zones are characterized by low and erratic rainfall, periodic droughts, and soils with different plant covers. The annual rainfall varies from up to 350 mm in arid zones to 700 mm in semiarid areas.

Either due to human-induced actions or natural conditions, the loss of soil organic C is linked to soil erosion, degradation, and desertification in arid and semiarid areas, and causes a decline in agronomical productivity and in soil ecosystem services.

As already mentioned arid and semiarid soils contain less organic C than soils under other regions (Lal, 2004). Therefore, any further loss of organic C in arid and semiarid regions may have devastating ecological and societal impacts. For instance, it is known that agriculture practices, such as tillage, can reduce the amount of soil organic C by increasing mineralization (Dick, 1992; Zuber and Villamil, 2016). Furthermore, the decrease of water availability can reduce microbial activity and mineralization in arid and semiarid soils (Bastida et al., 2017). The degradation of these soils may impede their ecosystem functioning, as acting as a sink for atmosphere C through improving plant growth. The degradation of soils due to C losses in many arid and semiarid areas of the planet cannot be afforded in the future because of two reasons. First, many of these areas are located in extensive agricultural zones, which must provide enough food for a growing population. Second, the global change mitigation by C sequestration by these soils can play a key role.

It is well established that microbial communities determine C dynamics (Fontaine et al., 2003; Kuzyakov, 2010). Microbial biomass in arid and semiarid soils is usually constrained by the low amounts of plant inputs and the water availability (Bastida et al., 2008a; Ben-David et al., 2011; Drenovsky et al., 2010; Hortal et al., 2013). Usually, microbial biomass is correlated with the amount of soil organic C (Bastida et al., 2016) and also depends on moisture of dry soils. Land use that affects organic C content is also reflected in a change of microbial biomass (Entry et al., 2004; Jia et al., 2010). Furthermore, microbial biomass responded to plant growth and the parallel changes in soil organic matter (Hortal et al., 2013).

Wet–dry cycles can initially enhance the mineralization of soil organic matter and usually microbial biomass decreases during drying, whereas it increases after rewetting (Fierer and Schimel, 2002; Huxman et al., 2004).

Accumulation of inorganic N usually occurs during dry periods because diffusion of ions is restricted in the thin water films of dry soil and both microbial growth and root nutrient uptake decrease (Stark and Firestone, 1995). A portion of the microbial biomass is killed under dry conditions (Bottner, 1985), and the relative necromasses are decomposed by surviving organisms when the soil is rewetted. This leads to high N mineralization and large pulses of CO_2 (Placella et al., 2012).

As suggested by Entry et al. (2004), the abundance of Gram-positive bacteria increase in desiccated or degraded soils due to their sporulation capacity under harsh conditions. Indeed, the ratio between Gram-positive and Gram-negative bacteria is a suitable marker of microbial stress (de Vries and Shade, 2013).

Usually enzyme activities are positively correlated to soil organic C contents (Nannipieri et al., 2012). The potential activities of enzymes that decompose proteins (e.g., aminopeptidase) and recalcitrant C compounds, such as lignin and humic substances (e.g., phenol oxidases) are higher in arid and semiarid than in more temperate soils (Stursova and Sinsabaugh, 2008). In arid and semiarid soils, pH can reach or be higher than 8 due to the great content of carbonates and strongly modulates enzyme activity. High pH has been suitable for phenol oxidase activity (Sinsabaugh et al., 2002). In contrast, the pH optima of glycosidases (e.g., cellulase, chitinase) generally range from 4 to 6.

STRATEGIES FOR BUILDING SOIL C

As already mentioned tillage practices generally cause loss of soil organic C with formation and evolution of CO_2 to the atmosphere. On the contrary, conservation tillage practices, which include minimum tillage and no-till, conserve soil C and reduce that amount of fossil fuel needed for tillage.

Kern and Johnson (1993) calculated that the amount of soil C that would either be lost or sequestered and the amount of fossil fuel required for agriculture by considering three different scenarios of conservation tillage in USA from 1990 to 2020 were: 27% (Scenario 1), 57% (Scenario 2), and 76% (Scenario 3). In the case of Scenario 1, the level of conservation tillage was held constant at the actual 1990 level of 27% for 30 years; for Scenarios 2 and 3, tillage was simulated to begin at 27% in 1990 and linearly increased up to 57% and 76%, respectively, over the first 20 years of the analysis and then held constant for the remaining 10 years.

Today, the scarce organic C (and N) in some soils of the planet limit soil functionality and the relative ecosystem services. Adequate practices such soil restoration, reduced tillage or afforestation are critical to promote C sequestration in soil.

Soil Restoration With Organic Amendments (Exogenous C Addition)

There is a strong link between the biomass and composition of the plant community, microbial diversity, and their ecosystemic functions, and the soil organic C content and type of arid and semiarid soils (Bastida et al., 2016; Delgado-Baquerizo et al., 2016). Overall, the decline of organic C content negatively impacts microbial biomass and physical structure of soils. Therefore, it is needed to increase the soil organic C content thus improving: (1) physical structure by promoting the formation of aggregates and macroporosity (Caravaca et al., 2002; Yazdanpanah et al., 2016); (2) the source of nutrients for plants and microbes (Ros et al., 2003; Tejada et al., 2006); (3) water retention and reducing soil erosion (Albaladejo et al., 2000) (Fig. 8.2).

Organic amendments are a source of organic matter and can compensate the scarce input from plant debris in arid and semiarid soils (Bastida et al., 2008a, 2012; García et al., 2002). Thus they can improve soil quality with beneficial effects on environmental perspectives (Bastida et al., 2007, 2008a, 2013, 2015; García et al., 1992; Pascual et al., 1999, 1997), agricultural productivity (Baldi et al., 2010; Jindo et al., 2016), and restoration of degraded soils (Luna et al., 2016; Pardo et al., 2016; Zornoza et al., 2016). In the case of arid and semiarid soils different organic amendments, such as crop residues, farm yard manure, wastewater sludges, olive mill wastes, can be used (Table 8.1).

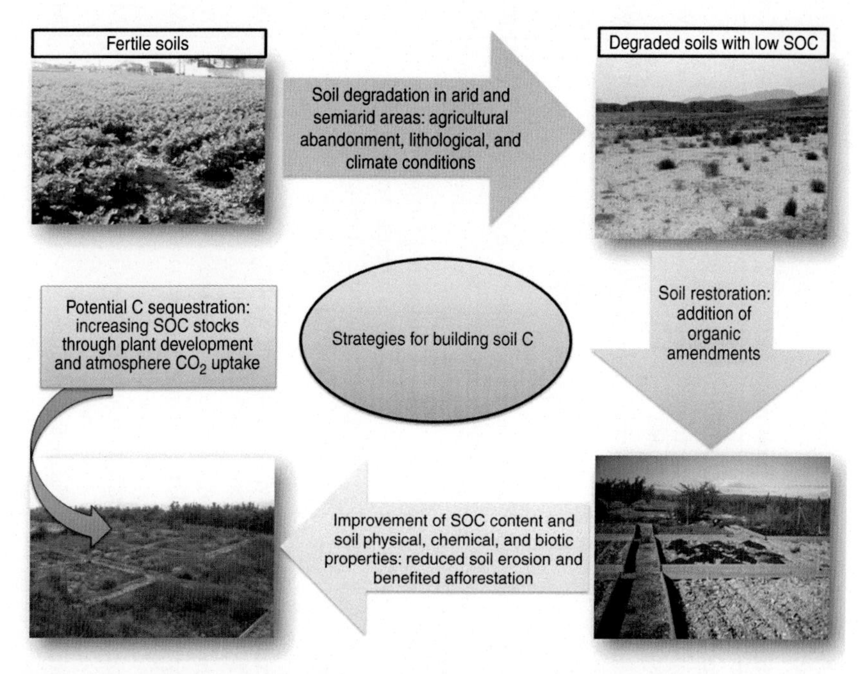

Figure 8.2 ***Strategies for building soil organic C (SOC) in arid and semiarid areas through the application of organic amendments.***

Table 8.1 Examples of long-term field experiments where organic amendments have been applied for increasing soil productivity and quality

Location	Organic materials	Crop	Time	References
Murcia, Spain	Fresh and composted vegetal wastes	None	7 years	Torres et al. (2015)
Bennett, CO, USA	Anaerobically digested biosolids	Wheat	12 years	Barbarick and Ippolito (2007)
Ravenna, Italy	Municipal-industrial wastewater sludge	Wheat–maize–sugarbeet rotation	12 years	Mantovi et al. (2005)
Toledo, Spain	Barley straw and crop waste; cattle manure	Barley, wheat, and sorghum	16 years	Dorado et al. (2003)
Punjab, India	Rice straw compost	Rice-wheat rotation	10 years	Sodhi et al. (2009)
Obere Lobau, Austria	Biowaste compost	Cereals and potatoes	10 years	Erhart et al. (2005, 2008)
Turin, Italy	Cattle slurry, Composted farmyard manure	Maize	11 years	Monaco et al. (2008)
Linz, Austria	Composts of: urban organic waste, green waste, cattle manure, or sewage sludge	Maize, wheat, and barley rotations	12 years	Ros et al. (2006)
Murcia, Spain	Sewage sludge and compost	None	Up to 11 years	Bastida et al. (2008a, 2015)
Murcia, Spain	Municipal solid waste	None	17 years	Bastida et al. (2008b, 2013)
Orange, VA, USA	Anaerobically digested sewage sludge	Barley, radish, and lettuce	19 years	Sukkariyah et al. (2005)
Tokyo, Japan	Dried sewage sludge; rice husk compost; sawdust compost	Maize, barley, and rye	19 years	Kunito et al. (2001)
Bologna, Italy	Cattle manure, cattle slurry, wheat or corn residues	Maize-wheat rotation	34 years	Triberti et al. (2008)
Central Sweden	Farmyard manure	Maize	47 years	Elfstrand et al. (2007)
Towada, Japan	Rice straw compost	Potato, maize, and soybean rotation	60 years	Takeda et al. (2005)

However, for safety reason, adequate composting of these residues is needed for their stabilization and sanitization. Usually, the surface application of these amendments is not always effective for improving soil quality (Rostagno and Sosebee, 2001) and ploughing should be used. However, the application of organic amendments for restoring soil quality must be done carefully and is not without risks (Plaza et al., 2004). For instance, Tejada et al. (2007) observed that the application of beet vinasse impacted negatively physical and biological soil properties due to the presence of sodium ions (Tejada et al., 2007). The application rate of organic amendments is also important even if a rate increase does not always improve soil quality (Bastida et al., 2007). Interestingly, a singular application of organic amendment is able to maintain long-term positive effects in soil quality (Bastida et al., 2007; Diacono and Montemurro, 2010; Torres et al., 2015). As already mentioned organic amendments improve the soil physical properties and reduce soil erosion (Albaladejo et al., 2008; Caravaca et al., 2001; Díaz et al., 1994). However, in dryland agroecosystems, the application of organic amendments must be accompanied with conservational management practices, in order to maintain C inputs and soil structure (Lal, 2004).

Despite metagenomics approaches showing that microbial diversity was not affected by organic amendments (Cherif et al., 2009; Crecchio et al., 2004), Cytryn et al. (2011) found that Bacteroidetes preferentially proliferated in compost-amended soil. The abundance of Bacteroidetes, copiotrophs, increases by increasing organic C availability (high C mineralization) (Bastida et al., 2013), indicating that this phylum plays an important role in decomposing organic materials in soil (Fierer et al., 2007). Accordingly, in southwest Sydney (Australia), Ge et al. (2010) found that the effect of organic waste on the structure of the soil microbial community depended on the residue type but also the addition rate according to Nicolás et al. (2017). Moreover, Chen et al. (2016a) found that the fungal and bacterial communities respond differently to the long-term organic and inorganic fertilization. Other studies have evaluated the long-term impacts of fertilization in the P cycle and observed that soil practices, including organic and inorganic fertilization, influenced the microbial composition of the alkaline phosphomonoesterase gene (Luo et al., 2017).

Recent advances in soil metaproteomics have permitted to identify proteins and their respective functionalities in arid and semiarid soils restored by the application of organic amendments (Bastida et al., 2015). Particularly organic amendments improved ecosystem functions and affected the cellular functionality of microbial populations through effects on the amount of proteins involved in translation, transcription, energy production, and C-fixation in soil.

Biochar is the C-rich solid material obtained from biomass through a pyrolysis process and can be used to enhance soil quality and mitigate climate change (Lehmann, 2007) being a porous and low density C rich material (Beesley et al., 2011). Due to its structure, biochar C is very stable in soil for hundreds to thousands of years (Kuzyakov et al., 2009). Biochar can stimulate the mineralization of native soil organic C (Wardle et al., 2008) but Sánchez-García et al. (2016) concluded that biochar alone did not alter N dynamics, but markedly increased soil organic C content under semiarid Mediterranean conditions.

Beesley et al. (2011) suggested that soil amendment with the combination of composts, manures, and biochar could be an effective strategy for revegetation, probably due to the positive effects of biochar on soil physical properties. Indeed, biochar can reduce bulk density and increase soil porosity and water-holding capacity of soil and the effects depend on biochar rates, as reviewed by Omondi et al. (2016). Therefore, biochar amendment can reduce soil erosion through the formation of aggregates (Jien and Wang, 2013). Furthermore, it has been observed that the use biochar either alone or combined with mineral and organic fertilizers did not increase abiotic and biotic CO_2 emissions from a semiarid calcareous soil (Fernández et al., 2014). It was also shown that after 182 days of incubation, soil organic and microbial biomass C contents increased by increasing the application rates of organic fertilizer, especially of compost, whereas increasing the rate of added mineral fertilizer decreased microbial biomass. Biochar addition increased both organic and inorganic contents of carbon.

Foster et al. (2016) also studied the impact of biochar and manure amendments on soil nutrients and enzyme activities in a semiarid irrigated maize cropping system. Biochar increased soil total C content but had little effects on microbial biomass and moisture. Manure increased soil total soil N and available P contents, microbial biomass, and moisture. Enzyme activities were affected by biochar but not by manure amendment. Limited irrigation maintained maize yield, while amendments showed no effect on maize yield.

Soil Afforestation and Organic C

As already mentioned climate change mitigation through C sequestration depends on the establishment of a stable plant cover. Land-use changes, such as those due to afforestation and management of fast-growing tree species, may affect the regional rate of C sequestration by incorporating carbon dioxide (CO_2) in plant biomass and soil (Jandl et al., 2007). In addition, the establishment of mixed species forest can avoid high rates of organic matter mineralization (Jandl et al., 2007).

The soil organic C content can decrease not only by converting natural to agricultural soils but also by abandoning agricultural soils under different climates (Bastida et al., 2006; Chen et al., 2016b). Therefore, the afforestation of agricultural or abandoned soils has been suggested as a suitable strategy for increasing the SOC content and thus improving C sequestration (Lal, 2005). Nevertheless, C sequestration in soils is complex and depends on multiple factors, such as climate, soil properties, plant species, and management of the soil–plant system (Kuzyakov, 2010; Lal, 2005; Schmidt et al., 2011). Therefore, soil restoration through organic amendments may be a preceeding practice before afforestation to improve soil quality.

The changes in plant cover after afforestation cause change litter inputs and thus the composition and activity of the belowground microbial community and this may promote the formation of aggregates with the consequent physical protection of soil organic C within soil aggregates in semiarid environments (García-Franco et al., 2015). The potential for C sequestration after afforestation is higher in sites with no large disturbance preparation (García-Franco et al., 2014). In Burkina Faso, the potential of the biofuel crop *Jatropha curcas* L. for C sequestration has been demonstrated in the long term (Baumert et al., 2016). A meta-analysis conducted by Deng and Shangguan (2017) evaluated soil C and N dynamics after afforestation and factors that affected soil organic matter in China. Precipitation, temperature, land use, tree species, soil depth, and forest age were the main factors that influenced SOC content after afforestation. Furthermore, Hu et al. (2016) demonstrated that root rather than leaf litter of *Populus simonii* Carr. controls C sequestration in soil and concluded that deep-rooted trees with large root biomass could be used for promoting C sequestration in soil.

A meta-analysis on the C accumulation in agricultural soils after afforestation showed that the main factors that affect SOC stocks after afforestation were previous land use, clay content, preplanting disturbance, and type of planted tree species (Laganière et al., 2010). It was also highlighted that: (1) the impact of afforestation in C stocks was more pronounced in cropland soils than in pasture or gasslands; (2) broadleaf tree species have a greater capacity to increase SOC than coniferous species; and (3) soils with a high clay content (>33%) have a greater capacity to accumulate SOC than soils with low clay content.

As already mentioned afforestation of pasture and agricultural systems can often increase C sequestration in soils, but the magnitude, timing, and direction of soil organic C dynamics, depend on site conditions, management practices, and previous land use (Hernández et al., 2016). The positive impacts of afforestation in SOC sequestration are sometimes contradictory due to the multiple factors controlling C dynamics (Laganière et al., 2010). For instance, Hernández et al. (2016) found no significant changes in SOC stocks after 8 years of afforestation of native pasture with either *Eucalyptus* or *Pinus* in a temperate region of Uruguay.

PERSPECTIVES

Past and current knowledge on soil erosion can surely be a great help in suggesting where and how future erosion might be a problem. Future erosion will depend on future rates of water and wind erosion, which are both controlled by climate change, as well as by land-use change. Rates of water erosion are likely to respond to increases in rainfall in a non-linear manner, with disproportionately greater increases occuring in wet years. There are, though, still knowledge gaps in this matter and about scenaries influencing soil C content. Therefore, it is a matter of debate how erosion will respond to climate change–related factors and how this combination will affect soil C stocks and ecosystem services. These controlling factors will ultimately influence the response of activity and composition of plant and microbial communities and, consequently, the CO_2 emissions from soil to the atmosphere. The promotion of C sequestration in arid and semiarid regions may involve afforestation strategies. For this purpose, an improvement of soil conditions through the application of organic amendments must be neccesary. However, the C balance of amended and afforested soils should be studied under the expected impacts of climate change.

ACKNOWLEDGMENTS

The authors thank the Spanish Ministry for the CICYT projects AGL2014-54636-R and AGL2014-55658-R. Felipe Bastida thanks the Spanish Government for his "Ramón y Cajal" contract and FEDER funds (RYC-2012-10666). Authors are grateful to "Fundación Séneca" for his grant to Excellence Research Groups.

REFERENCES

Abule, E., Smit, G.N., Snyman, H.A., 2005. The influence of woody plants and livestock grazing on grass species composition, yield and soil nutrients in the Middle Awash Valley of Ethiopia. J. Arid. Environ. 60, 343–358.

Albaladejo, J., Castillo, V., Díaz, E., 2000. Soil loss and runoff on semiarid land as amended with urban solid refuse. Land Degrad. Dev. 11, 363–373.

Albaladejo, J., López, J., Boix-Fayos, C., Barberá, G.G., Martínez-Mena, M., 2008. Long-term effect of a single application of organic refuse on carbon sequestration and soil physical properties. J. Environ. Qual. 37, 2093–2099.

Albaladejo, J., Ortiz, R., García-Franco, N., Ruiz-Navarro, A., Almagro, M., García-Pintado, J., Martínez-Mena, M., 2013. Land use and climate change impacts on soil organic carbon stocks in semi-arid Spain. J. Soil Sediments 13, 265–277.

Austin, A.T., Yahdjian, L., Stark, J.M., Belnap, J., Porporato, A., Norton, U., Ravetta, D.A., Schaeffer, S.M., 2004. Water pulses and biogeochemical cycles in arid and semiarid ecosystems. Oecologia 141, 221–235.

Austin, A.T., Vivanco, L., 2006. Plant litter decomposition in a semi-arid ecosystem controlled by photodegradation. Nature 442, 555–558.

Austin, A.T., 2011. Has water limited our imagination for aridland biogeochemistry? Trends Ecol. Evol. 26, 229–235.

Baldi, E., Toselli, M., marcolini, G., Quartieri, M., Cirillo, E., Innocenti, A., Marangoni, B., 2010. Compost can successfully replace mineral fertilizers in the nutrient management of commercial peach orchard. Soil Use Manage. 26, 346–353.

Barbarick, J.A., Ippolito, J.A., 2007. Nutrient assessment of a dry-land wheat agroecosystem after 12 years of biosolids applications. Agron. J. 99, 715–722.

Bastida, F., Moreno, J.L., Hernández, T., García, C., 2006. Microbiological degradation index of soils in a semiarid climate. Soil Biol. Biochem. 38, 3463–3473.

Bastida, F., Moreno, J.L., Garcia, C., Hernandez, T., 2007. Addition of urban waste to semiarid degraded soil: long-term effect. Pedosphere 17, 557–567.

Bastida, F., Kandeler, E., Moreno, J.L., Ros, M., García, C., Hernández, T., 2008a. Application of fresh and composted organic wastes modifies structure, size and activity of soil microbial community under semiarid climate. Appl. Soil Ecol. 40, 318–329.

Bastida, F., Kandeler, E., Hernández, T., García, C., 2008b. Long-term effect of municipal solid waste amendment on microbial abundance and humus-associated enzyme activities under semiarid conditions. Microb. Ecol. 55, 651–661.

Bastida, F., Nicolás, C., Moreno, J.L., Hernández, T., García, C., 2012. Microbial functionality and diversity in agroecosystems: a soil quality perspective. In: Benkeblia, N. (Ed.), Sustainable Agriculture and New Biotechnologies. CRC Press, Taylor & Francis Group, Boca Raton, FL, pp. 187–213.

Bastida, F., Hernández, T., Albaladejo, J., García, C., 2013. Phylogenetic and functional changes in the microbial community of long-term restored soils under semiarid climate. Soil Biol. Biochem. 65, 12–21.

Bastida, F., Selevsek, N., Torres, I.F., Hernández, T., García, C., 2015. Soil restoration with organic amendments: linking cellular functionality and ecosystem services. Sci. Rep. 5, 15550.

Bastida, F., Torres, I.F., Moreno, J.L., Baldrian, P., Ondoño, S., Ruíz-Navarro, A., Hernández, T., Richnow, H.H., Starke, R., García, C., Jehmlich, N., 2016. The active microbial diversity drives ecosystem multifunctionality and is physiologically related to carbon availability in Mediterranean semi-arid soils. Mol. Ecol. 25, 4660–4673.

Bastida, F., Torres, I.F., Romero-Trigueros, C., Baldrian, P., Vetrovský, T., Bayona, J.M., Alarcón, J.J., Hernández, T., García, C., Nicolás, E., 2017. Combined effects of reduced irrigation and water quality on the soil microbial community of a citrus orchard under semi-arid conditions. Soil Biol. Biochem. 104, 226–237.

Baumert, S., Khamzina, A., Vlek, P.L.G., 2016. Soil organic carbon sequestration in *Jatropha curcas* systems in Burkina Faso. Land Degrad. Dev. 27, 1813–1819.

Beesley, L., Moreno-Jiménez, E., Gomez-Eyles, J.L., Harris, E., Robinson, B., Sizmur, T., 2011. A review of biochar' potential role in the remediation, revegetation and restoration of contaminated soils. Environ. Pollut. 159, 3269–3282.

Ben-David, E.A., Zaady, E., Sher, Y., Nejidat, A., 2011. Assessment of the spatial distribution of soil microbial communities in patchy arid and semi-arid landscapes of the Negev Desert using combined PLFA and DGGE analyses. FEMS Microbiol. Ecol. 76, 492–503.

Berhe, A., Harte, A., Harden, J., Torn, W., M, S., 2007. The significance of erosion-induced terrestrial carbon sink. Bioscience 57, 3374-346.

Berhe, A.A., Harden, J.W., Torn, M.S., Harte, J., 2008. Linking soil organic matter dynamics and erosion-induced terrestrial carbon sequestration at different landform positions. J. Geophys. Res. 113, G04039.

Berhe, A., Arnold, A., Stacy, C., Lever, E., McCorkle, R., Araya, E., S, N., 2014. Soil erosion control son biogeochemical cycling of carbon and nitrogen. Nat. Educ. Knowl. 5 (2).

Black, C.K., Davis, S.C., Hudiburg, T.W., Bernacchi, C.J., DeLucia, E.H., 2017. Elevated CO_2 and temperature increase soil C losses from a soybean-maize ecosystem. Global Change Biol. 23, 435–445.

Booker, K., Huntsinger, L., Bartolome, J.W., Sayre, N.F., Stewart, W., 2013. What can ecological science tell us about opportunities for carbon sequestration on arid rangelands in the United States? Global Environ. Change 23, 240–251.

Bond-Lamberty, B., Thomson, A., 2010. Temperature-associated increases in the global soil respiration record. Nature 564, 579–582.

Borken, W., Matzner, E., 2009. Reappraisal of drying and wetting effects on C and N mineralization and fluxes in soils. Global Change Biol. 15, 808–824.

Borrelli, P., Paustian, K., Panagos, P., Jones, A., Schütt, B., Lugato, E., 2016. Effect of good agricultural and environmental conditions on erosion and soil organic carbon balance: a national case study. Land Use Policy 50, 408–421.

Bottner, P., 1985. Response of microbial biomass to alternate moist and dry conditions in a soil incubated with ^{14}C- and ^{15}N-labelled plant material. Soil Biol. Biochem. 17, 329–337.

Caravaca, F., Lax, A., Albaladejo, J., 2001. Soil aggregate stability and organic matter in clay and fine silt fractions in urban refuse-amended semiarid soils. Soil Sci. Soc. Am. J. 65, 1235–1238.

Caravaca, F., García, C., Hernández, M.T., Roldán, A., 2002. Aggregate stability changes after organic amendment and mycorrhizal inoculation in the afforestation of a semiarid site with *Pinus halepensis*. Appl. Soil Ecol. 19, 199–208.

Chen, A., Xie, X., Ge, T., Hou, H., Wang, W., Wei, W., Kuzyakov, Y., 2016a. Rapid decrease of soil carbon after abandonment of subtropical paddy fields. Plant Soil 415, 203–214.

Chen, C., Zhang, J., Lu, M., Qin, C., Chen, Y., Yang, L., Huang, Q., Wang, J., Zhenguo, S., Shen, Q., 2016b. Microbial communities of an arable soil treated for 8 years with organic and inorganic fertilizers. Biol. Fert. Soils 52, 455–467.

Cherif, H., Ayari, F., Ouzari, H., Marzorati, M., Brusetti, L., Jedidi, N., Hassen, A., Daffonchio, D., 2009. Effects of municipal solid waste compost, farmyard manure and chemical fertilizers on wheat growth, soil composition and soil bacterial characteristics under Tunisian arid climate. Eur. J. Soil Biol. 45, 138–145.

Collins, S.L., Sinsabaugh, R.L., Crenshaw, C., Green, L., Porras-Alfaro, A., Stursova, M., Zeglin, L.H., 2008. Pulse dynamics and microbial processes in aridland ecosystems. J. Ecol. 96, 413–420.

Conant, R.T., Klopatek, J.M., Klopatek, C.C., 2000. Environmental factors controlling soil respiration in three semiarid ecosystems. Soil Sci. Soc. Am. J. 64, 383–390.

Crecchio, C., Curci, M., Pizzigallo, M.D.R., Ricciuti, P., Ruggiero, P., 2004. Effect of municipal solid waste compost amendments on soil enzyme activities and bacterial genetic diversity. Soil Biol. Biochem. 36, 1595–1605.

Cytryn, E., Kautsky, L., Ofek, M., Mandelbaum, R.T., Minz, D., 2011. Short-term structure and functional changes in bacterial community composition following amendment with biosolids compost. Appl. Soil Ecol. 48, 160–167.

Daryanto, S., Eldridge, D.J., Throop, H.L., 2013. Managing semi-arid woodlands for carbon storage: grazing and shrub effects on above- and belowground carbon. Agr. Ecosyst. Environ. 169, 1–11.

de Baets, S., Meersmans, J., Vanacker, V., Quine, T.A., van Oost, K., 2012. Spatial variability and change in soil organic carbon stocks in response to recovery following land abandonment and erosion in mountainous drylands. Soil Use Manage 29, 65–76.

de Vries, F.T., Shade, A., 2013. Controls on soil microbial community stability under climate change. Front. Microbiol. 4, article 265.

Delgado-Baquerizo, M.D., Maestre, F.T., Reich, P.B., Jeffries, T.C., Gaitan, J.J., Encinar, D., Berdugo, M., Campbell, C.D., Singh, B.K., 2016. Microbial diversity drives multifunctionality in terrestrial ecosystems. Nat. Commun. 7, 10541. doi: 10.1038/ncomms10541.

Deng, L., Shangguan, Z.P., 2017. Afforestation drives soil carbon and nitrogen changes in China. Land Degrad. Dev. 28, 151–165.

Diacono, M., Montemurro, F., 2010. Long-term effects of organic amendments on soil fertility: a review. Agron. Sustainable Dev. 30, 401–422.

Díaz, E., Roldán, A., Lax, A., Albaladejo, J., 1994. Formation of stable aggregates in degraded soil by amendment with urban refuse and peat. Geoderma 63, 277–288.

Dick, R.P., 1992. A review: long-term effects of agricultural systems on soil biochemical and microbial parameters. Agr. Ecosyst. Environ. 40, 25–36.

Dorado, J., Zancada, M.C., Almendros, G., López-Fando, C., 2003. Changes in soil properties and humic substances after long-term amendments with manure and crop residues in dryland farming system. J. Plant Nutr. Soil Sci. 166, 31–38.

Dorrepaal, E., Toet, S., van Logtestijn, R.S.P., Swart, E., van der Weg, M.J., Callaghan, T.V., Aerts, R., 2007. Carbon respiration from subsurface peat accelerated by climate warming in the subartic. Nature 460, 616–619.

Drenovsky, R.E., Steenwerth, K.L., Jackson, L.E., Scow, K.M., 2010. Land Use and climatic factors structure regional patterns in soil microbial communities. Global Ecol. Biogeogr. 19, 27–39.

Elfstrand, S., Hedlund, K., Martensson, A., 2007. Soil enzyme activities, microbial community composition and function after 47 years of continuous green manuring. Appl. Soil Ecol. 35, 610–621.

Entry, J.A., Fuhrmann, J.J., Sojka, R.E., Shewmaker, G.E., 2004. Influence of irrigated agriculture on soil carbon and microbial community structure. Environ. Manage. 33, 363–373.

Erhart, E., Hartl, W., Putz, B., 2005. Biowaste compost affects yield, nitrogen supply during the vegetation period and crop quality of agricultural crops. Eur. J. Agron. 23, 305–314.

Erhart, E., Hartl, W., Putz, B., 2008. Total soil heavy-metal concentrations and mobile fractions after 10 years of biowaste-compost fertilization. J. Plant Nutr. Soil Sci. 171, 378–383.

Evans, S.E., Wallenstein, M.D., 2012. Soil microbial community response to drying and rewetting stress: does historical precipitation regime matter? Biogeochemistry 109, 101–116.

FAO, 2015. Status of the world's soil resources. Main report. Food and Agriculture Organization of the United Nations (FAO), Rome, pp. 607.

Fernández, J.M., Nieto, M.A., López-de-Sa, E.G., Gascó, G., Méndez, A., Plaza, C., 2014. Carbon dioxide emissions from semi-arid soils amended with biochar alone or combined with mineral and organic fertilizers. Sci. Total Environ. 482 (483), 1–7.

Ferrari, A., Hagedorn, F., Niklaus, P.A., 2016. Experimental soil warming and cooling alters the partitioning of recent assimilates: evidence from a ^{14}C-labelling study at the alpine treeline. Oecologia 181, 25–37.

Fierer, N., Schimel, J.P., 2002. Effects of drying-rewetting frequency on soil carbon and nitrogen transformations. Soil Biol. Biochem. 34, 777–787.

Fierer, N., Bradford, M.A., Jackson, R.B., 2007. Toward an ecological classification of soil bacteria. Ecology 88, 1354–1364.
Fontaine, S., Mariotti, A., Abbadie, L., 2003. The priming effect of organic matter: a question of microbial competition? Soil Biol. Biochem. 35, 837–843.
Foster, E.J., Hanse, N., Wallenstein, M., Cotrufo, M.F., 2016. Biochar and manure amendments impact soil nutrients and microbial enzymatic activities in a semi-arid irrigated maize cropping system. Agric. Ecosyst. Environ. 233, 404–414.
García, C., Hernández, T., Costa, F., 1992. Variation in some chemical parameters and organic matter in soils regenerated by the addition of municipal solid-waste. Environ. Manage. 16, 763–768.
García, C., Hernández, T., Roldán, A., Martín, A., 2002. Effect of plant cover decline on chemical microbiological parameters under Mediterranean climate. Soil Biol. Biochem. 34, 635–642.
García, C., Roldán, A., Hernández, T., 2005. Ability of different plant species to promote microbiological processes in semiarid soil. Geoderma 124, 193–202.
García-Franco, N., Wiesmeier, M., Goberna, M., Martínez-Mena, M., Albaladejo, J., 2014. Carbon dynamics after afforestation of semiarid shrublands: implications of site preparation techniques. Forest Ecol. Manage. 319, 107–115.
García-Franco, N., Martínez-Mena, M., Goberna, M., Albaladejo, J., 2015. Changes in soil aggregation and microbial community structure control carbon sequestration after afforestation of semiarid shrublands. Soil Biol. Biochem. 87, 110–121.
Ge, Y., Changrong, C., Zhihong, X., Eldridge, S.M., Chan, K.Y., He, Y., He, J.Z., 2010. Carbon/nitrogen ratio as a major factor for predicting the effects of organic wastes on soil bacterial communities assessed by DNA-based molecular techniques. Environ. Sci. Pollut. Res. 17, 807–815.
Gebauer, R.L.E., Ehleringer, J.R., 2000. Water and nitrogen uptake patterns following moisture pulses in a cold desert community. Ecology 81, 1415–1424.
Giardina, C.P., Ryan, M.G., 2000. Evidence that decomposition rates of organic carbon in mineral soil do not vary with temperature. Nature 404, 858–861.
Girmay, G., Singh, B.R., Mitiku, H., Borresen, T., Lal, R., 2008. Carbon stocks in Ethiopian soils in relation to land use and soil management. Land Degrad. Dev. 19, 351–367.
Hernández, J., del Pino, A., Vance, E.D., Califra, A., Del Giorgio, F., Martínez, L., González-Barrios, P., 2016. *Eucalyptus* and *Pinus* stand density effects on soil carbon sequestration. Forest Ecol. Manage. 368, 28–38.
Hortal, S., Bastida, F., Armas, C., Lozano, Y.M., Moreno, J.L., García, C., Pugnaire, F.I., 2013. Soil microbial community under a nurse-plant species changes in composition, biomass and activity as the nurse grows. Soil Biol. Biochem. 64, 139–146.
Hu, Y.L., Zeng, D.H., Chang, S.X., Mao, R., 2013. Dynamics of soil and root C stocks following afforestation of croplands with poplars in a semi-arid region in northeast China. Plant Soil 368, 619–627.
Hu, Y.L., Zeng, D.H., Ma, X.Q., Chang, S.X., 2016. Root rather than leaf litter input drives soil carbon sequestration after afforestation on a marginal cropland. Forest Ecol. Manage. 362, 38–45.
Hueso, S., García, C., Hernández, T., 2012. Severe drought conditions modify the microbial community structure, size and activity in amended and unamended soils. Soil Biol. Biochem. 50, 167–173.
Huxman, T.E., Snyder, K.A., Tissue, D., Leffler, A.J., Ogle, K., Pockman, W.T., Sandquist, D.R., Potts, D.L., Schwinning, S., 2004. Precipitation pulses and carbon fluxes in semiarid and arid ecosystems. Oecologia 141, 254–268.
Jandl, R., Lindner, M., Vesterdal, L., Bauwens, B., Baritz, R., Hagedorn, F., Johnson, D.W., Minkkinen, K., Byrne, K.A., 2007. How strongly can forest management influence soil carbon sequestration? Geoderma 137, 253–268.

Jia, G.M., Zhang, P.D., Wang, G., Cao, J., Han, J.C., Huang, Y.P., 2010. Relationship between microbial community and soil properties during natural succession of abandoned agricultural land. Pedosphere 20, 352–360.

Jien, S.H., Wang, C.H., 2013. Effects of biochar on soil properties and erosion potential in a highly weathered soil. Catena 110, 225–233.

Jindo, K., Chocano, C., Melgares de Aguilar, J., González, D., Hernández, T., García, C., 2016. Impact of compost application during 5 years on crop production, soil microbial activity, carbon fraction, and humification process. Commun. Soil Sci. Plant Anal. 47, 1907–1919.

Jones, C., McConnell, C., Coleman, K., Cox, P., Fallon, P., Jenkinson, D., Powlson, D., 2005. Global climate change and soil carbon stocks; predictions from two contrasting models for the turnover of organic carbon in soil. Global Change Biol. 11, 154–166.

Kern, J.S., Johnson, M.G., 1993. Conservation tillage impacts on national soil and atmospheric carbon levels. Soil Sci. Soc. Am. J. 57, 200–210.

Koven, C.D., Ringeval, B., Friedlinstein, P., Ciais, P., Cadule, P., Khvorostyanov, D., Krinner, G., Tarnocai, C., 2011. Permafrost carbon-climate feedbacts accelerate global warming. Proc. Natl. Acad. Sci. U.S.A 108, 14769–14774.

Kraus, D., Weller, S., Klatt, S., Santabárbara, I., Hass, E., Wassmann, R., Werner, C., Kiese, R., Butterbach-Bahl, K., 2016. How well can we assess impacts of agricultural land management changes on total greenhouse gas balance (CO_2, CH_4 and N_2O) of tropical rice-cropping systems with a biogeochemical model? Agric. Ecosyst. Environ. 224, 104–115.

Kunito, T., Saeki, K., Goto, S., Hayashi, H., Oyaizyu, H., Matsumoto, S., 2001. Copper and zinc fractions affecting microorganisms in long-term sludge amended soils. Bioresour. Technol. 79, 135–146.

Kuzyakov, Y., Subbotina, I., Chen, H., Bogomolova, I., Xu, X., 2009. Black carbon decomposition and incorporation into soil microbial biomass estimated by ^{14}C labeling. Soil Biol. Biochem. 41, 210–219.

Kuzyakov, Y., 2010. Priming effects: interactions between living and dead organic matter. Soil Biol. Biochem. 42, 1363–1371.

Laganière, J., Angers, D.A., Paré, D., 2010. Carbon accumulation in agricultural soils after afforestation: a meta-analysis. Global Change Biol. 16, 439–453.

Lal, R., 2001. Soil degradation by erosion. Land Degrad. Dev. 12, 519–539.

Lal, R., 2003. Soil erosion and the global carbon budget. Environ. Int. 29, 437–450.

Lal, R., 2004. Soil carbon sequestration impacts on global climate change and food security. Science 304, 1623–1626.

Lal, R., 2005. Forest soils and carbon sequestration. Forest Ecol. Manage. 220, 242–258.

Lal, R., 2006. Enhancing crop yields in the developing countries through restoration of the soil organic carbon pool in agricultural lands. Land Degrad. Dev. 17, 197–209.

Lavee, H., Imeson, A.C., Sarah, P., 1998. The impact of climate change on geomorphology and desertification along a Mediterranean-arid transect. Land Degrad. Dev. 9, 407–422.

Lehmann, J., 2007. A handful of carbon. Nature 447, 143–144.

Li, Y., Awada, T., Zhou, X., Shang, W., Chen, Y., Zuo, X., Wang, S., Liu, X., Feng, J., 2012. Mongolian pine plantations enhance soil physico-chemical properties and carbon and nitrogen capacities in semi-arid degraded sandy land in China. Appl. Soil Ecol. 56, 1–9.

Luo, G., Ling, N., Nannipieri, P., Chen, H., Raza, W., Wang, M., Guo, S., Shen, Q., 2017. Long-term fertilisation regimes affect the composition of the alkaline phosphomonoesterase encoding microbial community of a vertisol and its derivative soil fractions. Biol. Fertil. Soils 53, 375–388.

Liu, W., Zhang, Z., Wan, S., 2009. Predominant role of water in regulating soil and microbial respiration and their responses to climate change in a semiarid grassland. Global Change Biol. 15, 184–195.

Luna, L., Pastorelli, R., Bastida, F., Hernández, T., García, C., Miralles, I., Solé-Benet, A., 2016. The combination of quarry restoration strategies in semiarid climate induces different responses in biochemical and microbiological soil properties. Appl. Soil Ecol. 107, 33–47.

Mantovi, P., Baldoni, G., Toderi, G., 2005. Reuse of liquid, dewatered, and composted sewage sludge on agricultural land: effects of long-term application on soil and crop. Water Res. 39, 289–296.

Monaco, S., Hatch, D.J., Sacco, D., Bertora, C., Grignani, C., 2008. Changes in chemical and biochemical soil properties induced by 11-yr prepeated additions of different organic materials in maize-based forage systems. Soil Biol. Biochem. 40, 608–615.

Nannipieri, P., Giagnoni, L., Renella, G., Puglisi, E., Ceccanti, B., Masciandaro, G., Fornasier, F., Moscatelli, M.C., Marinari S, 2012. Soil enzymology: classical and molecular approaches. Biol. Fert. Soils 48, 743–762.

Nicolás, C., Hernández, T., García, C., 2017. Type and quantity of organic amendments determine the amount of carbon stabilized in particle-size fractions of a semiarid degraded soil. Arid Land Res. Manage. 31, 14–28.

Nie, M., Pendall, E., Bell, C., Gasch, C.K., Raut, S., Tamang, S., Wallenstein, M.D., 2013. Positive climate feedbacks of soil microbial communities in a semi-arid grassland. Ecol. Lett. 16, 234–241.

Omondi, M.O., Xia, X., Nahayo, A., Liu, X., Korai, P.K., Pan, G., 2016. Quantification of biochar effects on soil hydrological properties using meta-analysis of literature data. Geoderma 274, 28–34.

Pardo, T., Bes, C., Bernal, M.P., Clemente, R., 2016. Alleviation of environmental risks associated with severely contaminated mine tailings using amendments: modelling of trace element speciation, solubility, and plant accumulation. Environ. Toxicol. Chem. 35, 2874–2884.

Pascual, J.A., García, C., Hernández, T., Ayuso, M., 1997. Changes in the microbial activity of an arid soil amended with urban organic wastes. Biol. Fert. Soils 24, 429–434.

Pascual, J.A., García, C., Hernández, T., 1999. Lasting microbiological and biochemical effects of the addition of municipal solid waste to an arid soil. Biol. Fert. Soils 30, 1–6.

Pérez-Quezada, J.F., Delpiano, C.A., Snyder, K.A., Johnson, D.A., Franck, N., 2011. Carbon pools in an arid shrubland in Chile under natural and afforested conditions. J. Arid. Environ. 75, 29–37.

Placella, S.A., Brodie, E.L., Firestone, M.K., 2012. Rainfall-induced carbon dioxide pulses result from sequential resuscitation of phylogenetically clustered microbial groups. Proc. Natl. Acad. Sci. U.S.A. 109, 10931–10936.

Plaza, C., Hernández, C., García-Gil, J.C., Polo, A., 2004. Microbial activity in pig slurry-amended soils under semiarid conditions. Soil Biol. Biochem. 36, 1577–1585.

Quine, T., Walling, D., Chakela, Q., Mandiringana, O., Zhang, X., 1999. Rates and patterns of tillage and water erosion on terraces and contour strips: evidence from cesium-137 measurements. Catena 36, 115–142.

Ringrose, S., Matheson, W., Vanderpost, C., 1998. Analysis of soil organic carbon and vegetation cover trends along the Botswana Kalahari Transect. J. Arid. Environ. 38, 379–396.

Ros, M., Hernández, M.T., García, C., 2003. Soil microbial activity after restoration of a semiarid soil by organic amendments. Soil Biol. Biochem. 35, 463–469.

Ros, M., Pascual, J.A., García, C., Hernández, M.T., Insam, H., 2006. Hydrolase activities, microbial biomass and bacterial community in a soil after long-term amendment with different composts. Soil Biol. Biochem. 38, 3443–3452.

Rostagno, C.M., Sosebee, R.B., 2001. Surface application of biosolids in the chihuahuan desert: effects on soil physical properties. Arid Soil Res. Rehabil. 15, 233–244.

Sánchez-García, M., Sánchez-Monedero, M.A., Roig, A., López-Cano, I., Moreno, B., Benítez, E., Cayuela, M.L., 2016. Compost vs biochar amendment: a two-year field study evaluating soil C build-up and N dynamics in an organically managed olive crop. Plant Soil 408, 1–14.

Scherr, S.J., Yadav, S., 1996. Land Degradation in the Developing World: Implications for Food, Agriculture and the Environment to 2020. IFPRI, Foof. Agril. and the Environment Discussion Paper 14, Washington, DC, 36.

Schimel, D.S., 1995. Terrestrial ecosystems and the carbon cycle. Global Change Biol. 1, 77–91.

Schlesinger, W.H., Raikks, J.A., Hartley, A.E., Cross, A.F., 1996. On the spatial pattern of soil nutrients in desert ecosystems. Ecology 77, 364–374.

Schmidt, M.W.I., Torn, M.S., Abiven, S., Dittmar, T., Guggenberger, G., Janssens, I.A., Kleber, M., Kögel-Knabner, I., Lehmann, J., Manning, D.A.C., Nannipieri, P., Rasse, D.P., Weiner, S., Trumbore, S.E., 2011. Persistence of soil organic matter as an ecosystem property. Nature 478, 49–56.

Seddaiu, G., Porcu, G., Ledda, L., Roggero, P.P., Agnelli, A., Corti, G., 2013. Soil organic matter content and composition as influenced by soil management in a semi-arid Mediterranean agro-silvo-pastoral system. Agric. Ecosyst. Environ. 167, 1–11.

Sukkariyah, B.F., Evanulo, G., Zelazny, L., Chaney, R.L., 2005. Cadmium, Copper, Nickel, and Zinc availability in a biosolids-amended piedmont soil years after application. J. Environ. Qual. 34, 2255–2262.

Shen, W., Reynolds, J.M., Hui, D., 2009. Responses of dryland soil respiration and soil carbon pool size to abrupt vs. gradual and individual vs. combined changes in soil temperature, precipitation, and atmospheric [CO_2]: a simulation analysis. Global Change Biol. 15, 2274–2294.

Sinsabaugh, R.L., Carreiro, M.M., Repert, D.A., 2002. Allocation of extracellular enzymatic activity in relation to litter composition, N deposition, and mass loss. Biogeochemistry 60, 1–24.

Six, J., Conant, R.T., Paul, E.A., Paustian, K., 2002. Stabilization mechanisms of soil organic matter: implications for C-saturation of soils. Plant Soil 241, 155–176.

Sodhi, G.P.S., Beri, V., Benbi, D.K., 2009. Soil aggregation and distribution of carbon and nitrogen in different fractions under long-term application of compost in rice–wheat system. Soil Till. Res. 103, 412–418.

Steffens, M., Kölbl, A., Totsche, K.U., Kögel-Knabner, I., 2008. Grazing effects on soil chemical and physical properties in a semiarid steppe of Inner Mongolia (PR China). Geoderma 143, 63–72.

Steffens, M., Kölbl, A., Kögel-Knabner, I., 2009. Alteration of soil organic matter pools and aggregation in semi-arid steppe topsoils as driven by organic matter input. Eur. J. Soil Sci. 60, 198–212.

Stark, J.M., Firestone, M.K., 1995. Mechanisms for soil moisture effects on activity of nitrifying bacteria. Appl. Environ. Microbiol. 61, 218–221.

Stursova, M., Sinsabaugh, R.L., 2008. Stabilization of oxidative enzymes in desert soil may limit organic matter accumulation. Soil Biol. Biochem. 40, 550–553.

Su, Y.Z., Wang, X.F., Yang, R., Lee, J., 2010. Effects of sandy desertified land rehabilitation on soil carbon sequestration and aggregation in an arid region in China. J. Environ. Manage. 91, 2109–2116.

Takeda, A., Tsukada, H., Nanzyo, M., Takaku, Y., Uemura, T., Hisamatsu, S., Inaba, J., 2005. Effect of long-term fertilizer application on the concentration and solubility of major and trace elements in a cultivated Andisol. Soil Sci. Plant Nutr. 51, 251–260.

Tejada, M., Hernández, M.T., García, C., 2006. Application of two organic amendments on soil restoration: effects on the soil biological properties. J. Environ. Qual. 35, 1010–1017.

Tejada, M., Moreno, J.L., Hernández, M.T., García, C., 2007. Application of two beet vinasse forms in soil restoration: effects on soil properties in an arid environment in southern Spain. Agric. Ecosyst. Environ. 119, 289–298.

Torres, I.F., Bastida, F., Hernández, T., García, C., 2015. The effects of fresh and stabilized pruning wastes on the biomass, structure and activity of the soil microbial community in a semiarid climate. Appl. Soil Ecol. 89, 1–9.

Triberti, L., Nastri, A., Giordani, G., Comellini, F., Baldoni, G., Toderi, G., 2008. Can mineral and organic fertilization help to sequestrate carbon dioxide in cropland? Eur. J. Agron. 29, 13–20.

UNCED, 1992. Report of the United Nations Conference on Environment and Development, Chapter 12, Managing Fragile Ecosystems: Combating Desertification and Drought (Rio de Janeiro, 3-14 June 1992), General A/CONF.151/26 (Vol. II), Chapter 12 (http://www.unccd.ch/).
van Oost, K., Quine, T.A., Govers, G., de Gryze, S., Six, J., Harden, J.C., Ritchie, G.W., McCarty, G., Heckrath, G., Kosmas, C., Giráldez, J.V., da Silva, M., Merckx, R., 2007. Science 318, 626–629.
von Lutzow, M., Koegel-Knabner, I., Ekschmitt, K., Matzner, E., Guggenberger, G., Marschner, B., Flessa, H., 2006. Stabilization of organic matter in temperate soils: mechanisms and their relevance under different soil conditions: a review. Eur. J. Soil Sci. 57, 426–445.
Wardle, D.A., Nilsson, M.C., Zackrisson, O., 2008. Fire-derived charcoal causes loss of forest humus. Science 321, 1295.
Yazdanpanah, N., Mahmoodabadi, M., Cerdá, A., 2016. The impact of organic amendments on soil hydrology, structure and microbial respiration in semiarid lands. Geoderma 266, 58–65.
Zhang, X., Johnston, E.R., Li, L., Konstantinidis, K.T., Han, X., 2016. Experimental warming reveals positive feedbacks to climate change in the Eurasian Steppe. ISME J., 180. doi: 10.1038/ismej.2016.180.
Zornoza, R., Acosta, J.A., Faz, A., Baath, E., 2016. Microbial growth and community structure in acid mine soils after addition of different amendments for soil reclamation. Geoderma 272, 64–72.
Zuber, S.M., Villamil, M.B., 2016. Meta-analysis approach to assess effect of tillage on microbial biomass and enzyme activities. Soil Biol. Biochem. 97, 176–187.

CHAPTER 9

The Future of Soil Carbon

Carlos Garcia*, Paolo Nannipieri, Teresa Hernandez***
*CEBAS-CSIC, Campus Universitario de Espinardo, Murcia, Spain
**University of Firenze, Firenze, Italy

Chapter Outline

WHAT IS KNOWN ABOUT SOIL ORGANIC C?

The Role of Organic C in Soil Fertility

Vegetal residues and rhizodeposition (root exudates, lysates, root cell material) are sources of soil organic matter (organic C) together with animal and microorganisms residues (Stevenson, 1986). The mineralization processes return the C to the atmosphere mainly as carbon dioxide, while a fraction of it is assimilated in microbial tissues (soil biomass) and part is transformed through humification processes in stable forms (humic substances; some doubt about the existence of humic materials in soil is discussed later), which can also be mineralized but more slowly (Stevenson, 1994). In this chapter we use the term organic matter and organic C although it must be taken into account that organic carbon represents about 58% of soil organic matter.

Conservation of soil organic C is crucial for ecosystem functioning because soil organic C plays a key role in regulating climate, water supplies, and biodiversity, providing the ecosystem services that are essential to human well-being. Soil organic matter influences soil physical, physical–chemical, chemical, biological and microbiological properties as well

The Future of Soil Carbon
http://dx.doi.org/10.1016/B978-0-12-811687-6.00009-2

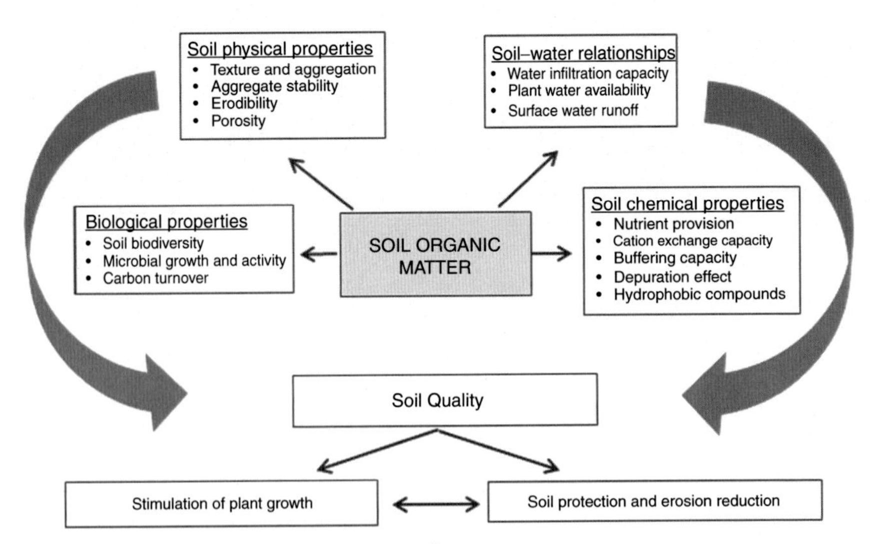

Figure 9.1 ***Effects of organic matter on soil.***

as plant physiological processes, playing a key role in the development and functioning of terrestrial ecosystems (Fig. 9.1). Organic matter is vital in the formation of stable aggregates (García-Orenes et al., 2005; Larchevêque et al., 2006; Lax et al., 1994) and a secondary pore system that allows water movement into the soil profile, better aeration, and soil moisture retention (Annabi et al., 2007; Ellies, 2004; Mabuhay et al., 2006); it also positively affects heat capacity, cation exchange capacity, and soil buffering power (Díaz, 1992; López-Piñeiro et al., 2007; Weber et al., 2007).

The presence of stable aggregates reduces the tendency of soil to compaction and increases soil resistance to deformation, because soil degradation is largely controlled by the presence and abundance of water-stable macroaggregates (Bayhan et al., 2005). From a biological point of view, organic matter favors microbial activity and the biochemical processes involved in the nutrient cycles, activating the edaphic fauna that, in turn, increases its porosity (O'Dell et al., 2007; Tejada et al., 2006).

Soil organic matter can be classified into different pools, depending on their resistance to microbial degradation (Smith and Read, 1997): (1) the labile pool is the most active part of organic matter with easily degradable C compounds, such as carbohydrates, amino acids, polysaccharides, lipids, and other compounds of low molecular weight; (2) the slowly degradable pool with organic compounds, such as cellulose, hemicellulose, chitin; and (3) the recalcitrant organic matter, with organic compounds more resistant to the degradation than those earlier mentioned; these compounds generally contain aromatic rings (lignin) and aliphatic chains (lipids). Therefore, soils differ not only in total soil organic C but also in the composition of the different pools of soil organic C (Stevenson, 1986).

The Dispute About the Existence of Humus in Soil

Part of the resistant soil organic matter is commonly termed humus and microorganisms are supposed to be involved in its formation (Dickinson et al., 1974). Humus is the product of the decomposition of animal, microbial, and plant residues in both aquatic and terrestrial ecosystems (Stevenson, 1994). Humus is one of the most abundant forms of organic matter present on the Earth's surface (Anderson, 1979) and it consists of a complex mixture of many different compounds of difficult definition, characterized by molecular weights ranging from 700 to 300,000 Da. The structure and reactivity of humic substances is poorly known due to their great chemical heterogeneity and geographical variability; they are considered as superstructures of only apparent high-molecular weight compounds self-assembled by relatively small heterogeneous molecules held together by mainly hydrophobic dispersive forces (Piccolo, 2002).

Humic substances exert a positive effect on soil properties, such as improving structure, biological activity, and C fixation of soil and stimulation of seed germination, roots, and plant growth with beneficial effects on crop yield (Chen et al., 2004; Jindo et al., 2011; Magdoff and Weil, 2004; Piccolo, 2002). Other positive effects of humic fractions on plant physiology include an increase in water and nutrient uptake, and stimulation of root permeability and respiration, and stimulation of photosynthesis. This physiological stimulation can protect plants from diseases and pests (El-Ghamry et al., 2009).

Nowadays, the existence of humus material in soil is questioned and Lehmann and Kleber (2015) have proposed that the humic material is not a distinct chemical category but a very complex mixture of microbial and plant biopolymers with their degradation products considering that NMR signals of humic substances resemble those of intact and degrading biopolymers. Even the use of advanced spectromicroscopic techniques, such as STXM/NEXAFS, can show the presence of a distinct chemical category. Future research should clarify the dispute on the existence of humic material in soil.

Soils as C Sinks

Atmospheric carbon dioxide fixed as plant residues or rhizodeposition as well as the C from exogenous organic matter reaching soil can be stored into soil. Soil is the main reservoir of organic C of terrestrial ecosystems, after oceans and geological sinks and soil C is at least 3 times greater than the pool of atmospheric carbon dioxide (Amundson, 2001). On average, soil contains in the upper 1 m more C than the atmosphere and vegetation together (Smith et al., 2010); two-thirds of soil C is present as organic matter and one-third as soil inorganic C. However, soils can lose C as CO_2 through mineralization of its organic matter due to unsustainable land management or to land use changes (Lal, 2010a,b).

On average inorganic C accounts for about 750 Pg C in the upper 1 m mainly present as $CaCO_3$ and $MgCO_3$ • $CaCO_3$, CO_2, HCO_3, and $CO_3^{=}$ (Batjes, 1996; FAO, 2001; Swift, 2001). These carbonates can be primary, which is inherited from the parent material or deposited as dust, or secondary if derived from precipitation of carbonate ions, resulting from respiration of root and microorganisms, or from the dissolution of carbonate of parent materials, with Ca and Mg ions, due to weathering (Mi et al., 2008; Ming, 2002; Wu et al., 2009). The weathering of calcium silicate of parent materials consumes 2 units of CO_2 whereas only 1 unit of CO_2 is released in the secondary carbonate deposition, and this leads to sequestration of atmospheric CO_2 (Wu et al., 2009). Soil inorganic C is also an active pool because free carbonates affect soil microbial activity and the rate of soil organic matter mineralization. Soil pH is affected by carbonates, whose concentration in soil depends on soil acidification, which is affected by climate change or agricultural practices; soil acidification can cause high C losses from soil carbonates (Emmerich, 2003; Mikhailova and Post, 2006). Soil inorganic C is more abundant under dry climates with a low mean annual rainfall and high evaporation rates, than under wet climates. Na et al. (2008) found a close correlation between the depths of secondary carbonate–mineral in soil profiles and mean annual rainfall indicating that a low annual rainfall prevents the rapid removal of carbonate and base cations through leaching (Rawlins et al., 2011).

The organic C content of soil is very wide, ranging from about 10% in the surface of alpine soils to less than 0.5% in the desert soils. It depends on the type of soil, climate, the organic inputs, and the mineralization of native organic matter and land management. Soils of cold areas can accumulate great amounts or organic C because microbial activity in these soils is very low due to the low temperatures and, consequently, mineralization processes are very low; thus, these soils constitute an important sink of organic C (Stevenson, 1986). Contrarily, in semiarid and arid areas, the existing high temperatures stimulate microbial activity and hence, organic matter breakdown. The richest soils in organic C are Cryosols/Gelisols and peatlands (Batjes, 2011), which are found mostly in northern Europe, particularly in the United Kingdom and Ireland. Grassland soils also store large amounts of C while soils of the warmest and driest areas of southern Europe contain less C because temperature and moisture regimes are the principal determinants of soil organic C dynamics (Lal, 2004; Prentice, 2001).

The study of the changes in the soil organic C pool is nowadays a hot topic due to the effects of global climatic change on the organic matter content of soil. To increase the C storage in soil, it is important to decrease the oxidation of organic C to CO_2 thus decreasing the greenhouse effect with beneficial effects on soil; this has beneficial effects on the terrestrial ecosystem functioning because, as discussed earlier, the organic C content plays an important role in regulating biological, chemical, and physical properties of soil, which participate to the regulation of climate, water supplies, and biodiversity.

Organic C Sequestration in Soil

Recently the concern about the increase of CO_2 and methane emissions into the atmosphere has increased interest on the potential role of soils acting as C sinks and on the study of the relationship between organic matter turnover in relation to C retention in soils. Increasing the capacity of soil to sequester C is a means of counteracting the increasing emission of CO_2 into the atmosphere in the medium term, thus contributing to alleviating the environmental impacts of the greenhouse effect (Swift, 2001). According to Chapman (2010) the term "sequestration" implies both capture and storage of C.

Although there is a limit on the amount of organic C that can be stored in soil, the high C losses of many agricultural and degraded soils indicate that these soils potentially can store organic C. Degraded soils can retain part of the C added as amendments when the added C or the relative microbial metabolites, derived from the use of these amendments by soil microflora, become associated with the mineral fraction, thus are protected against microbial decomposition (García et al., 2012; Larney et al., 2009; Nicolás et al., 2012; Six et al., 2000). Potential C accumulation is lower in semiarid areas than in temperate ecosystems, and under warmer and drier climates than in cold climates (Lal, 2009). Most of arid and Mediterranean soils also have high concentrations of inorganic C (Díaz-Fernández et al., 2003; Lal, 2009), whereas semiarid soils can potentially fix 200 Kg organic C ha^{-1} $year^{-1}$ on average (Lal, 2009).

Organic matter can be protected in soil against microbial degradation by: (1) physicochemical mechanisms due to the formation of aggregates. In this case organic matter is physically protected as it is inaccessible to microorganisms; the formation of microaggregates as well as the association of organic matter with silt and clay particles can protect organic matter from microbial degradation (Ladd et al., 1996); (2) biochemical stabilization through transformation of organic compounds into more resistant chemical compounds, as it occurs during the humification process (Fernández et al., 2009; Stevenson, 1994), which is questioned to occur by Lehmann and Kleber (2015) (Fig. 9.2). However, sooner or later each organic compound reaching soils, also the most recalcitrant one, is mineralized to CO_2 (Ladd et al., 1996; Schmidt et al., 2011).

The physical protection exerted by the macro- and/or microaggregates on organic matter can be attributed to: (1) compartmentalization of the substrate and soil microbial communities; (2) reduction of oxygen diffusion within macroaggregates and especially in microaggregates, with negative effects on microbial activity within microaggregates (Six et al., 2000).

Organic C sequestration can also depend on the hydrophobic moieties of soil organic compounds; indeed, microbial activity is higher under hydrophilic than hydrophobic conditions (Piccolo et al., 1999). However, Spacini et al. (2000a,b) observed that both hydrophilic and hydrophobic

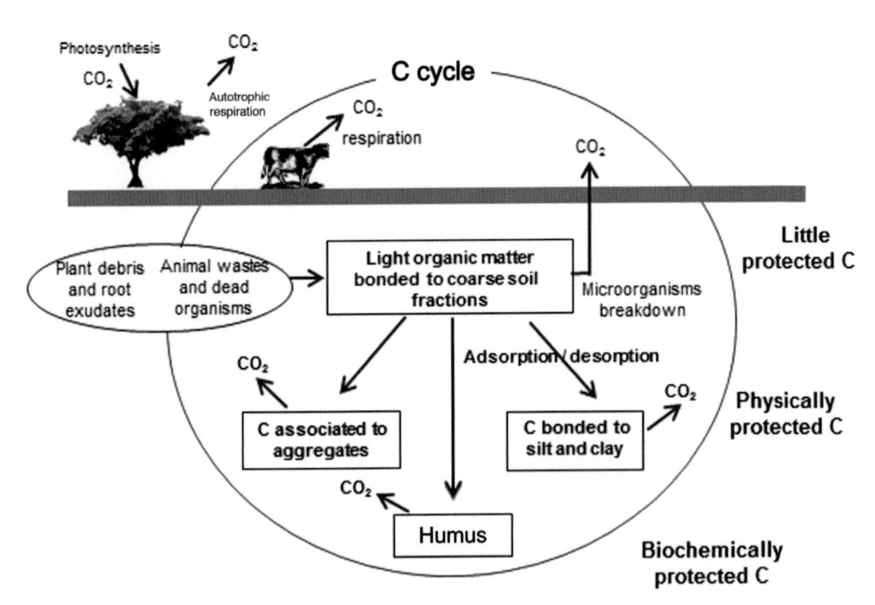

Figure 9.2 ***Organic matter dynamic in soil and carbon cycle.***

components of maize straw were incorporated into the humic pool in soils amended with maize straw and incubated for 1 year; it was also suggested that all humic fractions have short-term interactions with products of vegetal residue degradation. Spaccini et al. (2001) and Piccolo (2002) highlighted the importance of associations between humic fractions and fine textural fractions (silt- and clay-sized fractions), and suggested that the hydrophobic sequestration of C mainly occurs when hydrophobic compounds interact with finer soil particles, rather than with larger size fractions.

As mentioned earlier, the adsorption of organic compounds by surface-reactive mineral particles and the formation of aggregates are important processes for stabilizing organic C in soil, because both processes protect the target organic compound against microbial degradation. For example, the adsorption of organic compounds by soil minerals increases their average residence time in soil (Kaiser and Guggenberger, 2003; Swift, 2001). Organic C bound to the coarse soil fraction (>250 μm) is poorly protected (biochemically and physically), easily mineralized and thus rapidly disappear from the environment (Cambardella and Elliot, 1992; Solomon et al., 2000), while organic matter of free microaggregates (<250 μm) or microaggregates within macroaggregates is physically protected from decomposition. Both ^{14}C- or ^{13}C-labeled compounds or the metabolites, derived from their use by soil microflora retained in larger particles were more rapidly mineralized than those of the finer particles (Bird et al., 2002). Therefore, microaggregates contain older C than macroaggregates (Pulleman and Marinissen, 2004; Six and Jastrow, 2002).

Factors Influencing Soil C Levels: Pressures on Soil C Stocks

In a mature soil system, the organic matter content tends to reach a dynamic equilibrium where the C inputs to soil by vegetal, microbial, and animal residues are equal to the C outputs by organic C mineralization and leaching of the soluble organic C fractions (Schulze and Freibauer, 2005). In such equilibrated ecosystems, the soil organic matter content is maintained at a more or less constant level while the C cycle is active. Changes in land use or management may lead to a mismatch of this equilibrium state and lead to large losses of organic matter, thus decreasing soil fertility.

The changes in soil organic C contents are determined by the imbalances between C inputs and outputs. Worldwide the soils of agroecosystems have decreased their C content by 25%–75% depending on climate, type of soil, and the management of soil, and the losses of soil C can range from 10 to 50 t C ha^{-1} (Lal, 2011). This decrease in the C content has led to a decrease in the productive capacity of these soils and in the incorporation efficiency of the added C inputs and thus the soil capacity to sequester atmospheric CO_2.

Climate (temperature, rainfall), erosion, land-use management, vegetation cover, pollution, peatland excavation, climate change, and soil type can influence soil C contents (Laganière et al., 2010; Shi et al., 2013); these effects depend on the region; for example, clearing land in the tropics generally emits larger C amounts than clearing land in temperate areas (West et al., 2014).

Erosion

Erosion is one the main causes of soil C losses. For example, 132 Mha in the EU are affected by erosion (including wind erosion) (Lal, 2003; Scherr, 1999). In the Mediterranean area soil losses range between 20 and 40 t/ha after a storm, and more than 100 t ha^{-1} can be lost due to extreme events. This erosion, which is initially superficial and laminar, if not stopped in time produces gullies and ravines. Regeneration of these gullies is costly and sometimes impossible. An adequate soil management to prevent erosion avoiding these detrimental effects is necessary (ENSAP, 2012; Lal, 2004).

Accelerated erosion, closely linked to desertification, differs from geological or natural erosion, which is not always detrimental because it forms sediments and adds nutrients to rivers, maintains the sedimentary balance in channels and on beaches, and forms very fertile areas, such as those of deltas or alluvial plains (Lal, 2003). This type of erosion is slow and gradual, with erosion rates usually low and sustainable. These "natural" processes can be accelerated, for example, by man and the resultant anthropic erosion is characterized by high erosion rates, which can lead to desertification (Brown, 1984; Lal, 2003).

As a consequence of erosion, soil fertility, intended as crop production, markedly decreases due to the reduction of the effective soil thickness, losses of organic matter and nutrients, degradation of the physical structure, and reduction of water retention. This decrease in soil infiltration leads to less rainwater harvesting and greater surface runoff, which increases the erosive force of rainfall on the surrounding terrain, thus increasing degradation effects (Berhe et al., 2007).

These are the main factors causing erosion:

- *Climate*: Dry climate, such as that of the Mediterranean areas, with scarce but torrential rains, capable of generating runoff (with soil trawling and organic matter) is one of the most erosive climate. The greater the precipitation intensity, the greater the raindrops and, thus, the greater soil erosion (FAO, 2015).

- *Soil type and its erodibility*: The resistance to erosion and the transport rate of particles determine soil erosion rates and both depend on texture and structure of soil. The stability of the aggregates plays a fundamental role in soil erosion. A soil with good aggregation keeps its particles together and allows the flow of water inside soil pores, making the erosion process difficult (Barthès and Roose, 2002). On the other hand, the particles can easily disintegrate in soils with low aggregate stability, with clogging pores, and thus avoid water infiltration, favoring water flow onto surface. Sand-lime and fine sand particles and those with little organic matter are more erodible in this context (Colombo et al., 2010; Fang et al., 2015).
- *Degree and length of the slope*: Both degree and length of slope influence the rate of erosion. In soils water speed on surface depends on degree and length of the slope. Agricultural soils with slopes higher than 10% are vulnerable to erosion (Bagio et al., 2017).
- *Vegetation cover*: The vegetation cover can protect the soil against erosive processes by reducing the impact of raindrops and facilitating water infiltration. In addition, the vegetation can decrease the runoff speed and its erosive capacity, and also improves porosity and increases organic matter content of soil (Ross et al., 2001).
- *Agricultural management*: Excessive tillage, plowing according to the maximum slope, and burning of stubble, that eliminates organic inputs to soil, contribute to erosive processes (Poulenard et al., 2001).
- *Ecological disasters*: Forest fires, reducing organic matter content of soil and kill soil biota, leave the soil bare and thus exposed to erosion processes (Poulenard et al., 2001).

According to Lal (2003), soil erosion affects soil organic C dynamics by: (1) slaking or disrupting aggregates; (2) preferentially removing C through runoff water or dust storms; (3) increasing mineralization rate of soil organic C on-site; (4) increasing mineralization of soil organic C, which has been displaced and redistributed over the landscape after being transported by rivers and dust storms; (5) favoring reaggregation of soil through formation of organic-mineral complexes at the depositional/protected sites; and (6) deeply burying C-enriched sediment in depositional sites, flood plains, reservoirs, and ocean floor.

Land-Cover and Land-Use Changes

Land-use changes altering plant biomass, and above- and belowground biodiversity, can alter soil properties thus affecting C losses and storage (Deng et al., 2014; Powers et al., 2011; Smith et al., 2015). Changes in land use can alter the equilibrium between C inputs and outputs in soil, and soil can be a C source or C sink until a new equilibrium is reached (Deng et al., 2016).

Deforestation, conversion of grasslands to cropland, and grazing land are the main land cover changes. Forests cover 29% of soils and represent 60% of the terrestrial vegetation and these soils contain more C per unit area than other soils. The C content of 1 m depth of forest soils accounts for about 40% of the total soil C on average (FAO, 2002). The deforestation leads to a considerable C loss in these soils (estimated at about 25% of soil C) through CO_2 emissions into the atmosphere (Guo and Gifford, 2002). Therefore, when deforestation cannot be avoided, proper management is necessary to minimize C losses. Deforestation of organic soils and of steep slopes prone to erosion should be avoided. Reforestation, especially of degraded soils with low organic matter content, can be an important practice leading to long-term C sequestration in both plant biomass and soil.

Both forests and grasslands can favor the accumulation of organic C whereas agricultural practices, such as tillage, can increase C losses from soil (Foley et al., 2005). The land-use change from pasture to cropland increases C losses, whereas the conversion of cropland to secondary forest favors the accumulation of organic matter in soil (Guo and Gifford, 2002). In a meta-analysis involving 103 publications including 160 sites and 29 countries, Deng et al. (2016) observed that land-use conversions of these soils significantly reduced soil C stocks (0.39 Mg ha^{-1} $year^{-1}$). Soil C stocks increased significantly after conversion from farmland to grassland (0.30 Mg ha^{-1} $year^{-1}$) and forest to grassland (0.68 Mg ha^{-1} $year^{-1}$), but significantly decreased after conversion from grassland to farmland (0.89 Mg ha^{-1} $year^{-1}$), forest to farmland (1.74 Mg ha^{-1} $year^{-1}$), and changing forest type (0.63 Mg ha^{-1} $year^{-1}$).

Other changes in land use can increase C flux from land to the atmosphere. For example, urban development decreases terrestrial C storage; urban sprawl consumes between 10,000 km^2 and 20,000 km^2 of arable land per year in the developing world, mostly high-quality agricultural land (Turner et al., 2007). Loss of permanent grasslands and peatlands also represents a net loss of terrestrial C.

Grazing lands also play an important role in C sequestration. According to FAO, grazing soils store between 200 and 420 Pg of the total terrestrial ecosystem, a large part of it below the surface and, therefore, in a relatively stable state. Due to poor management, many grazing areas in tropical and arid areas are degraded and therefore offer varying possibilities for C sequestration. Grazing can alter soil C content through two different mechanisms: (1) changes in net primary production; (2) changes in N stocks and in organic matter decomposition rate (Piñeiro et al., 2010). The impact of increasing grazing on soil organic C varies; Leifeld and Fuhrer (2010) have reported an increase of soil organic C whereas Medina-Roldán et al. (2012) have reported no change.

Drainage and conversion of peatlands or wetlands to agricultural soils also contribute to important organic C losses; conversion of anaerobic to aerobic soils, stimulates organic matter oxidation (Hooijer et al., 2010). The CO_2 emissions from drained peatlands have increased from $1.^{06}$ Pg CO_2 year^{-1} in 1990 to 1.30 Pg CO_2 year^{-1} in 2008 (Joosten, 2010). Meersmans et al. (2009) reported that intensified drainage practices for agricultural purposes greatly increased organic C losses in Belgian soils initially poorly drained.

As already mentioned tillage generally increases organic C losses and for this reason no tillage or minimum tillage has been introduced in agricultural management; indeed, the tillage of croplands breaks soil aggregates and this exposes previously protected C to microbial attack, accelerating organic matter mineralization, and thus reducing the soil C content (Chapman, 2010; Mishra et al., 2010; Moussadek et al., 2014).

UNCERTAINTY ABOUT SOIL ORGANIC C SINKS

As already mentioned organic C plays a key role in affecting physical, chemical, biological, and microbiological properties of soil and acts as a reservoir of nutrients for plants. Understanding how quickly organic C can be produced and destroyed in soil is critical not only for the implications on CO_2 emissions from soil to atmosphere but also to sustain soil functionality and fertility. There is a great deal of uncertainty as to how the C stored in soil will respond to changes in temperature and other factors related to the foreseen climate change. Processes that affect organic C content in response to alterations, such as land use or global warming are poorly understood.

Soil C Sinks and Climate Change

Changes in climate and in the atmospheric CO_2 concentration can affect soil C cycle and C stocks. The steady increase in the emission of gases into the atmosphere (CO_2, N_2O, and CH_4) is leading to an increase in the average global temperature, an increase in rainfall in most of the northern hemisphere (at the rate of 0.5%–1% decade^{-1}), especially at middle and high latitudes, and a decrease in rainfall in subtropical areas (0.3% decade^{-1})

(Lal, 2004). These changes can affect C, water and nutrient cycles, increasing soil susceptibility to runoff and erosion, and thus negatively impacting biomass production, biodiversity, and the environment.

Nowadays we cannot predict how global warming will affect the huge amount of organic C stored in soil. Generally, decomposition rates of fresh organic inputs increase by increasing temperature but we cannot predict how C stabilized on mineral surfaces will respond to temperature changes. Indeed, the bibliography shows contradictory results on whether the soil capacity for storing C will decrease or even increase by increasing temperature (Schneider et al., 2007). The following processes are an example of the complexity of the plant–soil system and how it responds to changes in temperature due the increase in the CO_2 concentration: the increase in the atmospheric CO_2 concentration increases the photosynthetic activity and the amount of plant fixed C released by the rhizodeposition with stimulation of microbial activity in soil; however, an increase in plant N uptake due to the stimulation of the photosyntehtic activity can cause N-limiting conditions in soil and this may lead to the increase of soil organic matter mineralization by soil microflora for releasing inorganic N (Drigo et al., 2008). The C losses as result of the priming effect may be lower, equal, or higher than the increases in organic C inputs to soil by increased rhizodeposition promoted by the higher photosynthetic activity.

Climate change may produce contrasting effects depending on the region. In some regions, the increased temperature can increase C storage in plants and soil due to the stimulation of plant growth because of the increased concentration of CO_2 in the atmosphere (Schneider et al., 2007). However, the increase in temperature can also increase soil organic matter mineralization giving rise to higher emissions of CO_2 to the atmosphere. In the colder regions, at high latitudes, where microbial activity is low because of the low temperatures, massive stocks of C have been built up over thousands of years (Stevenson, 1986). Temperature increases in these cold and often frozen areas, due to climate change, will stimulate microbial activity, increasing mineralization processes and leading, consequently, to great amounts of C losses (Kirsckbaum, 1995). There is concern about these cold regions (Arctic and subArctic regions) because they are expected to warm the most under climate change. US Geological Survey scientists have estimated that at least 10 gigatons of soil C are stored in organic soils of Alaska and these soils are extremely vulnerable to fire and decomposition under warming conditions. In temperate regions where there are smaller C stocks than in cold areas, the C losses due to climate change will be less important.

It has been predicted that for one degree of warming, about 30 pentagrams of soil C will be released into the atmosphere. It is important to underline that the organic matter decompose slowly in peat soils due to the low oxygen level of soil water. If these areas dry out, organic matter will decompose rapidly releasing CO_2 into the atmosphere (Strack, 2008).

According to the European Environment Agency, soil moisture is being affected by rising temperatures and changes in precipitation patterns and significant decreases are expected between 2021 and 2050 in the Mediterranean regions with minor increases in Northeast Europe (European Environmental Agency, 2015). Despite the effects of the global climate, change still being poorly understood, it is clear that these effects depend on the region and that the atmosphere warming will increase the production of greenhouse gases with the consequent acceleration in the climate change (European Environmental Agency, 2015).

Agriculture and Climate Change

According to the Intergovernmental Panel on Climate Change, fossil fuels, land use, and agriculture are the three main causes of the increase in greenhouse gases for the last years (IPCC, 2010). Between 1980 and the end of the 21st century, global temperature is expected to rise between 1.8 and 4°C. The impacts on the environment will depend on the magnitude of the temperature rise. For example, some crops of high or middle latitudes will have higher productivity if the local temperature will increase by 1–3°C, but if the increase exceeds this value the effects may be negative.

Global warming is expected to have a significant impact on agriculture, temperature, carbon dioxide evolution, snow melt, precipitation, and the interaction between these elements, will determine the carrying capacity of the biosphere to produce sufficient food for all living beings. The consequences of the climate change and agriculture will depend on the overall balance of these effects. The study of these phenomena could help to anticipate and to plan adequately agriculture managements to maximize quality and quantity of the agricultural products and the conservation and storage of organic C in soil (Chanillor et al., 2008).

If there is significant overheating, the adaptive capacity of ecosystems will be exceeded, leading to negative consequences, such as an increased risk of species extinction (Schneider et al., 2007). The cost associated with the impacts caused by climate change is expected to grow over time with increasing temperatures. The expected increase in the hardness and frequency of droughts, heat waves, and other extreme weather events would cause major impacts over the course of this century (Battisti and Naylor, 2009).

At the same time, agriculture has also been shown to produce significant effects on the climate, mainly by producing and releasing greenhouse gases, such as carbon dioxide, methane, and nitric oxide and by altering the surface of the planet, with losses in its capacity to absorb or reflect heat and light. Agricultural activities in the EU-28 generated 470.6 million tons of CO_2 equivalent in 2012, corresponding to about 10% of total greenhouse gas emissions (Eurostat, 2016). These greenhouse gas emissions from agriculture came mainly from one of three sources: agricultural soils (accounting for about one half of agricultural emissions), enteric fermentation (about one third), and manure management (about one sixth). The other sources of agricultural greenhouse gas emissions—field-burning of agricultural residues and rice cultivation—were only minor contributors (Eurostat, 2016).

Climate change and agriculture are interrelated at a global scale. Climate change, which is occurring in a period of increasing demand for food, seeds, fiber, and fuel, could irreversibly damage the natural resources on which agriculture depends. The relationship between climate change and agriculture is, however, bidirectional: agriculture contributes to climate change in several important ways and climate change can negatively affect agriculture.

In conclusion, climate change could affect agriculture in the long term in several ways:

- Productivity, in terms of quantity and quality of crops.
- Agricultural practices, through changes in water use (irrigation) and inputs, such as herbicides, insecticides, and fertilizers.
- Environment, particularly the relationship between frequency and soil drainage system, and processes, such as erosion, reduction in crop diversity.
- Rural space, through the loss or gain of cultivated land, land speculation, and hydraulic services.
- Biological adaptation, living organisms could become more or less competitive, humans could also have the need to develop more competitive agricultural practices; for example, selection of rice varieties resistant to salt or flooding.
- The effects of global warming are already visible. In some areas, moderate warming has slightly improved yields. But, in general, the negative consequences are overcoming the positive ones (IPCC, 2014). Floods and droughts are becoming increasingly frequent and severe, seriously affecting agricultural productivity and the livelihoods of rural communities,

increasing the risk of conflict over land and water. In addition, climate change facilitates the spread of pests and invasive species and can increase the geographic spread of some diseases (IPCC, 2014).

- Phenological variations caused by Global warming can affect the final crop production (Oteros et al., 2015). For instance, the phenology and the flowering of the olive tree is also being seriously affected by the global change (Garcia-Mozo et al., 2014), and this can have serious effects on olive production because the flowering of the olive tree and the harvest production are closely related (Oteros et al., 2014).

Facing the Future of Soil C

Adapting to Climate Change

The reduction of the organic C pool of soil as a consequence of the predicted climate change, causes, as already mentioned, the degradation of soil functioning and this may accelerate the depletion of the C storage in soil. This urges the use of strategies mitigating CO_2 emissions and the threat of global warming.

The negative effects of climate change on human populations may be reduced through the promotion of sustainable activities. Practices reducing greenhouse gas emissions can prevent, reduce, or delay the negative impacts and should be applied so as to avoid overcoming the Earth's adaptive capacity. Restoring degraded ecosystems can help in sequestering C from the atmosphere. In addition to offsetting CO_2 emissions by C sequestration, soil with a good quality can also increase CH_4 oxidation and reduce N_2O emission by completing the denitrification processes.

The sustainable soil management is an important strategy for managing climate change risks because it adapts soil to climate variability (Lal, 2011). As already mentioned, soils with good quality can moderate climate change by reducing the concentration of CO_2 and other greenhouse gas in the atmosphere through, for example, CO_2 sequestration and CH_4 oxidation. Marginal soils should be restored with appropriate management practices to create a positive C balance with C inputs greater than C outputs. Because of the severe depletion of the C pool, degraded soils have a higher capacity to sequester atmospheric CO_2 (Baethgen, 2009), as already mentioned.

Forest and soil conservation practices can increase C storage (by restoring and establishing new forests, wetlands, and grasslands) and reduce CO_2 emissions (by reducing soil tillage and suppressing wildfires). It is crucial to increase organic C in all soils by: (1) restoring degraded areas; (2) controlling soil erosion; (3) increasing the use of recommended agricultural

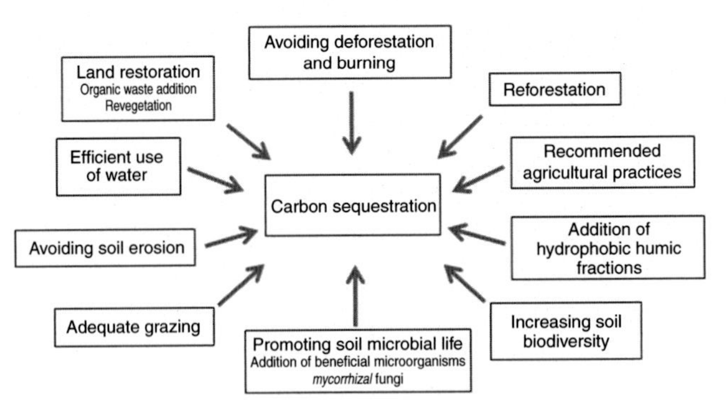

Figure 9.3 ***Management for increasing C sequestration in soil.***

practices (so as to avoid deforestation and the farming of peat lands, increasing mulching instead of burning of biomass, increasing large-scale use of biochar and organic amendments); and (4) improving the management of grazing (Fig. 9.3).

Restoration of Degraded Areas

The addition of new organic matter, and the maintenance of native organic matter in soil, will undoubtedly decrease the levels of CO_2 in the atmosphere. In this sense, the report on organic matter and biodiversity by the European Strategy for Soil Protection, which is the preamble to a new directive, identifies the loss of organic matter as one of the main causes of soil degradation, and it underlines that exogenous organic matter is nowadays an inestimable organic source to soil, favoring the implantation of a stable vegetal cover. The incorporation of organic elements into soil is considered as one of the most effective strategy for combating erosion and associated degradation processes (Van Camp et al., 2004). This is particularly important in Mediterranean countries, where many soils are exposed to semiarid climates associated with degradative processes. The recovery of degraded soils and ecosystems, therefore, has a high potential for C sequestration in soil. Although most of the degraded soils have lost a large part of their original C, they can be recovered by appropriate rehabilitation strategies and by their proper use. As already mentioned, the use of organic amendments to restore degraded and eroded soils can counteract the effect of global warming.

Soil Erosion Control

In order to avoid soil erosion, it is necessary to carry out a sustainable use of soil and to perform actions that do not degrade it. The United Nations Conference on Environment and Development (UNCED), also known as the "Earth Summit" held in Rio de Janeiro in 1992, agreed with the definition of sustainable development by the Brundlandt Report (United Nations, 1987); in this report sustainable development is defined as "Development that meets the present needs of people without compromising the ability of future generations to meet theirs." In any case soil should be always covered with stable vegetation that protects it. It is clear that any action to combat erosion will also combat desertification, as erosion is the main cause of soil degradation and desertification.

Measures to prevent erosion include:

- Maintenance of the vegetation covers of soil. This vegetation will improve the porosity, increase the organic matter content, and stimulate the activity and growth of microbial populations of soil, and act as a physical barrier against water and wind erosion. Plant cover also reduces soil temperature and water evaporation, increases the amount of water available to plants, and contributes to reduce soil compaction and surface crusting.
- Construction of terraces on crops on mountain slopes. These terraces will favor the use of rainwater and will prevent the loss of soil and organic matter due to runoff phenomena.
- Reforestation as support of soil since trees act as a barrier to wind and protect against rain.
- Maintenance and construction of retaining walls to slow the progress of dunes and runoff.
- Use of machines adapted to the conditions of the slope, following the slope borders.
- Use of sustainable farming techniques that do not damage soil structure.

Conservation Agriculture: Implementation of Recommended Agricultural Practices

As already mentioned, all land uses and changes in land use can influence the increase in C stocks and the reduction of GHG emissions. The conservation and enhancement of C stocks through sustainable use of land must be promoted. Improved agricultural practices can increase C storage in plant biomass and soils, helping to mitigate climate change. The improvement of land-use systems and land-management practices will oppose land degradation and the losses of C due to erosion and soil degradation processes.

Some beneficial mitigation measures have already been identified, such as lower rates of agricultural expansion in natural habitats, afforestation, reforestation, intensified efforts to prevent deforestation, introduction of agroforestry, and agroecological systems, restoration of underused or degraded land and meadows, promotion of C sequestration, reduction, and more efficient use of N inputs, efficient management of fertilizers (Bindraban et al., 2015), the use of feed increasing the digestive efficiency of cattle, the use of complex cropping systems, and the use of livestock farming that involve forage and agroforestry (Table 9.1).

Conservation agriculture is a system of sustainable agriculture that includes a set of agricultural practices adapted to the local conditions and demands of cultivation. These practices prevent erosion and degradation, improve soil quality and biodiversity, and contribute to the good use of natural resources, such as water and air, without undermining farm production levels. According to the FAO, this type of agriculture comprises a series of techniques, whose main objective is to conserve, improve, and make more efficient use of natural resources through integrated management of soil, water, biological agents, and external inputs.

To avoid soil C losses, adoptive techniques are needed, such as reduction and minimization of work (plowing and tillage), use of crop rotation, rational use of chemical fertilizers, and use of crop residues as a natural means of protection and fertilization of soil, being able to increase the soil organic matter content, improve soil structure, and maintain crop yield.

Table 9.1 Recommended agricultural practices

1. Conservation measures that reduce soil erosion
2. Direct sowing or reduced tillage
3. Use of crops with high proportion of residues, such as maize, sorghum, and wheat
4. Minimal mechanical alteration of soil (no-tillage)
5. Maintenance of a plant cover on soil
6. Implementation of multifunctional margins and retention structures
7. Implantation of strategies of optimal irrigation and deficit
8. Increasing the intensity of crop rotation by eliminating summer fallow
9. Use of buffer strips
10. Use of advanced technologies (decision support systems, precision farming, and fleet management)
11. Selection of varieties and hybrids that accumulate more organic C
12. Measures to promote biodiversity
13. Agronomic, technical, and economic practices optimized for the improvement of irrigation water management
14. Optimization of the use of agrochemicals

This type of agriculture should be characterized by: (1) direct sowing of annual crops with machines capable of sowing on residues of the previous crop and at least 30% of the cultivated area should be protected by vegetable remains (Ding et al., 2000); (2) minimum or zero tillage with annual crops with only vertical alteration of soil profile and at least 20% to 30% of cultivated area should be protected by vegetal remains (Farkas et al., 2009; Ugalde et al., 2007); (3) the use of a continuous and protective plant cover; use of woody crops with at least 30% of the crown-free soil surface protected by living or inert cover (Ding et al., 2000); (4) the restoration of permanent vegetation (Glover et al., 2012, 2010; Thompson et al., 2006); (5) the complete addition of not-harvested to soil (Turmel et al., 2015); (6) the elimination of bare fallow; and (7) the improvement of water management (Rosenzweig and Hillel, 2000; Smith and Conen, 2004).

The conversion of conventional till to no-till restores some of the depleted soil C and N pools and the gains in the organic C contents depend on soil type and texture, drainage, previous tillage intensity, and duration of no-till practices (Mishra et al., 2010). Soane et al. (2012) indicated that soils with no-till increased C content and aggregate stability, especially near the surface.

This sustainable production concerns:

- More rational and efficient use of nutrients, water, space, and energy in all land-use systems, ending with excessive exploitation of natural resources
- More effective measures for soil and water conservation
- Increased nutrient recycling by reducing fertilizers and pesticides rates and by increasing native nutrient use efficiency
- Better use of biological resources to increase and sustain crop and livestock yields
- Increased recognition and use of indigenous knowledge in the local historical farming practices adapted to environmental conditions
- Recovery of native species and varieties, better adapted to local ecosystems by promoting seed banks and cattle breeds and their development in the field to increase biodiversity

These are the general recommendations for fertilization suggested by the conservation agriculture:

- The use of manure, compost, crop residues, and so on, can have beneficial effects on edaphic biodiversity and on the physical and chemical conditions of soil. This practice provides nutrients to culture in a balanced way, stimulating plant physiology, increasing the root growth, and controlling the activity of certain phytopathogens.

- The use of green manures (crops with a fast growth, cut at the time of flowering, and buried in the same place where they have been planted) improves the physical properties of soil, enriches soil with a fresh organic material to be rapidly mineralized with release of nutrients, and stimulates activity and growth of soil microbial populations.
- The use of associated crops (spatial biodiversity) and rotations (temporal biodiversity) increases the nutrient dynamics in soil, as different depths of the profile can be explored with the different root systems. It also improves infiltration and retention of water and protect soil against erosive agents.
- The implementation of agroforestry and agroforestry systems, provided that the socioeconomic and cultural conditions of the agrosystem can permit it, is a good strategy for recycling nutrients and closing nutrient cycles within the farm itself, with the consequent reduction in external inputs, while preserving the agricultural landscape, reducing erosion, diversifying local microclimates, and improving water conservation (Bugayong, 2003).

The strategy is not providing nutrients in a soluble or directly available form, as done in conventional fertilization, but providing nutrients that plant needs at the right time and through the action of edaphic life; this can be done by using organo–mineral complexes (e.g., the clay–humic complex) that also contribute to soil stability.

One promising strategy for increasing soil C stocks is to stimulate soil microbial activity by inoculating beneficial microbes to stimulate soil nutrient cycles in soil where they have been interrupted by inadequate agricultural practices (e.g., excessive use of agrochemical) (Chaparro et al., 2012; Trivedi et al., 2013). The use of rotations instead of monoculture may also maximize C storage (Bell et al., 2003). The addition of biochar to soil due to its ability of increasing soil organic C, while improving soil fertility, has also been proposed for some authors (Lehmann et al., 2011).

The addition of mycorrhizal fungi has also been described as a good strategy for increasing C sequestration (Clemmensen et al., 2013). These fungi favor the access of plants to nutrients and water throughout their hyphae, which are coated with glomalin, a sticky substance involved in soil structure and C storage (Treseder and Holden, 2013). It has been suggested that mycorrhized plant transfer to soil up to 15% more C than their nonmycorrhizal counterpart, and that fungi compete for N with soil microorganisms and decompose organic matter releasing C (Clemmensen et al., 2013).

Promoting the hydrophobicity of native organic matter by, for example, nontillage practices or by the addition of hydrophobic materials, such as organic wastes (compost) or *humic* substances, can increase the stability of soil aggregates and thus favoring the C sequestration (Piccolo, 2002).

Sustainable Management of Grazing Lands

Grazing can alter soil organic C content through changes in net primary production, changes in N stocks, and changes in organic matter decomposition (Piñeiro et al., 2010). Pasture soils can store stable C and in general terms, C content in pasture soils is higher than in agricultural soils. However, it has been indicated that about 7.5% of grassland soils have been degraded because of overgrazing (Conant, 2012).

Suitable management of grazing lands (e.g., optimized stocking density) can be used as a tool for large-scale land restoration (Table 9.2). The intensity, frequency, and seasonality of grazing must be controlled; animals must be moved in order to avoid overgrazing, which is one of the main causes of soil degradation; grazing stimulates biological activity in soil, animal wastes increase soil fertility, and as animals move in a herd their trampling aerates soil, presses in seeds, and pushes down dead plant matter so it can be degraded by soil microorganisms (Marshall et al., 2009; McSherry and Ritchie, 2013).

A best control of fire management, used for controlling wood species is also recommended. The sustainable management of grazing lands also included a reduction in applying mineral fertilizer rates by using more ecological and sustainable N fertilization, and by introducing leguminous species fixing N. Another important measure is to improve pasture quality by introducing more productive plant species with deeper root systems and with the relative root debris being slowly degraded by soil microflora (Kell, 2012).

Table 9.2 Measures for increasing C storage in grazing soils

- Improvement of forage quality
- Regular use of fire to increase forage productivity
- Reduction of overgrazing
- Water conservation

CONCLUSIONS

Soils are finite natural resources and the increasing degradation of soil resources caused by population pressures, environmental conditions, and inadequate governance over this valuable resource has become a worrying reality. However, it is well known that soil resources play a key role in assuring food security and also in climate change adaptation and mitigation. Consequently, with the threat to this valuable resource urgent actions have to be implemented in order to avoid soil degradation. In this sense, soil organic C is the basis of soil fertility and should be stored in soil; soil management should assure no C loss from soil and the necessary sustainability of food production. However, we are not using soil to its full potential. Due to loss of soil organic C we are not mitigating climate change like it could be, and we are not following sustainable management practices for soils.

Soil organic C is an important part of the natural C cycle (around twice the amount of C that is found in the atmosphere and in vegetation); this cycle is key in arid and semiarid soils where due to the erosion process, the loss of soil organic C has a direct role on soil fertility. The dynamics of organic C in soil is complex and subjected to strong and complex physical, chemical, biochemical, and biological processes that are ultimately responsible for organic C stabilization and mineralization; an alteration of such equilibriums due to land use (i.e., tillage) and climate pressures may alter the organic C stocks in soil and potentially cause soil degradation, hence affecting the sustainability of this natural resource. In addition, organic C can be protected in soil against microbial degradation by physicochemical mechanisms due to the formation of aggregates, or by biochemical stabilization through transformation of organic compounds into more resistant chemical compounds; in this sense, some aspects, such as the existence of humus material in soil is now questioned. Some aspects, such as the role of clay minerals in protecting organic C, or the importance of microbial residues in the soil organic matter formation, should be considered. However, in the coming decades, a reduction of soil organic matter as a consequence of climate change would negatively affect soil quality, productivity, and sustainability, but if more C is stored in the soil as organic C, it will reduce the amount present in the atmosphere, and help reduce global warming.

The prediction to the future of what is going to happen with soil organic C, due to the previous questions, is not simple. Studies combining molecular techniques for better studying the composition and activity of microbial communities, along with the measurement of parameters on physical and chemical properties of soil, as well as studies based on the composition and nature of the existing organic matter of the soil, could help to promote possible models on the future of soil C. Some of the practices that increase soil organic C include conservation farming (reducing or eliminating tillage and retaining plant debris from previous crops), improving crop management (e.g., through better rotation), maintaining and improving tree/forestry management, improving grazing management and adding organic materials, such as composts and manures; these practices should be adopted in an obligatory way in order to protect and increase soil organic C, and soil fertility and productivity.

ACKNOWLEDGMENTS

Authors thank the Spanish Ministry of Science and Innovation for the CICYT project AGL2014-55269-R. Authors also thank the Fundación Séneca of the Region of Murcia for its financial support within the program "Research Groups of Excellence of the Region of Murcia (19896/GERM/15)."

REFERENCES

Amundson, R., 2001. The carbon budget of soils. Annu. Rev. Earth Planet Sci. 29, 535–562.

Anderson, D.W., 1979. Processes of humus formation and transformation in soils of the Canadian great plains. J. Soil Sci. 30, 77–84.

Annabi, M., Houot, S., Francou, C., Poitrenaud, M., Le Bissonnais, Y., 2007. Soil aggregate stability improvement with urban composts of different maturities. Soil Sci. Soc. Am. J. 71, 413–423.

Baethgen, W.E., 2010. Climate risk management for adaptation to climate variability and change. Crop Sci. 50 (2 Suppl. 1), S70–S76, CGIAR Science Forum 2009, Wageningen, Netherlands, Jun 16–17.

Bagio, B., Bertol, I., Wolschick, N.H., Schneiders, D., Santos, M.A.N., 2017. Water erosion in different slope lengths on bare soil. Rev. Bras Cienc. Solo 41doi: 10.1590/18069657rbcs20160132, p. e0160132.

Barthès, B., Roose, E., 2002. Aggregate stability as an indicator of soil susceptibility to runoff and erosion; validation at several levels. Catena 47, 133–149.

Batjes, N.H., 1996. Total carbon and nitrogen in the soils of the world. Eur. J. Soil Sci. 47, 151–163.

Batjes, N.H., 2011. Soil organic carbon stocks under native vegetation: revised estimates for use with the simple assessment option of carbon benefits project system. Agr. Ecosyst. Environ. 142, 365–373.

Battisti, D., Naylor, R.L., 2009. Historical warnings of future food insecurity with unprecedented seasonal heat. Science 323, 240–244.

Bayhan, K., Isisdar, A., Akgul, M., 2005. Tillage impacts on aggregate stability and crop productivity in a loam soil of a dryland in Turkey. Soil Plant 55, 214–220.

Bell, J.M., Smith, J.L., Bailey, V.L., Bolton, Jr., H., 2003. Priming effect and C storage in semi-arid no-till spring crop rotations. Biol. Fertil. Soils 37, 237.

Berhe, A., Harte, J., Harden, W., Torn, M.S., 2007. The significance of erosion-induced terrestrial carbon sink. Bioscience 57, 3374–4346.

Bindraban, P.S., Dimkpa, C., Nagarajan, L., Roy, A., Rabbinge, R., 2015. Revisiting fertilisers and fertilization strategies for improved nutrient uptake by plants. Biol. Fert. Soils 51, 897–912.

Bird, S.B., Herrick, J.E., Wander, M.M., Wright, S.E., 2002. Spatial heterogeneity of aggregate stability and soil carbon in a semi-arid rangeland. Environ. Pollut. 116, 445–455.
Brown, L.R., 1984. The global loss of topsoil. J. Soil Water Conserv. 39, 162–165.
Bugayong, L.A., 2003. The contribution of plantation and agroforestry to rural livelihoods. Socioeconomic and Environmental Benefits of Agroforestry Practices in a Community-Based Forest Management Site in the Philippines. International Conference on Rural Livelihoods. Forests and Biodiversity, Bonn, Germany, May 19–23, 2003.
Cambardella, C.A., Elliot, E.T., 1992. Particulate soil organic matter changes across a grassland cultivation sequence. Soil Sci. Soc. Am. J. 56, 777–783.
Chaparro, J.M., Sheflin, A.M., Manter, D.K., Vivanco, J.M., 2012. Manipulating the soil microbiome to increase soil health and plant fertility. Biol. Fert. Soils 48, 489–499.
Chapman, S.J., 2010. Carbon sequestration in soils. Hester, R.E., Harrison, R.M. (Eds.), Carbon Capture: Sequestration and storage. Issues in Environmental Science and Technology, 29, RSC Publishing, Cambridge, UK, pp. 179–202.
Chanillor, A.J., Ewert, F., Arnold, S., Simelton, E., Fraser, E., 2008. Crop and climate change: progress, trends, and challenges in simulating impacts and informing adaptation. J. Exp. Bot. 60, 2775–2789.
Chen, Y., Clapp, C.E., Magen, H., 2004. Mechanisms of plant growth stimulation by humic substances: the role of organo iron complexes. Soil Sci. Plant Nutr. 50, 1089–1095.
Clemmensen, K.E., Bahr, A., Ovaskainen, O., Dahlberg, A., Ekblad, A., Wallander, H., et al., 2013. Roots and associated fungi drive long-term carbon sequestration in boreal forest. Science 339, 1615–1618.
Colombo, C., Palumbo, G., Aucelli, P.P.C., de Angelis, A., Rosskopf, C.M., 2010. Relationship between soil properties, erodibility and hillslope features in Central Apennines southern Italy. In: Proceedings of the 19th World Congress of Soil Science: Soil solutions for a changing world. August 1–6, Brisbane, Australia. , Published on DVD.
Conant, R.T., 2012. Grassland soil organic carbon stocks: status, opportunities, vulnerability. In: Lal, R., Lorenz, K., Hüttl, R.F., Schneider, B.U., von Braun, J. (Eds.), Recarbonization of the Biosphere. Springer, Dordrecht, pp. 275–302.
Deng, L., Liu, G.B., Shangguan, Z.P., 2014. Land use conversion and changing soil carbon stocks in China's Grain-for-Green Program: a synthesis. Global Change Biol. 20, 3544–3556.
Deng, L., Zhu, G.Y., Tang, Z.S., Shangguan, Z.P., 2016. Global patterns of the effects of land-use changes on soil carbon stocks. Global Ecol. Conserv. 5, 127–138.
Díaz, E., 1992. Efecto de la adición de residuos urbanos en la regeneración de suelos degradados como medio de control de la desertificación. Tesis Doctoral, Universidad de Murcia.
Díaz-Fernández, J.L., Barahona, E., Linares, J., 2003. Organic and inorganic carbon in soils of semiarid regions: a case study from the Guadix-Baza basin (Southeast Spain). Geoderma 114, 65–80.
Dickinson, C.H., Wallance, B., Given, P.H., 1974. Microbial activity in Florida-everglades peat. New Phytol. 73, 107–113.
Ding, G., Novak, J.M., Amarasiriwardena, D., Hunt, P.G., Xing, B., 2000. Soil organic matter characteristics as affected by tillage management. Soil Sci. Am. J. 66, 421–429.
Drigo, B., Kowalchuk, G.A., Van Veen, J.A., 2008. Climate change goes underground: effects of elevated atmospheric CO_2 on microbial community structure and activities in the rhizosphere. Biol. Fert. Soils 44, 667–679.
El-Ghamry, A.M., Abd El-Hai, K.M., Ghoneem, K.M., 2009. Amino and humic acids promote growth, yield and disease resistance of faba bean cultivated in clayey soil. Aust. J. Basic Appl. Sci. 3, 731–739.
Ellies, A., 2004. Efecto de la materia orgánica en el suelo. Residuos orgánicos, su uso en sistemas, agroforestales: simposio de la Sociedad Chilena de la Ciencia del Suelo. Universidad de La Frontera, Temuco.

Emmerich, W.E., 2003. Carbon dioxide fluxes in a semiarid environment with high carbonate soils. Agr. Forest Meteorol. 116, 91–102.
ENSAP, 2012. Nile Basin Initiative. Eastern Nile Subsidiary Action Program (ENSAP). Eastern Nile Technical Regional Office (ENTRO). Eastern Nile Watershed Management Project. A Field Guide on Gully Prevention and Control.
European Environmental Agency, 2015. The Soil and the Climatic Change. www.eea.europa.eu.
Eurostat, 2016. Oficina Europea de Estadística. , ec.europa.eu/eurostat.
Fang, H., Sun, L., Tang, Z., 2015. Effects of rainfall and slope on runoff, soil erosion and rill development: an experimental study using two loess soils. Hydrol. Process 29, 2649–2658.
FAO, 2001. Soil carbon sequestration for improved land management. World soil reports 96. Rome, 58.
FAO, 2002. Capture of carbon in soils for a better land management. Word soil reports. ISBN 92-5-304690-2. Rome, 83.
FAO, 2015. Status of the world's soil resources. Main report. Food and Agriculture Organization of the United Nations (FAO), Rome, 607.
Farkas, C., Birkas, M., Varallyay, G., 2009. Soil tillage systems to reduce the harmful effect of extreme weather and hydrological situations. Biologia 64, 624–628.
Fernández, J.M., Senesi, N., Plaza, C., Brunetti, G., Polo, A., 2009. Effects of composted and thermally dried sewage sludge on soil and soil humic acid properties. Pedosphere 19 (3), 281–291.
Foley, J.A., de Fries, R., Asner, G.P., Barford, C., Bonan, G., Carpenter, S.R., et al., 2005. Global consequences of land use. Science 309, 570–574.
García, E., García, C., Hernández, T., 2012. Evaluation of the suitability of using large amounts of urban wastes for degraded arid soil restoration and C fixation. Eur. J. Soil Sci. 63, 650–658.
Garcia-Mozo, H., Yaezel, L., Oteros, J., Galan, C., 2014. Statistical approach to the analysis of olive long-term pollen season trends in southern Spain. Sci. Total Environ. 473, 103–109.
García-Orenes, F., Guerrero, C., Mataix-Solera, J., Navarro-Pedreño, J., Gómez, I., Mataix-Beneyto, J., 2005. Factors controlling the aggregate stability and bulk density in two different degraded soils amended with biosolids. Soil Till. Res. 82, 65–76.
Glover, J.D., Reganold, J.P., Bell, L.W., Borevitz, J., Brummer, E.C., Buckler, E.S., et al., 2010. Increased food and ecosystem security via perennial grains. Science 328, 1638–1639.
Glover, D., Reganold, J.P., Cox, C.M., 2012. Plant perennials to save Africa's soils. Nature 489, 359–361.
Guo, L.B., Gifford, R.M., 2002. Soil carbon stocks and land use change: a meta-analysis. Global Change Biol. 8, 345–360.
Hooijer, A., Page, S., Canadell, J.G., Silvius, M., Kwadijk, J., Wosten, H., Jauhiainen, J., 2010. Current and future CO_2 emissions from drained peatlands in Southeast Asia. Biogeosciences 7, 1505–1514.
IPCC, 2010. WG1 AR4 Report of the Intergovernmental Panel on Climate Change.
IPCC, 2014. In: Barros, V.R., Field, C.B., Dokken, D.J., Mastrandrea, M.D., Mach, K.J., Bilir, T.E., Chatterjee, M., Ebi, K.L., Estrada, Y.O., Genova, R.C.-., Girma, B., Kissel, E.S., Levy, A.N., MacCracken, S., Mastrandrea, P.R., White, L.L. (Eds.), Contribution of Working Group II to the Fifth Assessment Report of the Intergovernmental Panel on Climate Change. Cambridge University Press, Cambridge, UK and New York, NY, USA, p. 688.
Jindo, K., Martin, S.A., Navarro, E.C., Pérez-Alfocea, F., Hernandez, T., García, C., et al., 2011. Root growth promotion by humic acids from composted and non-composted urban organic wastes. Plant Soil 353 (1–2), 209–220.

Joosten, H., 2010. The Global Peatland CO_2 Picture: Peatland Status and Drainage Related Emissions in All Countries of the World. Wetlands International, Wageningen, The Netherlands.
Kaiser, K., Guggenberger, G., 2003. Mineral surfaces and soil organic matter. Eur. J. Soil Sci. 54 (2), 219–236.
Kell, D., 2012. Large-scale sequestration of atmospheric carbon via plant roots in natural and agricultural ecosystems: why and how. Philos. Trans. Roy. Soc. 367, 1589–1597.
Kirsckbaum, M.U.F., 1995. The temperature dependence of soil organic matter decomposition, and the effect of global warming on soil organic C storage. Soil Biol. Biochem. 27, 753–760.
Ladd, J.N., Foster, R., Nannipieri, P., Oades, J.M., 1996. Soil structure and biological activity. Stotzky, G., Bollag, J.M. (Eds.), Soil Biochemistry, 9, Marcel Dekker, New York, pp. 23–78.
Laganière, J., Angers, D.A., Paré, D., 2010. Carbon accumulation in agricultural soils after afforestation: a meta-analysis. Global Change Biol. 16, 439–453.
Lal, R., 2003. Soil erosion and the global carbon budget. Environ. Int. 29, 337–450.
Lal, R., 2004. Soil carbon sequestration impacts on global climate change and food security. Science 304, 1623–1627.
Lal, R., 2009. Sequestering carbon in soils of arid ecosystems. Land Degrad. Dev. 20 (4), 441–454.
Lal, R., 2010a. Managing soil and ecosystems for mitigating anthropogenic carbon emissions and advancing global food security. Bioscience 60, 708–721.
Lal, R., 2010b. Carbon sequestration in saline soils. J. Soil Salin. Water Qual. 1, 30–40.
Lal, R., 2011. Soil health and climate change: an overview. Singh, B.P., Cowie, A.L., Chan, K.Y. (Eds.), Soil Health and Climate Change, Soil Biology, 29, Springer-Verlag, Berlin, Heidelberg, pp. 1–24, 3–24.
Larchevêque, M., Ballini, C., Korboulewsky, N., Montes, N., 2006. The use of compost in afforestation of Mediterranean areas: effects on soil properties and young tree seedlings. Sci. Total Environ. 369, 220–230.
Larney, F.J., Janzen, H.H., Olson, B.M., Olson, A.F., 2009. Erosion-productivity soil amendment relationships for wheat over 16 years. Soil Till. Res. 103, 73–83.
Lax, A., Díaz, E., Castillo, V., Albadalejo, J., 1994. Reclamation of physical and chemical properties of a salinized soil by organic amendment. Arid Soil Res. Rehabil. 8, 9–17.
Lehmann, J., Kleber, M., 2015. The contentious nature of soil organic matter. Nature 528, 60–68.
Lehmann, J., Rillig, M.C., Thies, J., Masiello, C., Hockaday, W.C., Crowley, D., 2011. Biochar effects on soil biota: a review. Soil Biol. Biochem. 43, 1812–1836.
Leifeld, J., Fuhrer, J., 2010. Organic farming and soil carbon sequestration: what do we really know about the benefits? AMBIO J. Human Environ. 39, 585–599.
López-Piñeiro, A., Murillo, S., Barreto, C., Muñoz, A., Rato, J.M., Albarrán, A., García, A., 2007. Changes in organic matter and residual effect of amendment with two-phase olive-mill waste on degraded agricultural soils. Sci. Total Environ. 378, 84–89.
Mabuhay, J.A., Nakagoshi, N., Isagi, Y., 2006. Microbial responses to organic and inorganic amendments in eroded soil. Land Degrad. Dev. 17, 321–332.
Magdoff, F., Weil, R.R., 2004. Soil Organic Matter in Sustainable Agriculture. CRC Press, Upper Saddle River, NJ.
Marshall, M.R., Francis, O.J., Frogbrook, Z.L., Jackson, B.M., McIntyre, N., Reynolds, B., et al., 2009. The impact of upland land management on flooding: result from an improved pasture hillslope. Hydrol. Process 23, 464–475.
McSherry, M.E., Ritchie, M.E., 2013. Effects of grazing on grassland soil carbon: a global review. Global Change Biol. 19, 1347–1357.
Medina-Roldán, E., Paz-Ferreiro, J., Bardgett, R.D., 2012. Grazing exclusion affects soil and plant communities, but has no impact on soil carbon storage in an upland grassland. Agr. Ecosyst. Environ. 149, 118–123.

Meersmans, J., Van wesemael, B., De Ridder, F., Dotti, M.F., De Baets, S., Van molle, M., 2009. Changes in organic carbon distribution with depth in agricultural soils in northern Belgium, 1960–1990. Global Change Biol. 15, 2739–2750.
Mi, N., Wang, S.Q., Liu, J.Y., Yu, G.R., Zhang, W.J., Jobbágy, E., 2008. Soil inorganic carbon storage pattern in China. Global Change Biol. 14, 2380–2387.
Mikhailova, E.A., Post, C.J., 2006. Effects of land use on soil inorganic carbon stocks in the Russian Chernozem. J. Environ. Qual. 35, 1384–1388.
Ming, D.W., 2002. Carbonates. In: Lal, R. (Ed.), Encyclopedia of Soil Science. Marcel Dekker Inc, New York, pp. 139–141.
Mishra, U., Ussiri, D.A.N., Lal, R., 2010. Tillage effects on soil organic carbon storage and dynamics in corn belt of Ohio USA. Soil Till. Res. 107, 88–96.
Moussadek, R., Mrabet, R., Dahan, R., Zouhari, A., El Mourid, M., Van Ranst, E., 2014. Tillage system affects soil organic carbon storage and quality in Central Morocco. Appl. Environ. Soil Sci., dx.doi.org/10.1155/2014/654796.
Na, M.I., Shaoqiang, W., Jiyuan, L.I.U., Guirui, Y.U., Wenjuan, Z., Esteban, J., 2008. Soil inorganic carbon storage pattern in China. Global Change Biol. 14, 2380–2387.
Nicolás, C., Hernández, T., García, C., 2012. Organic amendments as strategy to increase organic matter in particle-size fractions of a semi-arid soil. Appl. Soil Ecol. 57, 50–58.
O'Dell, R., Silk, W., Green, P., Claassen, V., 2007. Compost amendment of Cu–Zn minespoil reduces toxic bioavailable heavy metal concentrations and promotes establishment and biomass production of *Bromus carinatus* (Hook and Arn.). Environ. Pollut. 148, 115–124.
Oteros, J., Garcia-Mozo, H., Botey, R., Mestre, A., Galan, C., 2015. Variations in cereal crop phenology in Spain over the last twenty-six years (1986–2012). Climatic Change 130, 545–558.
Oteros, J., Orlandi, F., Garcia-Mozo, H., Aguilera, F., Dhiab, A.B., Bonofiglio, T., Galan, C., 2014. Better prediction of Mediterranean olive production using pollen-based models. Agron. Sustain. Dev. 34, 685–694.
Piccolo, A., 2002. The supramolecular structure of humic substances: a novel understanding of humus chemistry and implications in soil science. Adv. Agron. 75, 57–134.
Piccolo, A., Spaccini, R., Haberhauer, G., Gerzabek, M.H., 1999. Increased sequestration of organic carbon in soil by hydrophobic protection. Naturwissenschaften 86, 496–498.
Piñeiro, G., Paruelo, J.M., Oesterheld, M., Jobbágy, G., 2010. Pathways of grazing effects on soil organic carbon and nitrogen. Rangeland Ecol. Manage. 63, 109–119.
Poulenard, J., Podwojewski, P., Janeau, J.L., Collinet, J., 2001. Runoff and soil erosion under rainfall simulation of andisols from Ecuadorian Páramo: effect of tillage and burning. Catena 45, 185–207.
Powers, J.S., Corre, M.D., Twin, T.E., Veldkamp, E., 2011. Geographic bias of field observations of soil carbon stocks with tropical land-use changes precludes spatial extrapolation. Proc. Natl. Acad. Sci. USA 108, 6318–6322.
Prentice, I.C., 2001. The carbon cycle and the atmospheric carbon dioxide. Climate Change: 2001: The Scientific Basis: Intergovernmental Panel on Climate Change. Cambridge University Press, Cambridge, UK.
Pulleman, M.M., Marinissen, J.C.Y., 2004. Physical protection of mineralizable C in aggregates from long-term pasture and arable soil. Geoderma 120, 273–282.
Rawlins, B.G., Henrys, P., Breward, N., Robinson, D.A., Keith, A.M., Garcia-Bajo, M., 2011. The importance of inorganic carbon in soil databases and stock estimates: a case study from England. Soil Use Manage. 27, 312–320.
Rosenzweig, C., Hillel, D., 2000. Soil and global climate change: challenges and opportunities. Soil Sci. 165, 47–56.
Ross, M., García, C., Hernández, T., 2001. The use of urban organic wastes in the control of erosion in a semi-arid Mediterranean soil. Soil Use Manage. 17, 292–293.

Scherr, S., 1999. Soil degradation: a threat to developing country food security by 2020. IFRI Food Agr. Environ., 63, Washington, DC.
Schmidt, M.W.I., Torn, M., Abiven, S., Dittmar, T., Guggenberger, G., Janssens, I.A., et al., 2011. Persistence of soil organic matter as an ecosystem property. Nature 478, 49–56.
Schneider, S.H., Semenov, S., Patwardhan, A., Burton, I., Magadza, C.H.D., Oppenheimer, M., et al., 2007. Assessing key vulnerabilities and the risk from climate change: Climate change 2007: impacts, adaptation and vulnerability. In: Parry, M.L., Canziani, J.P., Palutikof, P.J., van der Linden, P.J., Hanson, C.E. (Eds.), Contribution of Working Group II to the Fourth Assessment Report of the Intergovernmental Panel on Climate Change. Cambridge University Press, Cambridge, UK, pp. 779–810.
Schulze, E.D., Freibauer, A., 2005. Environmental science: carbon unlocked from soils. Nature 437 (7056), 205–206.
Shi, S.W., Zang, W., Zang, P., Yu, Y.Q., Ding, F., 2013. A synthesis of change in deep soil organic carbon stores with afforestation of agricultural soils. Forest Ecol. Manage. 296, 53–63.
Six, J., Elliott, E., Paustian, K., 2000. Soil macroaggregate turnover and microaggregate formation: a mechanism for C sequestration under no-tillage agriculture. Soil Biol. Biochem. 32, 2099–2103.
Six, J., Jastrow, J.D., 2002. Organic matter turnover. In: Lal, R. (Ed.), Encyclopedia of Soil Science. Marcel Dekker, New York, pp. 936–942.
Smith, K.A., Conen, F., 2004. Impacts of land management on fluxes of trace greenhouse gases. Soil Use Manage. 20, 255–263.
Smith, P., Bhogal, A., Edgington, P., Black, H., Lilly, A., Barraclough, D., et al., 2010. Consequences of feasible future agricultural land-use change on soil organic C stocks and greenhouse gas emissions in Great Britain. Soil Use Manage. 26, 381–398.
Smith, P., House, J.I., Bustamante, M., Sobocká, J., Harper, R., Pan, G., et al., 2015. Global change pressures on soils from land use and management 22 (3), 1008–1028.
Smith, S.E., Read, D.J., 1997. Mycorrhizal Symbiosis. Academic Press, Inc, San Diego, CA, ISBN 0-12-652840-3.
Soane, B.D., Ball, B.C., Arvidsson, J., Basch, G., Moreno, F., Roger-Strade, J., 2012. Non-till in northern, western and south-western Europe: a review of problems and opportunities for crop production and the environment. Soil Till. Res. 118, 66–87.
Solomon, D., Lehmann, J., Zech, W., 2000. Land use effects on soil organic matter properties of chromic luvisols in semi-arid northern Tanzania: carbon, nitrogen, lignin and carbohydrates. Agr. Ecosyst. Environ. 78, 203–213.
Spaccini, R., Piccolo, A., Conte, P., Haberhauer, G., Gerzabek, M.H., 2001. Increased soil organic carbon sequestration through hydrophobic protection by humic substances. Soil Biol. Biochem. 34 (12), 1839–1851.
Spacini, R., Piccolo, A., Haberhauer, G., Gerzabek, M.H., 2000a. Transformation of organic matter from maize residues into labile and humic fractions of three European soils as revealed by ^{13}C distribution and CPMAS-NMR spectra. Eur. J. Soil Sci. 51, 583–594.
Spacini, R., Zena, A., Piccolo, A., 2000b. Carbohydrates distribution in size-aggregates of three European soils in a climate gradient. Fres. Environ. Bull. 9, 468–476.
Stevenson, F.J., 1986. Cycles of soil. Carbon, Nitrogen, Phosphorus, Sulphur, Micronutrients. John Wiley and Sons, New York, 448.
Stevenson, F.J., 1994. Humus chemistry: genesis, composition. Reactions. John Wiley and Sons, New York, p. 512.
Strack, M. (Ed.), 2008. Peatlands and Climate Change. University of Calgary, Canada, p. 223, Copyright (c) 2008 International Peat Society A5, ISBN: 978-952-99401-1-0.
Swift, R., 2001. Sequestration of carbon by soil. Soil Sci. 166, 858–871.

Tejada, M., Hernández, M.T., García, C., 2006. Application of two organic amendments on soil restoration: effects on the soil biological properties. J. Environ. Qual. 35, 1010–1017.

Thompson, A.M., Izaurralde, R.C., Rosenberg, N.J., He, X., 2006. Climate change impacts on agricultural and soil carbon sequestration potential in the Huang-Hai Plain of China. Agr. Ecosyst. Environ. 114, 195–209.

Treseder, K.K., Holden, S.R., 2013. Fungal carbon sequestration. Science 339, 1528–1529. doi: 10.1126/science.1236338.

Trivedi, P., Anderson, I.C., Singh, B.K., 2013. Microbial modulators of soil carbon storage: integrating genomic and metabolic knowledge for global prediction. Trends Microbiol. 21, 641–651, 10.1016/j.tim.2013.09.005.

Turmel, M.S., Speratti, A., Frédéric Baudron, F., Verhulst, N., Govaerts, B., 2015. Crop residue management and soil health: a systems analysis. Agr. Syst. 134, 6–16.

Turner, II, B.L., Lambin, E.F., Reenberg, A., 2007. The emergence of land change science for global environmental change and sustainability. Actas Acad. Nacion. Cienc. 104, 20666–20671.

Ugalde, D., Brungs, A., Kaebernick, M., McGregor, A., Slattery, B., 2007. Implications of climate change for tillage practice in Australia. Soil Till. Res. 97, 318–330.

United Nations, 1987. Our common future. Report of the World Commission on Environment and Development.

Van Camp, L., Bujarrabal, B., Gentile, A.R., Jones, R.J.A., Montanarella, L., Olazabal, C., Selvaradjou, S.K., 2004. Reports of the Technical Working Groups Established under the Thematic Strategy for Soil Protection. Office for Official Publications of the European Communities, Luxembourg, Vol. III, Organic Matter. EUR 21319 EN/3 European Commission.

Weber, J., Karczewska, A., Drozd, J., Licznar, M., Licznar, S., Jamroz, E., Kocowicz, A., 2007. Agricultural and ecological aspects of a sandy soil as affected by the application of municipal solid waste composts. Soil Biol. Biochem. 39, 1294–1302.

West, P.C., Gerber, J.S., Engstrom, P.M., Mueller, N.D., Brauman, K.A., Carlson, K.M., et al., 2014. Leverage points for improving global food security and the environment. Science 345, 325–328.

Wu, H.B., Guo, Z.T., Gao, Q., Peng, C.H., 2009. Distribution of soil inorganic carbon storage and its changes due to agricultural land use activity in China Agric. Ecosyst. Environ. 129, 413–421.

INDEX